W9-APD-621

as presented to the
XIV General Assembly of the
INTERNATIONAL UNION OF PURE AND APPLIED PHYSICS
on the occasion of the Union's
Fiftieth Anniversary, September 1972

Edited by
SANBORN C. BROWN

NATIONAL ACADEMY OF SCIENCES
WASHINGTON, D.C.
1973

Library of Congress Cataloging in Publication Data
Main entry under title:

Physics 50 years later.

Includes bibliographical references.
1. Physics–Addresses, essays, lectures. 2. International Union of Pure and Applied Physics. I. Brown, Sanborn Conner, 1913- , ed. II. International Union of Pure and Applied Physics.
QC71.P46 530 73-11320
ISBN 0-309-02138-3

Available from
Printing and Publishing Office
National Academy of Sciences
Washington, D.C. 20418

Printed in the United States of America

THE WHITE HOUSE

WASHINGTON

September 21, 1972

On behalf of all my fellow citizens, I welcome the distinguished physicists who have come to the United States to participate in the Fourteenth General Assembly of the International Union of Pure and Applied Physics. We are especially honored that you have chosen to celebrate your fiftieth anniversary in our country, and that you have given us this opportunity to pay special tribute to the achievements you have realized for all humanity.

Never before has there been such an urgent need for global cooperation in scientific affairs. It is a need that is positively and effectively served by meetings such as this. Governments have often lagged behind organizations such as yours in promoting scientific and technological cooperation. Your leadership has been an inspiration toward this end and has given incentive to scientific progress that will serve all mankind and promote the cause of world peace and prosperity.

I hope that your sessions will be rewarding and productive for all who attend, and that your deliberations will benefit the millions of people who, on this milestone in your history, gratefully acknowledge your considerable accomplishments.

Richard Nixon

Contents

Historical Introduction

To celebrate the Fiftieth Anniversary of the founding of the International Union of Pure and Applied Physics, leaders of various branches of physics were asked to present papers on the present state of their specialties and to speculate on the direction physics is taking. This memorial volume is the result.

In addition to the several days spent on the General Assembly business, the anniversary was marked by ceremonial functions and a historical exhibit—"Physics in 1922"—arranged by the American Institute of Physics' Center for History of Physics to commemorate the founding of the Union. The exhibit was planned and produced under the direction of Joan Nelson Warnow, librarian of the AIP Center's Niels Bohr Library.

Excerpts from letters written in 1922 are major elements in the exhibit. Most of the letters are from the Niels Bohr Library's microfilm collections of the George Ellery Hale papers, the Niels Bohr papers, and the Ernest Rutherford papers. The original documents are preserved at the California Institute of Technology Archives, the Niels Bohr Institute, and the Cambridge University Library, respectively.

A few highlights of this exhibit set a historical background for the 50 years of physics that began in 1922.

Physics in 1922

The International Union of Pure and Applied Physics was founded in 1922 to encourage and aid international cooperation in the field of physics. The historical materials in this exhibit provide glimpses of some of the highlights of physics activity in 1922, emphasizing the close ties among physicists at research centers throughout the world in their common efforts to probe new frontiers in several fields. As a special tribute to Niels Bohr's lifelong efforts to foster international cooperation in science, the exhibit also shows the background of Bohr's early impact on international science through his atomic concepts and the establishment of the Copenhagen Institute.

Radioactivity, atomic theory, and relativity were among the chief subjects of interest in 1922, and new developments in these fields provided the basis for lively discussions among the world's physicists through personal visits, meetings, and correspondence. 1922 was not only a year of fruitful results and discoveries, it was also a year in which international ties were being restored and research centers in several countries were establishing new facilities and productive environments. All of these activities stimulated the emergence of the new quantum mechanics during the next few years, an event that is now seen as one of the greatest intellectual achievements in the history of physics. The year ended with the award of the 1921 Nobel Prize in Physics to Albert Einstein and the 1922 Prize to Niels Bohr.

Au Conseil International des Recherches

Une assemblée générale

On se rappellera qu'en 1919 un congrès de savants décidait l'installation à Bruxelles d'un Conseil international de Recherches, organisme permanent destiné à rassembler des renseignements d'ordre scientifique de manière à coordonner les études scientifiques faites dans les différents pays.

Cet organisme devait tenir tous les trois ans une assemblée générale, que les délégués américains voudraient voir tenir successivement dans différentes villes.

Une assemblée générale a lieu actuellement à Bruxelles, au Palais des Académies. La séance inaugurale, mardi matin, était présidée par M. Emile Picard, secrétaire perpétuel de l'Académies des Sciences de Paris, ayant à ses côtés MM. Lameere, président de la section scientifique de l'Académie Royale de Belgique, remplaçant M. Lecointe, empêché; George Hale, secrétaire de l'Académie Nationale des Etats-Unis, directeur de l'Observatoire de Mount Wilson; V. Volterra, professeur, membre de l'Académie Royale « dei Lincei » de Rome; sir Schuster, secrétaire de la « Royal Society » de Londres.

Après que M. Emile Picard eut déclaré la séance ouverte, M. Lameere a souhaité la bienvenue aux congressistes. Il a dit la reconnaissance que la capitale belge leur voue pour avoir été choisie comme siège légal du conseil.

Puis M. Emile Picard a fait un bref historique du conseil. En 1919, dit-il, une des préoccupation de l'assemblée fut de recueillir l'adhésion de pays qui n'avaient pas encore délégué de représentants au Conseil, et les espoirs exprimés alors ne furent pas vains. Des craintes exprimées au sujet de la tutelle du Conseil vis-à-vis de certaines unions se prouvèrent non fondées. L'activité scientifique des unions définitivement constituées a été féconde; l'Union Internationale de Chimie, qui se réunit annuellement, témoigna d'une particulière activité. Certaines sciences ne se prêtent cependant pas au travail collectif. Tel est le cas des mathématiques. Un congrès international des mathématiciens eut cependant lieu à Strasbourg, en 1920.

Sir A. Schuster a fait connaître que vingt pays faisaient maintenant partie du Conseil; il a donné ensuite certains détails sur l'organisation administrative du Conseil. Sur la proposition de l'Angleterre, l'Egypte sera invitée à en faire partie.

L'assemblée a procédé ensuite à la nomination d'une commission financière, chargée d'arrêter le budget des diverses unions, le nombre de parts unitaires étant dans chacune de celles-ci, proportionnel à la population des Etats.

Elle a ensuite nommé une commission pour l'examen des amendements proposés aux articles des statuts du Conseil international, amendements qui seront discutés au cours de séances ultérieures.

Les délégations [1922?]

Le Conseil tiendra ses assises à Bruxelles, du 25 au 29 juillet. La France y est représentée par 42 délégués, parmi lesquels MM. Baillaud, directeur de l'Observatoire de Paris; E. Picard et R. Lacroix, secrétaires perpétuels de l'Académie des Sciences; le général Ferrié, inspecteur des services de télégraphie militaire; le colonel Bellot, directeur des services géologiques de l'armée; le prince Roland Bonaparte; Charles Mouren; le général Bourgeois; Deslandres; Grandidier; Lallemand, etc.

Les Etats-Unis ont envoyé huit délégués, parmi lesquels, outre M. Hale, membre du comité exécutif, les physiciens R.-A. Millikan et Townbridge, les chimistes Washburn et Nogès, etc.

La Belgique est représentée par quatorze délégués de l'Académie des Sciences et de Médecine et des diverses universités.

Le Danemark, la Pologne, la Norvège, la Hollande, le Japon, l'Italie, l'Angleterre, le Canada, la Grèce sont également représentés.

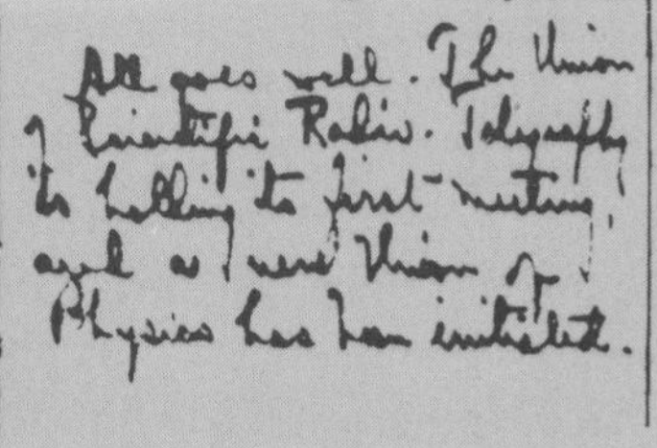

All goes well. The Union of Scientific Radio-Telegraphy is holding its first meeting, and a new Union of Physics has been initiated.

"AU CONSEIL INTERNATIONAL DES RECHERCHES," undated newspaper clipping, July 1922, with handwritten comments by George Ellery Hale. *Hale Papers.*

Founding of the IUPAP

All goes well. The Union of Scientific Radio-Telegraphy is holding its first meeting, and a new Union of Physics has been initiated.

These comments, marking the birth of IUPAP, were handwritten by George Ellery Hale at the bottom of a newspaper clipping reporting the July 1922 meeting in Brussels of the General Assembly of the International Research Council. Hale was a founder of the Council in 1919, which soon created International Unions in the various scientific disciplines and became known officially in 1931 as the International Council of Scientific Unions.

The Bohr Atom

Niels Bohr's research at the University of Manchester in the spring of 1912, which led to a Nobel Prize in 1922, was undertaken as part of the tradition of international postdoctoral studies in physics. There he formed a lifelong friendship with Ernest Rutherford and developed concepts of atomic structure that became central to all of twentieth-century physics. When Bohr arrived, the Manchester group was following up Rutherford's recent discovery of the concept of the nuclear atom. Soon Bohr was devoting all of his time to the study of atomic structure and, before he returned to Denmark in July 1912, had prepared a first draft of his ideas for discussion with Rutherford.

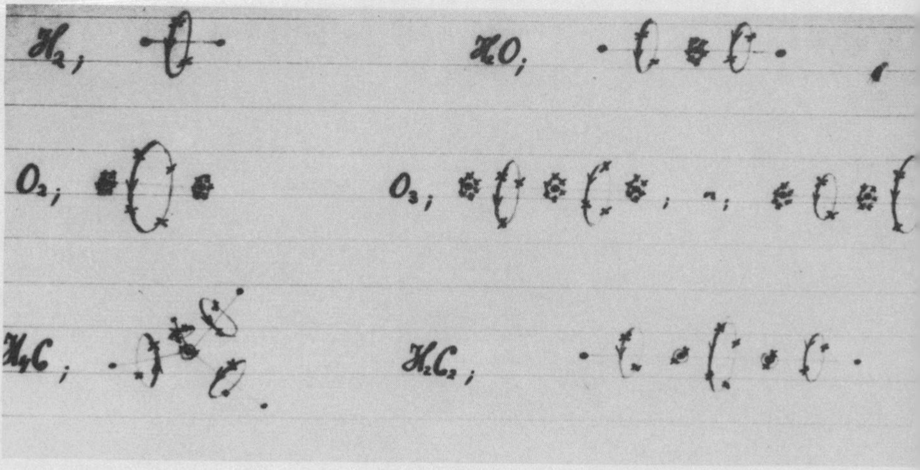

. . . Enclosed I send the first chapter of my paper on the constitution of atoms. I hope that the next chapters shall follow in a few weeks. In the latest time I have had good progress with my work, and hope to have succeeded in extending the considerations used to a number of different phenomena. . . .

. . . I have tried to show that it, from such a point of view, seems possible to give a simple interpretation of the law of the spectrum of hydrogen, and that the calculation affords a close quantitative agreement with experiments. . . .

I hope that you will find that I have taken a reasonable point of view as to the delicate question of the simultaneous use of the old mechanics and of the new assumptions introduced by Planck's theory of radiation. I am very anxious to know what you may think of it all. . . .

NIELS BOHR to ERNEST RUTHERFORD
March 6, 1913, *Rutherford Papers*

. . . Your ideas as to the mode of origin of [the] spectrum and hydrogen are very ingenious and seem to work out well; but the mixture of Planck's ideas with the old mechanics make it very difficult to form a physical idea of what is the basis of it. There appears to me one grave difficulty in your hypothesis, which I have no doubt you fully realize, namely, how does an electron decide what frequency it is going to vibrate at when it passes from one stationary state to the other? It seems to me that you would have to assume that the electron knows beforehand where it is going to stop. . . .

ERNEST RUTHERFORD to NIELS BOHR
March 20, 1913, *Bohr Papers*

◀ Niels Bohr's drawings of his models of electronic structures, from a memorandum he prepared in June and July 1912 as the basis for discussion of his new ideas with Ernest Rutherford.

These concepts were subsequently included in Parts II and III of Bohr's trilogy, "On the Constitution of Atoms and Molecules" (*Philosophical Magazine*, Vol. 26, 1913). *Bohr Papers.*

The Birth of the Copenhagen Institute

The Institute for Theoretical Physics at the University of Copenhagen was emerging as an international center in physics in 1922. Its origins went back to 1916 when Niels Bohr became the first occupant of the Chair of Theoretical Physics at the University. He had visions of a research institute for atomic physics where experiment and theory would be closely tied and where collaboration of scientists from many countries would be stressed. In 1917, funds were sought from individual Danish citizens to provide a site for the proposed institute. The solicitation letter emphasized that a major purpose of the institute would be to restore international scientific collaboration, which had broken down during the war. The Danish Government provided funds for erection of a building on the site, the Danish Carlsberg Foundation provided research grants, and stipends for foreign scientists were provided by the Rask-Oersted Foundation, set up by the Danish state to foster international cooperation in research. When the Institute opened in 1921 its international staff was already at work on a number of research problems.

Photograph of George Gamow's 1931 drawings starring Mickey Mouse in highlights of atomic physics, Bohr's career, and the Copenhagen Institute, 1900–1931. 14 drawings. *Margrethe Bohr Collection, AIP Niels Bohr Library.* ▸

Professor Harald Bohr wrote me that his brother, Professor Niels Bohr, is applying to the great Danish Carlsberg Foundation for support of his Institute, and that he feels that a favorable statement from me might have influence. . . .

But I want to treat this question from a particular historical view. The war burdens and unbearable peace terms have made scientific efforts in Germany impossible for a long time to come. Previously, Germany's numerous universities and institutes of technology were able to further experimental research with good financial support. Together with Germany, almost the whole European continent has become impoverished. But happy Denmark can step into the breach here. Denmark will enjoy doing this the more as such an act would honor the name of one of its most outstanding sons. The Institute of Mr. Bohr should not only serve the upcoming scientific generation of Denmark, it will also be an international place of work for foreign talent whose own countries are no longer in a position to make available the golden freedom of scientific work. Just as in the past at the Radium Institute of Vienna, future researchers of all countries should meet one another in Copenhagen for special studies and to pursue common cultural ideals at the Bohr Institute for Atomic Physics.

ARNOLD SOMMERFELD to THE CARLSBERG FOUNDATION
October 1919, *Bohr Papers*

Research at the Cavendish

The Cavendish Laboratory at Cambridge University, which had long been renowned as a major center of experimental physics, was flourishing in 1922 under Rutherford's leadership. The number of foreign students, visitors, and guest lecturers was increasing steadily, and the Laboratory served as an international communications center in the rapidly developing field of radioactivity and the study of the properties and structure of the nucleus.

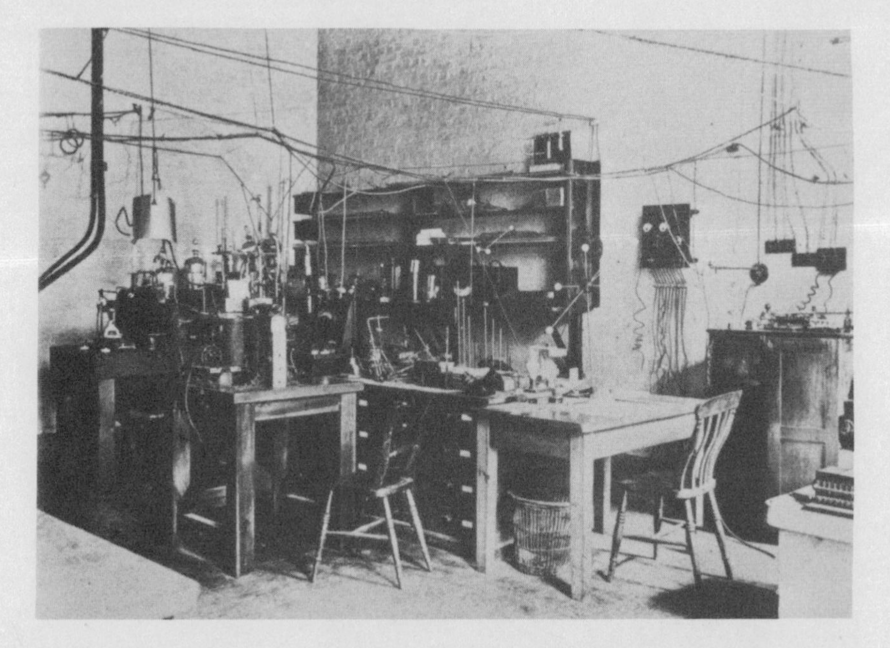

Photograph of Rutherford's research room at the Cavendish Laboratory in the early 1920's. *AIP Niels Bohr Library.*

ERNEST RUTHERFORD to NIELS BOHR, June 5, 1922. ▸
Photograph of original letter: TLS, English, 2 pp. *Bohr Papers.*

Cavendish Laboratory,

Cambridge.

June 5th 1922

ear Bohr

I am glad to hear that you are feeling fit and active.
ou must have been rather tired after your strenuous time in
ngland. I need hardly tell you how much your lectures were
ppreciated by the men in the laboratory. They would all like
o see you over here again next year to talk over matters
ith them, if not to give a formal course of lectures.

I note what you suggest about Kramers. I should like
o have him for a short timeif I could afford it, but you
robably know that finances are very tight in this University
nd I must be just before I am generous.

I have got the paper of Coster's which I will get attended
o and sent to the Philosophical Magazine. I have not yet had
ime to read it.

Ellis is making steady progress and I hope ultimately
hat we will get a great deal of information about nuclear
tructure from his experiments.

You will have seen Miss Meitner's theory on the same
ubject. Ellis is quite convinced that her theory is untenable.

Blackett is going strong, and by a statistical study of
he deflections has deduced the relation between range and
elocity of the α particles within 1 cm of the end. The
ange appears to be connected with a much lower power of
he velocity than the old Geiger relation, but of course
is results are only as yet in the preliminary stage, but he
opes to connect on directly with the velocity measured by
he photographic method. He is at present engeged in making
xperiments in Helium and Argon.

I spoke to Henderson about the matter you mentioned and
o doubt his paper will be published in due course. He has
een tackling the velocity of α particles by the old method,
sing Hilger plates and it looks as if some very interesting
esults will come out of it. However his experiments are only
n a preliminary stage.

The laboratory has been in a great state of excitement the
ast week, due to trying certain new types of photographic
lates , which one makes in the laboratory,Aston finds them
bout six times faster and very much clearer for his positive
ays. He has examined Tin Methide and finds seven isotopes
nd has still further extended Xenon ; by good fortune he
ppears to have fairly definite evidence of the break down of
he whole number rule, between Tin and Xenon, but of course
ther experiments will be required to make sure of it.

Chadwick and I have been trying out, Lithium, Beryllium,
agnesium, Silicon, and Chlorine for H particles, but certainly
hlorine gives nothing and we have no evidence that any of the others
o either. It is a very surprising result that only the elements
f atomic numbers 5,7,9,11,13,15, show this effect.

We have had wonderful weather in England, but rather
et. We gave a garden partyon Saturday night where we had
bout a hundred people.

We are all fit and well

Yours very sincerely

E Rutherford

Telegram from W. W. Campbell to Albert Einstein, April 12, 1923. English, 1 p. *Einstein Papers.* Institute for Advanced Study, Princeton, New Jersey.

Blatt Nr. 037

einstein akademie wissenchaften
berlin

Telegramm Nr. ....

Angenommen den 12 4 1923 — Telegraphie des Deutschen Reichs.

um ... Uhr ... Min. vorm. nachm.

von ... Antw S — Antw S

durch ... — Berlin, Haupt-Telegraphenamt — Leitung Nr. ...

Befördert den ...

BERLIN NW 12.4.23. 8 V.

Telegramm aus - lickobservaory calif 129/12 50 11 19/16.- westernunion =

= three pairs australia tahiti eclipse plates measured by campb
trumpler sixty two to eighty four stars each five of six
measurements completely calculated give einstein deflection
between one point fifty nine and one point eighty six seconds
arc mean value one point seventy four seconds = campbell .+

1,24 — 1,59 / 86

Relativity Makes the News

Worldwide interest in relativity, which was heightened by Eddington's 1919 eclipse expedition, was further enhanced in 1922 when an American expedition, specially equipped to test the Einstein theory, encountered ideal observing conditions. Einstein's predictions were confirmed through recorded images of 147 stars, 20 times more than had been recorded in 1919. During 1922, Einstein gave a series of scientific lectures in France and made a popular lecture tour in Japan, where his collected works were published in Japanese translation that year.

Discussions at Göttingen

The University of Göttingen, long one of the world's leading centers of physics and mathematics, invited Niels Bohr to give a series of lectures in 1922. Bohr knew the Göttingen mathematicians Felix Klein, David Hilbert, and Richard Courant through his brother Harald, who had studied mathematics there. Niels Bohr had also corresponded with Max Born and with James Franck, who was one of the first guest lecturers at the Copenhagen Institute. The Göttingen scientists, and others throughout Germany, were intensely interested in Bohr's interpretation of quantum theory, and they flocked to his lecture series, which later was known as the "Bohrfestspiele." The lectures had an important impact on subsequent physics research in Germany and in Denmark, not only because they brought the current ideas and problems of physics to a focus but also because the occasion established personal ties among the physicists who in the next few years would make outstanding contributions to the solution of these problems.

. . . I am busy calculating your model for the ground state of He and do hope it will be possible to arrive at a definite result.

I am immensely looking forward to my stay in Copenhagen and the work at your Institute.

. . . Finally, I would also like to thank you very much for the great friendliness with which you have given me information in Göttingen about the greatest variety of physical questions; for me it was a quite incalculable gain.

WOLFGANG PAULI to NIELS BOHR
July 7, 1922, *Bohr Papers*

Photograph of Enrico Fermi, Werner Heisenberg, and Wolfgang Pauli at the 1927 Como Conference. *Segrè Collection. AIP Niels Bohr Library.*

The Nobel Prize

At the end of 1922, Einstein was awarded the Nobel Prize in Physics for 1921 and Bohr the Prize for 1922. During his Nobel Address in Stockholm, Bohr announced that two scientists working in his Copenhagen laboratory–George de Hevesy from Hungary and Dirk Coster from The Netherlands–had just discovered element 72, named hafnium after the ancient name of Copenhagen. The discovery provided additional support for Bohr's atomic theory through which he accomplished a reordering of the periodic system of the elements in 1922. It also demonstrated the links between theory and experiment and the international collaboration that were the aims of the Copenhagen Institute.

Dear–or rather, beloved–Bohr!

Your heartfelt letter arrived shortly before I left Japan. I can say without exaggeration that it made me as happy as the Nobel Prize. I found especially charming your fear that you might have received the prize before me. That is genuinely Bohr-like (bohrisch) . . . .

ALBERT EINSTEIN to NIELS BOHR
January 10, 1923, *Bohr Papers*

. . . how glad I was when I read of the Nobel Prize report in the newspapers. It reminded me vividly of that beautiful day in Cambridge in 1913 when you set forth your ideas for me in the quadrangle of Trinity. Thanks to prior suggestion by Harald [Bohr], who had so often told me wonderful things about his brother, I was at that point immediately ready to believe that you must be right. But when I then reported of these things here in Göttingen, they laughed at me that I should take such fantasies seriously. Thus I became, so to speak, a martyr of the Bohr model. . . .

RICHARD COURANT to NIELS BOHR
December 8, 1922, *Bohr Papers*

International Communication

Scientific journals, personal visits, scientific meetings, and personal correspondence kept physicists abreast of new research and enabled them to collaborate with colleagues throughout the world, aiding current work and enriching the education of a new generation of young physicists.

The Indian physicist, C. V. Raman, at the University of Leningrad during a visit in the early 1920's: Professor Rogdestvenski (at left) greets Raman (center). Just behind Raman to the right is George Gamow, a graduate student at that time. In 1922, Raman published "On the Molecular Scattering of Light in Water and the Colour of the Sea" (*Proceedings of the Royal Society*, A, vol. 101, p. 64, 1922), the first in a series of papers that led to his important discovery subsequently known as the Raman effect. *AIP Niels Bohr Library.*

Physics
50 YEARS
LATER

PROFESSOR PIERRE FLEURY *was Secretary-General of the Union from 1947 to 1963. During these fifteen years he carried the major responsibility of the day-to-day operation of IUPAP, and his faithfulness and wisdom did much to rekindle the spirit of the Union after World War II. The organizing committee of the XIV General Assembly asked Professor Fleury to prepare an historical summary of the Union's activities, and the following are extracts from his printed paper.*

The International Union of Pure and Applied Physics from 1923 to 1972 / P. FLEURY

The International Union of Pure and Applied Physics is now entering maturity. Its development, which was rather inhibited by World War I, accelerated rapidly after World War II and was greatly assisted by the support then given by the United Nations Education, Scientific and Cultural Organization (UNESCO). More than 50 years of cordial international collaboration have characterized IUPAP's contributions to the extraordinary progress of physics.

In 1919, an International Research Council was created to assist in re-establishing the relationship between scientists which had been broken by World War I. It was during a meeting of this Council in Brussels in 1922 that an "Executive Committee" decided to take the necessary steps to constitute an International Union of Pure and Applied Physics. This Committee included Sir William Bragg, President; Messrs. M. Brillouin, O. M. Corbino, M. Knudsen, M. Leblanc, H. A. Lorentz, R. Millikan, H. Nagaoka, E. vanAubel, who became Vice-President of IUPAP, and Henri Abraham, who became the Union's Secretary-General.

A *Charter General Assembly* was held in Paris in 1923; sixteen countries were represented: Belgium, Canada, Czechoslovakia, Denmark, France, Italy, Japan, Netherlands, Norway, Poland, Spain, Sweden, Switzerland, Union of South Africa, United Kingdom, and the United States of America. Following the adoption of provisional statutes, it was decided that the members of the Steering Committee should remain in office. Mr. Rateau replaced Mr. Leblanc, who had died.

The history of IUPAP has been marked by General Assemblies held in principle every three years. The General Assemblies decide on the administration of the Union and pass resolutions concerning commissions and the support to be given to scientific conferences.

The *Second General Assembly* (Brussels, 1925) saw admission to the Union of Australia and Mexico and the election of four new Vice-Presidents: B. Cabrera, C. E. Guye, W. Natanson, and M. Siegbahn.

The *Third General Assembly* (Brussels, 1932). China was admitted. R. A. Millikan was elected President. A. Cotton, Sir Richard Glazebrook, M. Keeson, V. Posejpal, and L. Vegard were elected Vice-Presidents. An international Commission on Bibliography and Publications was created, as well as a Commission for Symbols, Units and Nomenclature (SUN).

The *Fourth General Assembly* (London, 1934). At this Assembly, the Union joined the International Council of Scientific Unions (ICSU), which replaced the International Research Council. Niels Bohr being unable to accept the presidency of the Union, this office was filled until 1937 by Manne Siegbahn. Enrico Fermi was elected Vice-President.

A SUN Commission report was adopted which included units in thermodynamics (including the choice of the joule as unit of quantity of heat), electrical and mechanical units (maxwell, gauss, gilbert, oersted), and thermodynamical symbols. B. Cabrera presented a report (prepared in collaboration with a special committee of the League of Nations) on the coordination of vocabulary in the various scientific disciplines.

This Assembly was marked by an International Congress organized jointly by IUPAP and the Physical Society; more than 700 delegates participated. Résumés of the papers and discussions were published in three books under the titles *Cosmic Rays*, *Transmutations*, and *The Solid State of Matter.*

The advent of World War II delayed preparation for the next general assembly and prevented the meeting of a second congress scheduled to be held in Paris in 1940. However, in cooperation with the International Institute for Intellectual Cooperation, which published the proceedings, IUPAP organized three study groups: New Theories in Physics (Warsaw, 1938); Physicochemical Determinations of the Molecular Weights of Gases (Neuchatel, 1938); and Magnetism (Strasbourg, 1939). A meeting devoted to the measurement of ionizing radiations, organized in cooperation with the International Union of Biological Sciences, was held at Groningen (Netherlands) in 1939.

The *Fifth General Assembly*, which was held in Paris in 1947, provoked interesting discussions on the nature of the Union's work. The meeting was organized by P. P. Ewald, who had served as Secretary-General since 1946. The Assembly gave a moving tribute to the memory of Henri Abraham, Secretary-General of IUPAP from the time it was established until his tragic death in a German concentration camp in 1943. It was with great emotion that President Siegbahn mentioned "this dignified and lovable colleague, always ready to help others, who during the entire period between two World Wars was the permanent center of the Union about which revolved its affairs."

The Assembly accepted the membership of Finland and elected a new Executive Committee, designating as Vice-Presidents C. Bialobrezeski, K. K. Darrow, Sir Charles Darwin, P. P. Ewald, C. J. Gorter, J. C. Jacobsen, and P. Scherrer, with P. Fleury as Secretary-General. The choice of the new President, H. A. Kramers, was a particularly happy one. This great physicist gave much of his time to the Union in spite of the illness to which he was subject. His wide knowledge of problems and of men, as well as his discreet authority which all acknowledged, made him a most efficient leader. His broad views and his attention to detail were particularly valuable during the difficult period following the war.

The fifth General Assembly approved the agreement that had been concluded between ICSU and UNESCO: grants that IUPAP subsequently received from ICSU were of considerable help in organizing conferences and congresses: from 3 in 1948 and 5 in 1949, their number increased progressively to 35 in 1972.

The General Assembly reconstituted the SUN Commission and also created commissions for Optics, Thermodynamic Notations, Radioactive Units, and Cosmic Rays. The Union also agreed to participate in various interunion commissions.

The *Sixth General Assembly* (Amsterdam, 1948). Hungary was admitted, the Statutes were modified and provision was made for "affiliated commissions." The Assembly approved the status of affiliation for the International Commission for Optics and created the Very Low Temperature Commission. E. Amaldi and J. C. Slater were elected Vice-Presidents.

The *Seventh General Assembly* (Copenhagen, 1951). Argentina, Brazil, Egypt, India, and Israel were admitted. N. F. Mott was elected President. Vice-Presidents elected were G. Borelius, J. Heyrovsky, P. Huber, K. S. Krishnan, M. L. Oliphant, and J. A. Wheeler. New Commissions created were Acoustics, Publications (which has functioned since 1949), and Solid State Physics.

The *Eighth General Assembly* (London, 1954). At this Assembly, the Federal Republic of Germany, Yugoslavia, and New Zealand were admitted to the Union. New Vice-Presidents elected were J. deBoer, R. B. Brode, M. Kotani, and J. H. Van Vleck.

The *Ninth General Assembly* (Rome, 1957). New members admitted were Austria, Bulgaria, and the U.S.S.R. Elected to office were the following: President, E. Amaldi; Vice-Presidents, G. Herzberg, A. Joffe, E. Rasmussen, H. H. Staub, F.C.A. Trandelenburg, and J. Weyssenhoff. The Commission on High Energy Physics was created, as well as that of a Subcommission on Semiconductors and Magnetism of the Commission on Solid State.

The *Tenth General Assembly* (Ottawa, 1960). Admitted to the Union were German Democratic Republic, People's Republic of China (not confirmed by Peking), Republic of China (Taiwan), Pakistan, and Romania. Elected to office were the following: President, H. J. Bhabha; Vice-Presidents, D. I. Blokhintsev, C. J. Gorter, L. Néel, H. H. Nielsen, E. Rudberg, and F. Seitz. C. C. Butler was elected Associate Secretary-General. New Commissions established were Nuclidic Masses and Atomic Constants, Physics Education, Low Energy Nuclear Physics.

The *Eleventh General Assembly* (Warsaw, 1963). Bolivia was admitted to the Union. Elected to office were the following: President, L. Néel; Vice-Presidents, M. Danysz, J. Jauch, Sir Gordon Sutherland, and H. Yukawa. C. C. Butler replaced P. Fleury as Secretary-General. Larkin Kerwin was elected Associate Secretary-General.

The *Twelfth General Assembly* (Basel, 1966). Ireland was admitted to the Union. Elected to office were the following: President, D. I. Blokhintsev; Vice-Presidents, R. Bacher, G. Bernardini, S. Bhagavantam, W. Boas, W. Dekeyser, and S. Rozental. The International Commission on Atomic and Molecular Physics and Spectroscopy was created.

The *Thirteenth General Assembly* (Dubrovnik, 1969). South Korea and Cuba were admitted to the Union. Elected to office were the following: President, R. F. Bacher; Vice-Presidents, A. Kastler, H. Maier-Leibnitz, L. Pal, and V. Weisskopf. The names of the Commissions on High and Low Energy Physics were modified. The Commission on the Physics of Plasmas was created.

Since 1947, the rapid developments in physics have prevented holding general conferences, as the vast number of delegates would have greatly complicated the organization of the meeting and much diminished its usefulness. However, the IUPAP Commissions have met frequently, and a large number of specialized colloquia and conferences have been organized by local committees with the cooperation of the Commissions and the approval of the Executive Committee. The Union has given careful study to the frequency, geographic location, and timeliness of these conferences before extending them official status. A large number of conferences that did not fall within the domain of a particular Commission were also held following examination by the Executive Committee.

IUPAP COMMISSIONS

1. *Commission for Symbols, Units and Nomenclature* This Commission maintains close relations with the corresponding commission of the International Union of Pure and Applied Chemistry, with the International Commission for Optics, and with the other commissions of IUPAP. It is also in communication with the Commission on Lighting, the International Electrotechnical Committee, the International Standards Organization (ISO), and the International Commission on Weights and Measures (CIPM). Its recommendations, approved by IUPAP, have included among others the definition of the absolute scale of temperatures, based on the triple point of water as a single fixed point; on the use (subsequently approved by CIPM) for international purposes of the practical system of units based on the meter, the kilogram-mass, the second, and the ampere; on the designation of the *newton* as the unit of force in this system; on the need for four principal units (and not three) in the CGS system; on the rationalization of electromagnetic equations; on the inconvenience of the use of the word *billion* in scientific literature; and on the adoption of the expression *nuclide*.

The booklet, *Symbols and Units Recommended by the Physics Union*, published in 1955 and reproduced in numerous scientific journals, was completed and re-edited in 1961 and 1965. It has had wide distribution (5000 copies of the third edition).

2. *Commission on Thermodynamics and Statistical Mechanics* Founded in 1947 under the name "Grandeurs et notations thermodynamiques," its purview was enlarged and its present name given in 1948.

3. *Commission on Cosmic Rays* Formed in 1947, it has sponsored conferences at two-year intervals ever since. This Commission recommended the terms *positron*, *negatron*, and *meson*; it also prepared a complete directory of specialists in cosmic rays (1949); studied projects concerning the creation or development of high-altitude research stations; and set up a subcommission for assembling and publishing the results of cosmic-ray monitoring in various parts of the world, as well as notes on standardization of equipment.

4. *Commission on Very Low Temperatures* This Commission cooperates with the International Institute on Cold; it worked toward an agreement for the use of the single temperature scale based on the vapor pressure of helium.

5. *Commission on Publications* The work of this Commission concerns placement of review articles into physics journals, rapid exchange of information between periodicals, problems of classification, use of uniform abbreviations, letters to the editor, critical reviews, projects of new specialized international journals, translation problems, and bibliographies of papers published in Slavic languages. A meeting of editors of physics journals took place in 1967.

6. *Acoustics Commission* There has been some question of transforming this Commission, which works in cooperation with 19 national acoustical societies, into an "affiliated commission."

7. *Solid State Commission* Created in 1951, this Commission initially concerned itself with matters relating to semiconductors and magnetism. These sections became subcommissions in 1957 and subsequently independent commissions.

8. *Semiconductors Commission*

9. *Magnetism Commission*

10. *Commission on Particles and Fields* (formerly High Energy Nuclear Physics) Created in 1959 as the Commission on High Energy Nuclear Physics, the Commission was given its present name in 1969.

11. *Nuclear Physics Commission* (formerly Low Energy Nuclear Physics) This Commission replaced in 1960 that concerned with radioactive units, whose secretary was Madame Joliot.

12. *Atomic Masses and Related Constants Commission* This Commission was created in 1960 under the name "Nuclidic Masses and Atomic Constants" to continue the work of a Committee formed in 1958 to prepare the adoption of a unified nuclidic mass scale. This scale, based on the choice of the number 12 as the relative nuclidic mass of the isotope of carbon containing 12 nucleons, was recommended by both IUPAC and IUPAP. It replaced both the old physics scale for nuclidic masses and the old chemical scale for atomic weights whose differences presented many inconveniences.

13. *Commission on Physics Education* The seriousness of the problems created by the great number of students (both young and adult), the evolution of curricula, the shortage of professors, and the training of physicists in mathematics all pointed to the need for this Commission, which was created in 1960.

14. *Atomic and Molecular Physics and Spectroscopy Commission* A mixed Commission on Spectroscopy had been created in 1948 by

the ICSU. In 1960, IUPAP created its own Commission in which IUPAC and IAU were represented. The present Commission dates from 1966.

15. *Plasma Physics Commission* This Commission was detached from the preceding one in 1969.

International Commission for Optics (CIO) The International Commission for Optics is the only "affiliated commission" of IUPAP. Under the Union's direction, it groups the optics committees of the following 17 countries: Australia, Austria, Belgium, Canada, Czechoslovakia, France, Federal Republic of Germany, Hungary, Italy, Japan, Netherlands, Poland, Spain, Sweden, Switzerland, United Kingdom, and United States of America. Following preparatory meetings in Paris (1946) and Prague (1947), the Commission was constituted in 1948. A history of its first 20 years of activity was published in 1968 (CIO Booklet No. 4).

The Commission instigated publication of the European journal, *Optica Acta* and worked to standardize certain symbols and conventions, particularly those relating to the design of apparatus. It has facilitated the exchange of scientists and extends its sponsorship to Summer Schools (in Paris in 1960 on the formation of images and in 1968 on the optical treatment of information; in Marseilles in 1970 on space optics).

As a member of the International Council of Scientific Unions, IUPAP has been regularly represented at the meetings of this Council. According to the agreement between ICSU and UNESCO, it has given this organization information and advice in its field of competence, for example, on the choice of experts and technicians, the organization of exchange of scientists and students, the choice of scientific films.

IUPAP welcomes the many members designated by other unions to several of its commissions (e.g., Solid State, Spectroscopy), and, in the same spirit, has cooperated in the work of the International Bureau of Physical Chemical Standards (attached to IUPAC), Commissions on Macromolecular Chemistry and on Polymers and of Crystal Growth (also of IUPAC), Radiobiology Commission of the International Union of Pure and Applied Biophysics, and the Upper Mantle Committee of the International Union of Geodesy and Geophysics.

IUPAP has also participated in the following "mixed commissions" created by ICSU: Physical Chemistry Data (1947–1953); Rheology (1947–1953); Radiometeorology (1947–1954); Ionosphere (1947–1958); High Altitude Stations (1948–1953); Applied Radioactivity (1948–1966); Electronic Microscopy (1951–1954); International Geophysical Committee (1959–1968).

IUPAP also has delegates on the following ICSU bodies (the date is that of the founding of the committee): Inter-Union Commission on Spectroscopy (ICSU, 1948); ICSU Abstracting Board (IAB, 1953); Scientific Committee on Oceanographic Research (SCOR, 1957); Committee on Space Research (COSPAR, 1958); ICSU Commission on the Teaching of Sciences (ISTC, 1960); Committee on Data for Science and Technology (CODATA, 1966); Inter-Union Commission on Solar-Terrestrial Physics (IUCSTP, 1966). Details on the constitution and on the activities of all these committees (old as well as new) have been published by ICSU. The minutes of the General Assemblies of IUPAP, the numerous bulletins and letters issued by the Secretariat, as well as the proceedings of the various conferences add up to an impressive and detailed account of the services rendered to physicists by the Union over the past 50 years. We find in all of it much promise for the future.

EDOARDO AMALDI *is Professor of General Physics at the University of Rome and holds many positions of importance in Italy. From 1957 to 1960 he was President of the International Union of Pure and Applied Physics. As an internationally known expert in nuclear physics, he has been honored by the Royal Academy of Sciences, Uppsala; the U.S.S.R. Academy of Sciences; the Royal Institution of Great Britain; the United States National Academy of Sciences; and the Royal Society of London.*

1 *The Unity of Physics*/EDOARDO AMALDI

1. THE ESTABLISHMENT OF WORLDWIDE INTERNATIONAL BODIES

The establishment of the International Union of Pure and Applied Physics that took place in Brussels in 1922[1] should be seen in the frame of the atmosphere prevailing among the scientists of all countries immediately after the end of the first World War.

Three years before, i.e., in 1919, the Constitutive Assembly of the International Research Council (ICR) had been held in Brussels, and the Unions of Astronomy, Geodesy and Geophysics, Pure and Applied Chemistry, and Scientific Radio had been established the same year.[2] The passions aroused by the war had not yet died down, and one had to wait until 1926 for a unanimous decision of the General Assembly of the International Research Council to delete from the statutes any restriction to the participation of all countries in the new international bodies.[3] Yet the establishment of these bodies in the period 1919–1922 was a consequence of the recognition, by larger and larger scientific and cultural circles, of the increasing necessity for an international collaboration, which could no longer be left completely to the initiative of single scientists, national bodies, or private foundations.[4]

The aims of these bodies—the Research Council and the Unions—as they appear from the corresponding statutes, are of purely organizational and practical nature and refer to aspects that, at first sight, seem to be marginal with respect to the great flow of science in its continuous development. But in reality the International Council and the Unions reflect, in their very structures, the stage of development reached in our century; they have been planned to provide a supporting frame to the development of sciences and to favor the general trend of research. In some way, they mirror certain fundamental conceptions of a general nature that in a more or less explicit form stay at the base of our very image of the world.

The existence of the International Council of Research—transformed in 1931 into the International Council of Scientific Unions (ICSU)—reflects the idea of the unity of science, which, in turn, is an aspect of the unity of human knowledge.

The existence of the different Unions reflects the conviction of a unity of the corresponding sciences from a practical point of view, which, however, has its roots in some kind of deeper unity that, although not mentioned explicitly, lies constantly behind our way of thinking.

The ideas of unity of human knowledge, of unity of science as a whole, and of the single sciences probably have their prehistoric roots in the monotheistic conception. The structure of the world reflects the rationality of God, so that from the study of Nature man is bound to discover the unity of the Creation. Essentially the same ideas persist in modern times, but with different justifications, which have also led to different specifications of the meaning of these various forms of unity.

II. THE UNITY OF HUMAN KNOWLEDGE AND THE UNITY OF SCIENCE

The expression "unity of human knowledge" is used to indicate at least two different things. It is used in connection with the origin and nature of knowledge, which, according to some currents of thought, is composed of layers that differ in their origin and nature.

For example, some people assert that, besides the knowledge deriving from the empirical data, there is a separate knowledge originating from the mystical–religious experience. The discussion of this problem would bring me too far away from my main subject, and, therefore, I will not deal with it, although my personal point of view is that the only origin of human knowledge can and should be looked for in the empirical data.

The other meaning refers to the unity of knowledge with respect to the cultural, social, and geographical differences between the various groups of human beings inhabiting the earth. It is usually considered to be a direct consequence of the unity of mankind.[5] The last expression means that the constant factors in our genetic heritage and our upbringing imply similarities in our aspirations and in the process of our lives. While the latter similarities may be somewhat masked by obvious differences imposed by life in diverse human environments, many features of our common humanity stand out above the variations.

The phrase "unity of mankind" also describes some aspects of the long-term evolution of societies, which takes place in two ways. With the improvement of transportation and communication the differences between the different societies that still exist tend to diminish so that within a foreseeable future they will have become so small that there will be in effect a single society inhabiting the world. Furthermore, a process of equalization has been going on within the existing societies. In most societies today an increased equality among their members is considered a theoretically desirable goal, and, by most standards—distribution of income, amount of education, typical daily activity, etc.—members of industrially developed societies have actually been approaching such equality. This does not mean that differences do not exist or will not continue to exist between people in a society, but rather that most options available to any one member of the group will become available to any other member, and that the decision between such options will be a matter of choice rather than of predetermination.

From this point of view, the striving for universal understanding as a means of elevating human culture can hardly be separated from the problem of the unity of human knowledge.[6]

Apart from these arguments of a general nature, one finds only very rarely attempts to substantiate the very reasonable idea of a unity of human knowledge, with some specific notion of general validity.

One of the very few examples—to my knowledge the only one—is found in some of the lectures of epistemological character of Niels Bohr,[6] where it is suggested that the notion of complementarity could find application not only in biology but also in psychology and perhaps even in arts and religion. Bohr noticed that in biology, for example, any experimental arrangement that would permit control of biological "functions" to the extent required for their well-defined description in physical terms would prohibit the free display of life. He also noticed that the notion of complementarity could be further extended to psychology, where, in a behavioristic approach, one has to deal with mutually exclusive applications of the words "instinct" and "reason," illustrated by the degree to which instinctive behavior is repressed in human societies.

I shall not try to recall other examples of application of the notion of complementarity suggested by Niels Bohr, because today the Bohr position appears overcome insofar as it concerns the life conceived as a primitive global phenomenon and the complementary between "living" and "analyzing" its atomic and molecular foundations.

Passing on to say a few words about the unity of sciences, I will recall that the problem was discussed by Kant in the *Streit der Facultäten*, where he examined the conflict between the different sciences—also from the practical point of view—and stressed the necessity of inserting them in a unitary superior scheme. In the *Metaphysische Anfangsgründe der Naturwissenschaft*, the unity of the scientific descriptions of nature is a consequence of Kant's philosophical conception deriving from the results of his transcendental analysis.

The necessity of a unification of human knowledge in general and of all sciences in particular was one of the most important theses of Comte and was the subject of many writings of Spencer. It was taken up, almost a century later, by the Movement for the Unification of Science with the publication of the *International Encyclopedia of Unified Science*, edited by R. Carnap, O. Newrath, and C. Morris.[7]

It may be pointed out that in this conception of a unified science no distinction was made between sciences, like physics and chemistry, based only on Galilean experiments and expressed by laws valid for any type of initial conditions, and sciences, like biology and cosmology, which are in part based on historical witness of unique and unrepeatable events.[8,9] We shall return later to this distinction.

The problem of the unification of sciences is considerably less general than that of the unification of human knowledge and is usually seen from a different point of view: scientists active in different fields and persons interested in sciences, conscious of the importance of a universal scientific attitude, try to promote what has been called a unified science.[10]

This term is usually quoted in connection with the unification of the results attained in the sciences. The problem of coordinating the scattered immense body of specialized findings into a systematic whole is a real one and cannot be neglected. It includes a comparison of the argumentation in cosmology, geology, physics, biology, behavioristics, history and social sciences in different ages, and, first of all, a unification of the scientific language. Some difficulties in sciences, even within a special discipline, arise from the fact that one cannot always decide whether two scientists (for example, psychologists, but also physicists) speak about the same or different problems or whether they state the same or different opinions by means of different scientific languages.[11]

There is also the question of unifying the efforts of all those who apply the scientific method to collective and social problems, so that these efforts may gain the force that comes from united effort. Very often the efforts to apply the scientific approach to certain problems are hampered—sometimes defeated—by obstruction due not merely to ignorance but to active opposition to the scientific attitude on the part of those influenced by prejudices, dogma, class interest, external authority, and nationalistic and social sentiments. From this point of view, the problem of the unity of science constitutes a fundamentally important social problem.

III. THE UNITY OF PHYSICS IN ITS NAIVE FORM

Coming to the central subject of my talk, the unity of physics, I will try to introduce a historical perspective by recalling two lectures given by Max Planck in Leiden. The first of these took place on December 9, 1908, at a time when Antoon Lorentz and Kamerlingh Onnes were still active, and its title was "Die Einheit des Physikalischen Weltbildes."[12] The second lecture, delivered 20 years later, on February 18, 1929, had the title "Zwanzig Jahre Arbeit am Physikalischen Weltbild."[13]

The first lecture begins with the remark that the science of nature had, from its beginning, as its last and highest goal that of succeeding in summarizing the extreme variety of the physical phenomena in a unitary system, possibly in a single formula. Thus the water of Thales, the energy of Wilhelm Oswald, and the principle of least curvature of Heinrich Hertz were considered in turn as the center and the essence of the physical image of the world, in which all physical processes should be framed and should find their explanation.

Planck argues that, in order to understand the direction of the development of physics, one should compare the present situation with that prevailing in a previous epoch. But the best indication of the stage reached by the development of a science is provided by the way in which its fundamental concepts are defined and its main parts delimited. The point made by Planck is that the subdivision of the matter of study and its definitions, when rigorous and appropriate, very often contain implicitly the latest and more mature results of the scientific investigation. He notices that the edifice of physics in 1908 was completely different from the primordial one, when the various parts of physics originated from immediate practical necessities and from particularly conspicuous phenomena.

In 1908 physics appears—always, according to Planck—to have a much more unitary character; the number of its chapters is considerably diminished because some of them have amalgamated. Thus acoustics has become a part of mechanics, magnetism and optics parts of electrodynamics.

This simplification is accompanied by an impressive disappearance of historical and human elements from all definitions. For example, in the study of electricity, nobody thinks anymore of the rubbed amber; and in the study of acoustics, optics, and heat, the physicist does not take into account the corresponding sense perceptions but refers to frequencies or wavelengths and to the absolute thermodynamic scale.

The general trend of the development of physics seems to be toward a unitary system, independent of anthropomorphic elements, in particular of sensorial judgments. The principal parts of physics in 1908 were reduced to two–mechanics and electrodynamics–or, as Planck says, the physics of matter and the physics of ether. Moreover, the limits between these two fields are not completely clear, since, for example, it is difficult to state whether the emission of light by atoms belongs to one or the other of these two broad chapters.

After these considerations, Planck moved on to discuss the general principles that will certainly play an essential role in the process of further unification. He concentrated his attention on the first and second principles of thermodynamics, dwelling upon the relation between probability and entropy and the conceptual difference between these two fundamental laws, the first of which forbids absolutely the *perpetuum mobile* of the first kind, while the second shows only that a *perpetuum mobile* of the second kind has exceedingly small probability.

The same general subject was again dealt with by Planck in his lecture of 1929: he noticed, of course, that in the meantime the situation had changed completely, mainly because of the advent of quantum mechanics. His attention was now concentrated mainly on the probabilistic description of the atomic phenomena and on the meaning of the expression "sensible world" and its distinction from the "real world" that–in Planck's opinion–should exist in itself independently of man. Planck recognized that the existence of a "real world" is certainly not imposed by logical considerations, proper to the intellect, but he claimed that it is imposed by the reason (*Vernunft*) that, together with the intellect, governs physics as any other science.

He noticed that with the passing of time the image of the "sensible world" becomes more and more abstract, and interpreted this tendency as being due to the progressive approach to the "real world" (which, however, remains unknowable, in principle). Then he argued that in this process of approach the image of the world should become increasingly free of all anthropomorphic elements. Therefore, any concept connected with the human technique of measurement cannot be accepted as part of the physical "image of the world." For this reason, Planck concluded, the probabilistic interpretation of quantum mechanics, and in particular the uncertainty principle, cannot be accepted as definitive parts of the edifice of physics.

I have devoted a rather large fraction of the space at my disposal to a review of the two Planck lectures, not in order to take a critical attitude with respect to some of his views but because of the lesson that we can learn from reading them. It is very instructive to consider, about half a century later, how the problem of the unity of physics was formulated and treated by the scientist who introduced in physics the quantum of action and thus opened the door to one of the most important upheavals of modern scientific thought. I will not enter into a discussion of the Copenhagen interpretation of quantum mechanics, which was clearly unacceptable to Planck. He insisted on the existence of an unknowable real world separated from the world of our sensory perceptions, which, as he recognized explicitly, is the only one knowable to man.

The same problem is still a matter of discussion today. But I will certainly not embark on a review or summary of the various attempts either to change the language used for interpreting quantum mechanics in order to give it a closer resemblance to classical physics or to introduce "hidden parameters," which escape observation but determine the outcome of the experiment in the causal way typical of classical physics.[14] Such a discussion would bring me too far away from my main subject. I will limit myself to stressing the following fact, which in my opinion is by far stronger than any argument of principle, even though its enunciation is a pure platitude. Today any physicist, or, more generally, any pure or applied scientist, uses in his daily work the formalism of quantum mechanics, discusses his problems and expresses his results in a language typical of the probabilistic interpretation of this theory, irrespective of his political or religious opinion and of the society in which he lives. This widespread consensus provides a kind of measure of the objectivity of the quantum-mechanical description of phenomena.

Of course, quantum mechanics also will most probably be superseded in a more or less distant future, but we do not yet have any idea of when, how, and in which direction this will take place. What seems today to be improbable is that such a step forward will provide a description of the observed phenomena closer to the classical conceptions than that suggested by quantum mechanics.

Going back to Planck's lectures, one should recognize that the fact that his arguments, and, here and there, even his language, appear so naive and old-fashioned to us, reminds us how ephemeral are all considerations of a general nature, especially when compared with new basic physical concepts and formal procedures such as the concept of quantum of action and the quantization of the harmonic oscillator.

IV. A FEW EXAMPLES OF HORIZONTAL AND VERTICAL PARTIAL UNIFICATION

Looking backward, one has the impression that the historical development of the physical description of the world consists of a succession of layers of knowledge of increasing generality and depth. Each layer has a well-defined field of validity beyond whose limits one has to pass to the successive one, characterized by more general and more encompassing laws and by discoveries constituting a deeper penetration into the structure of the universe than the layers recognized before.[15]

In this descriptive frame, a number of partial unification processes do take place, some of which have a vertical, others a horizontal, nature. The adjective "horizontal" is used to specify the unification of different chapters or parts of physics as in the cases mentioned by Planck in his 1908 lecture. The vertical unification refers to relationships between the descriptions of the same phenomena provided by theories belonging to layers of knowledge of different depth. In many cases, a horizontal unification carries with it, or derives from, a vertical one, and vice versa.

Processes of partial unification of both types have been going on at an extraordinary pace during the last 50 years. The most important steps in this direction are connected with the advent of quantum mechanics and its successive applications.

Among these, one should recall the study of the structure of matter in general, including an adequate description of vast categories of atomic and molecular, liquid, and solid-state phenomena. These developments, begun in 1926–1927, are still going on and have opened the door to a large number of applications, many of which are of paramount importance.

The same should be said about the application of quantum mechanics to the nuclear processes and the nuclear structure, where often the same basic concepts appear, as in the study of liquid and solid states.

Other important processes of unification took place during the 1950's and 1960's. Among these, one should mention a kind of vertical unification resulting from the recognition that the macroscopic parameters, describing the observed properties of matter, can be expressed in terms of space and time correlation functions. They are embedded in the quantum-statistical description of the system, without any recourse to specific models.[16]

While space correlation functions were used in 1927 by Zernike and Prins[17] and by Debye and Menke,[18] the first example of space and time correlation functions was given by Van Hove in the problem of the scattering of slow neutrons by matter.[19] Subsequently, the various macroscopic parameters, such as the dielectric constant, the magnetic permeability, and the electric and thermal conductivity, were all obtained by the same two-step procedure[20]: the first one consists of expressing these parameters in terms of two-particle time-correlation functions; the second of a Fourier transformation of these correlation functions and of taking its limit for infinite wavelength.

An interesting example of horizontal unification is provided by the recognition that the phenomena occurring very near the critical points show quite marked similarities. The molecular-field approach brings in the concept of an "order parameter" and suggests that there are close relationships between different phase-transition problems. A different theoretical approach, known as a "scaling law," predicts relationships between the critical indices used to describe singularities in the various correlation functions and thermodynamic derivatives.[21]

A well-known theory that has been created and developed as a unifying thinking frame, is the field theory, which attempts to describe all known types of interaction between subnuclear particles by analogy with the case of the electromagnetic field. In the latter case, the interaction between two charged particles is mediated by photons, and the use of the perturbative methods is fully justified by the smallness of the coupling constant and has allowed the computation of all observed purely electromagnetic phenomena with very high accuracy. But also in the case of the electromagnetic field, the theory is affected by a few constitutional faults, whose influence on the results to be compared with the experiments is eliminated by mathematical devices (such as renormalization), which cannot be considered as fully satisfactory without a proof of the convergence of the perturbative expansions.

As one proceeds from the theory of the electromagnetic field to the theory of other interactions, the situation becomes much worse. It is true that in the case, for example, of the strong interactions, the first steps in this direction were marked by a few fundamental discoveries such as the existence of mesons suggested by Yukawa. But the successive developments have met unsurpassable obstacles originating from the fact that the use of perturbative methods is no longer justified because of the very large value of the coupling constant. Thus, in this case, one no longer has at one's disposal a computational procedure capable of providing numerical results sufficiently accurate to allow a significant comparison with the experiments.

In the case of the weak interactions, the coupling constant is much smaller than that of the electromagnetic interaction, so that, at first sight, one would expect the application of the perturbative methods to be fully justified. But, since the weak interaction involves the product of the amplitudes of four particles, the coupling constant has the dimension of an energy to the power minus 2 (M^{-2}), and the theory is not renormalizable.

Furthermore, the W meson, which has been hypothesized as the mediator of the weak interactions (by analogy with mediators of the electromagnetic and strong interactions, i.e., the photon and the π- and K-mesons), has not yet been observed in spite of various experimental efforts aiming at its discovery. There are arguments suggesting that its mass should be so large ($\gtrsim$37 GeV)[22] that it cannot be produced by existing accelerators nor by those now under construction. It could be observed in cosmic rays as a secondary product of very-high-energy mu-mesons. But for the moment it remains a purely mathematical device for describing the weak interactions in the language of field theory.

In conclusion, we cannot today assert with certainty that field theory provides a satisfactory unification of the description of the different interactions. It is based on analogies and extensions that appear very reasonable but that derive after all from the presupposition of a uniformity of structure of the observed world, which, *a priori*, is not justified.

An interesting historical precedent of a similar uniformity of structure that was looked for unsuccessfully for many years is provided by the theory of gravitation of Einstein.[23] At its appearance in 1919, this theory conquered the scientific world by its elegance, simplicity, and amplitude of conception. Consequently, for about ten years, many of the greatest theoreticians of the time regarded this theory as the model to which the future developments of physics had to conform. Many physicists were convinced that a further extension of the two Einstein principles, the representation of physical reality as geometrical and the invariance with respect to general coordinates, would have led to an understanding of the chief phenomena that remained outside the original theory of Einstein, i.e., the electromagnetic field and the structure of matter. These attempts, however, were unsuccessful, since the theories of matter and electromagnetic field were developed along completely different lines with the advent of quantum mechanics in 1925.

The possibility that a similar situation might arise in the case of quantum electrodynamics should be kept in mind. After the first steps, made in the period 1928–1930, mainly by Heisenberg, Pauli, Dirac, and Fermi, and the great successes obtained in 1946–1948 by Tomonaga, Schwinger, Feynman, Dyson, and others, many physicists arrived at the conviction that quantum electrodynamics had to represent the model for the construction of the field theory, in particular for the case of the mesonic field. But today there are indications that some experimental observations of subnuclear particles may require theoretical concepts that fall outside the framework of field theory.

One should, however, recognize that the formalism developed for the field theory, in its nonrelativistic approximation, has found wide and important applications in nuclear physics and in solid- and liquid-state studies. The concepts of the quasi-particle, of the phonon, roton, or magnon represent stones of these constructions which are essential parts of what we call the observable world, or reality, no less so than the concepts of the electron, neutron, or neutrino.

Among the adaptations of the methods of quantum field theory to quantum many-body systems, one should recall the theory of nuclear matter and nuclei initiated by Bruckner and collaborators[24] and developed by Bethe, Goldstone, Hugenholtz, and others.[25] Although not yet satisfactory from the point of view of providing the observed values of binding energies and density distributions, this approach is of considerable methodological interest as being one of the few that aims at deriving the properties of many-nucleon systems from those observed for the two-nucleon system.

Also, the statistical mechanics of irreversibility and the development of the mathematical technique for treating dynamical problems, known as the Liouville representation of quantum mechanics,[16,26] originate from the reformulation, made by Prigogine and collaborators,[27] of results obtained by adapting the methods of quantum field theory to quantum many-body systems with irreversible behavior, such as solids.[28]

In the domain of elementary particles, there are other attempts at unification, again formulated in the language of field theory, that, although in their infancy, should be mentioned here.

The first one regards the unification of the strong interaction and gravitational field.[29] Its basic assumption is similar to the mixing of the photon with the ρ–ω–φ complex ($\Xi\ \rho^0$), which had been previously postulated[30] in an attempt to stress that hadronic electrodynamics can, to a good approximation, be separated from lepton electrodynamics, with the result that photons interact directly with leptons but only indirectly with hadrons via ρ^0–γ mixing.

By analogy, the attempt at unifying strong interactions and gravitation is based on the assumption that a mixing takes place of the gravitation—i.e., the quantum of the Einstein field—with some mixture of known, massive, strongly interacting, spin-2 particles. In such a theory, a gravitation would interact directly with leptons but only indirectly with hadronic matter.

The second of the above-mentioned attempts refers to the unification of the electromagnetic and weak interactions.[31] The postulated Lagrangian is invariant with respect to a non-Abelian group, a condition that may make it possible to explain the universality of the charge and, at the same time, achieve the renormalizability of the theory.

V. AN ATTEMPT AT AN OVERALL UNIFICATION OF PHYSICS

These, as well as many other examples that can be taken from different chapters of physics, as well as from various interdisciplinary subjects, obviously illustrate one of the most important mechanisms of development of the theory in the various fields of research. They do not, however, refer to the unification of physics on the largest scale, i.e., the organization of all our present knowledge of the observable world in a single deductive logical system.[32] The existence of such a logical structure was a basic assumption of the Laplace description of the world and was clearly accepted by Planck, at least in 1908.

The problem of such a unification has been re-examined in recent years by Weizsäcker,[33,34] who organizes our present knowledge of the physical world in five interlinked fundamental theories as follows:

1. A theory of space-time structure (special or perhaps general relativity);
2. General mechanics (quantum theory);
3. A theory of the possible species of objects (elementary-particle theory);
4. A theory of irreversibility (statistical thermodynamics); and
5. A theory of the totality of physical objects (cosmology).

Theories of special objects like nuclei, atoms, molecules, wavefields, and stars do not appear in Weizsäcker's list since they in principle can be derived from fundamental theories. Weizsäcker notices that we are today inclined to consider the theories 1, 2, and 4 as more or less final, while much work is being done in order to find 3 and perhaps 5.

He noticed that these five theories seem to arrange themselves like parts of a systematic unity of physics that is still seen only rather confusedly. The principle of this unity can be expressed, by saying: There are objects in space-time. Hence an account of space and time must be given (1). Being in space and time means for an object that it can move. Hence there is a set of general laws that govern the motion of all possible objects (2). All objects can be classified in more or less distinct species. Hence there must be a theory telling what species of object are possible (3). This theory describes objects as composites of more elementary objects. The composition can be described in detail, leading to the higher species (atoms, molecules, etc.). It can also be described in a statistical manner (4). All known objects somehow interact, or else we would not know about them. Hence some theory about all existing objects may be needed (5).

This preliminary account of a possible unity of physics shows, however, a number of shortcomings when the interlinkage of the theories and the problems connected with the concepts used in their description are analyzed more closely. Thus, for example, the interlinkage between theories 1 and 2 was discussed years ago by E. P. Wigner and H. Salecker.[35] From an analysis of their basic concepts, Wigner arrived at the conclusion that "there is hardly any common ground between these two theories." The concepts used in quantum mechanics, like measurement of positions, momenta, etc., do not appear to be significant if the postulates of the theory of general relativity are adopted. Among these there is the premise that coordinates are only auxiliary quantities, which can be given arbitrary values for every event.

Many other shortcomings are listed in the 1971 paper of Weizsäcker.[34] Just to give an idea of the nature of his considerations, I will recall his discussion of the interlinkage of theories 1 and 3, where the problem is faced that, according to general relativity, the space-time structure is described by gravitation, which on the other hand seems to be a field that one would like to deduce from elementary-particle theory.

Regarding theories 2 and 3, Weizsäcker notices that quantum theory is described as stating the general laws of motion of all possible objects and elementary-particle theory as hoping to describe all possible species of objects. It is not clear what this distinction means. Either these two theories will "in the end turn out to be coextensive and then, probably, identical, or objects will be thought of which would be possible according to general quantum theory but which are excluded by the additional information of elementary particle physics. The second alternative expressed the conventional view. But then the quantum theory of rejected objects turns out to be physically meaningless: should we reject it at all?"

After this very interesting analysis, Weizsäcker proceeds to examine what can be done to change the situation. He recognized that the search for the unification of physics is a program far transcending the work of any one individual or even of any one generation. He considered, however, that such a search can get support from what he calls philosophical guidelines whose detailing is his main task.

He noticed that certain basic concepts are common to the five theories listed before, such as object, space, interaction, time, and probability, and believes that their thorough analysis helps to prepare the tools for the construction of a unified physics.

Concerning "time and probability" Weizsäcker argues that all science is based on experience, and experience means that we learn from the past for the future: physical laws set up on the basis of past experience are used to predict future experiments, which are verified in the present. Thus "time" is a presupposition of "experience," and a new logic of temporal propositions must be developed.

The term "object" presupposes "time": an object is something that remains identical with itself in time, though its contingent properties may change. In Weizsäcker's view, the simplest and most general object is one characterized by a single twofold "alternative." The term "alternative" is introduced by him to indicate the possible outcomes of an experiment.

The concept of "object" is closely linked to the concept of "interaction," and interaction in turn is closely linked to "space."

Thus one of the philosophical guidelines of Weizsäcker for the construction of a unified physics is that the mathematical structures of "space" and "interaction" should be developed jointly. In present-day physics, the mathematical description of space, provided by the Lorentz group, is disconnected from interactions, and we do not have a general theory of "interactions" but only a promise of a beginning.

In Weizsäcker's scheme, the analysis of "time" and "probability" (and of the two related terms "reversibility" and "indeterminism") leads to a theory of the probabilities with which changes in the observable state of any object can be predicted—i.e., to quantum mechanics, although not necessarily in its present form.

Finally, in collaboration with M. Drieschner, Weizsäcker tried to set down the foundations of a new form of quantum theory, conceived to provide a possible core to unified physics. This construction, however, appears, at least to me, rather arbitrary, so I will not try to summarize it nor to present its main implications. I will only recall that Weizsäcker's conclusions unavoidably remind one of those reached about 40 years ago by Eddington and by Milne, both of whom formulated general theories constructed to embrace all physical phenomena.

Not only the conclusions but also the program set up at the beginning can be questioned.

VI. CLOSED AND OPEN THEORIES

The scheme for a unified physics sketched by Weizsäcker clearly has been conceived as an attempt to set down the foundations of a new "closed theory." This term was introduced by Heisenberg, who described the past progress of theoretical physics as a series of distinct closed theories (*abgeschlossene Theorien*). While the piling up of empirical data and of their explanation by existing well-established theories seems to take place smoothly, the basic theories advance in rare great steps or jumps.

These jumps are certainly historically prepared for, but in many cases with no accompanying feeling of growing clarity but rather with increasing awareness of unresolved problems. This historical phenomenon is most clearly seen in the years preceding the formulation of special relatively and quantum mechanics, which represent the latest examples of closed theories. These are generally characterized by an intrinsic simplicity, although we are not able to define what the word "simplicity" means in such a context. In any case, closed theories show a remarkable ability to answer those questions that can be clearly formulated within their own framework and to give their followers the feeling that questions that cannot be so formulated may be altogether meaningless. In the historical sequence of closed theories, each new one usually reduces its predecessors to some "limited" or "relative truth," assigning them the role of approximations or limiting cases. Thus we have learned to speak of the field of applicability of a theory, the limits of which are not known in the beginning and are clearly defined only by later theories.[34]

The role and scope of theories is, however, not seen by all physicists in the same perspective. Many are of the opinion that a theory that pretends to comprehend everything is doomed to break down on this point. A closed theory, in particular, is often unnecessarily rigid since it does not have the possibility of incorporating new discoveries or concepts. Of course, new discoveries may always upset some theory and wreck it completely, but, in the opinion of Herman Bondi,[36] for example, physicists should aim at shaping theories in such a form that new discoveries will not upset every theory, and for this purpose plenty of open theories should be at our disposal. He attacks what he calls the type of heresy very popular about 40 years ago, when Eddington and Milne put forward their overall theories. Also today, says Bondi, there are attempts to find "the world equation" capable of telling us everything. Among the many objections that can be raised against this tendency, the following can be recalled. An equation that says everything says nothing, because if the enormous variety of things that we see in this remarkably variegated world all spring from one equation, then the way from the equation to the observed things must be awfully long and difficult to deal with.

Bondi, of course, is aware of the fact that this remark could be interpreted as a criticism of all fundamental works. But such an interpretation would be wrong; what he tries to stress is that fundamental work is not only fully justified but also very important, provided it is maintained within reasonable limits, trying in particular not to eliminate entirely the openness of the theory, so that its capacity to be adapted to new discoveries can be preserved. His view is that a theory is scientific only if it can be disproved, but the moment one tries to cover absolutely everything, the chances are that nothing is covered.

A number of remarks and questions appears to be in order at this point. First, there is a certain confusion due to the fact that the expression "scientific construct" is currently used in the literature to denote two different things.

It is used in a restricted sense to indicate the idealized description of reality that becomes concrete through observations and experiments. The word "reality" refers to the ensemble of our (present and possible) observations, and the word "experiment" is intended in the Galilean sense as a reproduction of natural phenomena under conveniently selected, and anyhow artificial, conditions. Such a construct refers to a well-determined piece of the observed reality. It is a construct *a priori* unhistorical; it can be built (though not necessarily) as a unique logical deductive system. It helps us in constructing even more general models.

The second meaning of "scientific construct" refers to a general image of the world; this is typically historical, since it involves the evolution of the universe (and of life), which are unique and unrepeatable. This construct should be unique and should have an ontological significance, but it is highly metaphysical and arbitrary.

A complete unification of science in general, and of physics in particular, would involve the organization of theoretical constructs of the first restricted type within a unique scientific general construction of the second kind. Such a unification certainly was impossible in the frame of the classical point of view because of the absence of any element of freedom. A way out of such a difficulty may be provided by quantum mechanics, which preserves the most powerful methods of classical physics, such as the use of differential equations, and, at the same time, liberates the single events from the determinism of classical type.

It should also be pointed out that such a unification, important for the physicist, is necessary for the biologist, since the various forms of life now present can hardly be conceived independently of, and separate from, their evolution through the past.[8,9,37]

In conclusion, the unity of physics, intended as the construction of a unique deductive logical system providing a satisfactory description of all observations and experiments, has too many facets to permit a simple, clear answer. Certainly, present theories do not constitute such a system, since they clearly have many points of mismatch or discordance.

A few scientists and philosophers tackle the problem of constructing a deductive logical system of this type, convinced that the problem should have a solution as a consequence of what may be briefly called the "unity of nature."[33] But the meaning of such an expression is not clear if it refers to something other than the totality of our possible observations.

These remarks should not be taken as criticism of those who have made or are making these attempts, which certainly are very interesting and in any case useful for clarifying and widening some deep aspects of the scientific construction. They help us to underline a few problems to which we are not yet able to give an answer.

If one accepts, however, the schematic distinction between closed and open theories, it is rather natural to ask oneself to which of these two categories would belong the unitary physical theory, envisaged by some authors today.

Should it be a closed theory, then one can consider two alternatives. The theory is, so to speak, a final theory that represents the final stage of our physical knowledge beyond which there is no further possible development. Or it represents one further layer of the physical knowledge that will be superseded, in a more or less distant future, by the construction of a deeper one. The first alternative seems very unlikely on the grounds of our past experience, while the second would imply the construction—not impossible but certainly not easily conceivable—of an overall deductive logical system that can be extended, without changing its basic postulates, to layers of the observable reality that originally were foreign to the theory.[38]

The last remark would obviously hold—only with minor changes—if the unique deductive logical system were an open theory, i.e., a theory capable of incorporating new discoveries.

One should, however, recognize that the very distinction between closed and open theories seems to be not very clear. If one examines all past and present theories from this point of view, only a few of them appear to conform to one or the other of these two extreme conceptions. Thus, for example, classical electrodynamics—summarized by the Maxwell equations—provides the best example of a closed theory.[39] On the contrary, the two Einstein principles mentioned above—the geometrization of the physical reality and the invariance with respect to general coordinates—constitute the basis for an open theory of gravitation.

In many other cases, however, it is not clear to which of these two categories a theory should be assigned, since, by adding convenient terms or introducing other modifications in the corresponding equations, at a later time it became possible to incorporate new sets of phenomena in frames of thinking that originally would have been considered clear examples of closed theories.[40]

VII. OTHER FORMS OF UNITY

There are, however, points of view different from an overall logical axiomatization from which the unity of physics should be considered.

These points of view are less important from a philosophical point of view but, perhaps, even more relevant for the development of physics. They are rather obvious but may still be usefully summarized under the present circumstances.

The development of physics proceeds along many different lines, which neither singly nor as a partial or global ensemble belong to a prearranged design. They are the result of the intelligent initiative of a large number of individuals whose logical strength and imagination may vary enormously but are anyhow determined by a few hereditary qualities and a great number of environmental factors.

This lack of general or partial plans coordinating the efforts of an appreciable fraction of the world physicists active in each specific field is of paramount importance: I would dare to say that it is even essential for the progress of our knowledge of the physical world. Plans exist and should exist for the development of the applications of scientific knowledge, to meet the needs of society and for its benefit; but wide plans directing the search for a deeper understanding of the physical reality would unavoidably orient the efforts and thereby limit the freedom of research.

The first of the other forms of unity that I mentioned before is the community of goals of experimental and theoretical physics. This unity is essential to the very construction of physics, but, unfortunately, sometimes one has the impression that the theory becomes a game closed in itself having nothing, or at least very little, to do with the observed (and observable) reality.

Furthermore, the community of experimental techniques and methods together with the community of the mathematical tools determines the unity of the language used by all physicists. Between the two, however, there is a great difference: The unification due to mathematics is, in general, only formal and often mainly due to the limitations of our imagination and ability in the use of complicated mathematical tools. The community of experimental techniques, on the contrary, has deep roots in the observed reality.

Both of these communities, however, open up the possibility of transferring ideas and procedures from one field of research to another, and we know from our past experience how often these transfers have yielded substantial progress. From this point of view, the increasing specialization taking place in our days constitutes a regrettable danger for the ready development of many parts of physics. Every effort should be made to reduce such an inconvenience by organizing conferences and publishing books and journals devoted to summarizing, in the proper form, the most recent developments and ideas taking place in the various fields of research.

Other elements of unity are found in the critical examination of concepts and in the return to the origin of basic physics. Critical analysis usually starts in a particular field, but, at the end, it influences all, or almost all, other branches of physics.

The interdisciplinary fields of research are based on knowledge from different parts of physics or even from different sciences and constitute at the same time multiple bridges among the parent disciplines. The unity typical of any Union is the unity with respect to any geographical and any political division. From time to time, the ideologists of some specific school of thought or creed assert that societies based on different principles produce different sciences.

Now it is true that the surrounding society with its many characteristic aspects and in particular its general culture has an influence on the way of thinking of its scientists and on their work's programs. An interesting example was pointed out, some years ago, by noting that the remarkable contributions to the field theory made by the Japanese school of theoretical physics may have been favored by the fact that their culture had never been under the influence of Aristotle's way of thinking. But once an idea or procedure has been put forward and its usefulness for improving our knowledge of the physical world has been proved, it is immediately accepted by the scientists, in particular by the physicists, belonging to any other cultural group or society. They immediately develop the new idea or apply the new procedure so that it is either universally rejected or adopted and amalgamated into the present description of the physical world as a more or less important part of it, irrespective of its place of origin.

Differences of opinion may remain, and do sometimes remain, in the philosophical interpretation, but "physics" as a final product—consisting of a certain number of definitions and mathematical relationships—is universal.

The Union of Pure and Applied Physics is one of the many places where the exchanges necessary for this type of unification are favored and promoted. It is even the only one or at least the first one from both the historical and the organizational point of view where the unity with respect to the different creeds is recognized and accepted as a sociological principle, even by the authorities of the adherent countries.[41]

REFERENCES AND NOTES

1. P. Fleury (Secretary General of IUPAP from 1947 to 1963), "L'Union International de Physique Pure et Appliquée," *ICSU Rev. World Sci. 6* (1), 28 (1964).
2. H. Spencer Jones (Secretary General of ICSU from 1956 to 1958), "The Early History of ICSU—1919–1946," *ICSU Rev. 2* (4), 169 (1960).

3. Because the feeling against the Central Powers was very bitter after the first World War, these countries were excluded from IRC and the Unions by a resolution passed at a conference of representatives of the principal national academies of the allied powers, held in London in October 1918, for establishing these new international bodies. By the time of the Third General Assembly of IRC (Brussels, July 7–9, 1925), there was a growing feeling on various sides that these bodies could not continue to be considered truly international as long as certain countries were excluded from participating in their activities. But only on the occasion of the General Assembly held on June 29, 1926, was a unanimous decision reached to delete from the statutes all references to the resolution adopted at the Conference in London in 1918 and to invite Germany, Austria, Hungary, and Bulgaria to join the Council and the Scientific Unions attached to it.
4. The first scientific conference is probably the one organized in 1815 in Geneva by the chemist H. A. Grosse on the Physical and Natural Sciences. A few other scientific conferences took place during the same century. Particularly important, also in retrospect, appears to have been the international conference on physics held in Paris in 1900, under the auspices of the Société Française de Physique. The proceedings were collected by Ch.-Ed. Guillaume and L. Poincaré, Secretary Generals of the Conference, and published by Gauthier-Villars, Paris, 1900. The first international exclusive conference on physics (only 20 official participants were foreseen) was held in Brussels in 1911 on the invitation of Ernest Solvay at the suggestion of Walter Nernst. Among the participants were H. A. Lorentz, P. Langevin, J. Perrin, Maurice and Louis de Broglie, M. Planck, A. Einstein, N. Bohr, W. Wien, E. Rutherford, Pierre and Marie Curie, and H. Kamerlingh Onnes.
5. These remarks on the unity of mankind and human knowledge lean heavily on a few pages by Gerald Feinberg, *The Prometheus Project* (Doubleday, Garden City, N.Y., 1968).

6. N. Bohr, *Atomic Physics and Human Knowledge* (Wiley, New York, 1958).
7. *International Encyclopedia of Unified Science*, O. Neurath, R. Carnap, and C. Morris, eds. (Univ. of Chicago Press, Chicago, Ill., 1955).
8. M. Ageno, *L'Origine Della Vita Sulla Terra* (Zanichelli, Bologna, 1971).
9. M. Eigen, *Naturwissenschaften 58*, 465 (1971).
10. O. Neurath, "Unified Science of Encyclopedic Integration," in *International Encyclopedia of Unified Science*, O. Neurath, R. Carnap, and C. Morris, eds. (Univ. of Chicago Press, Chicago, Ill., 1955), Vol. 1, p. 1.
11. J. Dewey, "Unity of Science as a Social Problem," in *International Encyclopedia of Unified Science*, O. Neurath, R. Carnap, and C. Morris, eds. (Univ. of Chicago Press, Chicago, Ill., 1955), Vol. 1, p. 29.
12. M. Planck, *Physikalische Abhandlungen und Vorträge* (Vieweg, Braunschweig, 1958), Vol. III, p. 6.
13. M. Planck, *Physikalische Abhandlungen und Vorträge* (Vieweg, Braunschweig, 1958), Vol. III, p. 179.
14. See, for example, W. Heisenberg, *Physics and Philosophy* (George Allen and Unwin, London, 1958); B. D'Espagat, *Conceptions de la Physique Contemporaine* (Herman, Paris, 1965); and a few articles by J. S. Bell (p. 447) and D. Bohm and J. Bub (pp. 453 and 470) that appeared in *Rev. Mod. Phys. 38* (1966).
15. See, for example, E. P. Wigner, *Commun. Pure Appl. Math. 13*, 1 (1960).

16. See, for example, U. Fano, "Liouville Representation of Quantum Mechanics with Application to Relaxation Processes," in *Lectures on the Many-Body Problem*, E. Caianiello, ed. (Academic Press, New York, 1964), Vol. 2, p. 217.
17. F. Zernike and J. Prins, *Z. Phys. 41*, 184 (1927).
18. P. Debye and H. Menke, *Ergeb. Tech. Röntgenk. 2*, 1 (1931).
19. L. Van Hove, *Phys. Rev. 25*, 249 (1954).
20. See, for example, R. Kubo, "Some Aspects of the Statistical Mechanical Theory of Irreversible Processes," in *Lectures in Theoretical Physics* (delivered at the Summer Institute for Theoretical Physics, University of Colorado, Boulder, 1958), W. E. Brittin and L. G. Dunham, eds. (Interscience, New York, 1959), Vol. 1, p. 120.
21. L. P. Kadanoff, W. Götze, D. Hamblen, R. Hecht, E. A. S. Lewis, V. V. Palciauskas, M. Rayl, J. Swift, D. Aspenes, and J. Kane, *Rev. Mod. Phys. 39*, 395 (1967).
22. J. Schechter and Y. Ueda, *Phys. Rev. D2*, 736 (1970); T. D. Lee, *Phys. Rev. D3*, 801 (1971).
23. F. J. Dyson, *Phys. Today 18* (6), 21 (1965).
24. K. A. Bruckner, C. A. Levinson, and H. M. Mahmond, *Phys. Rev. 95*, 217 (1954); K. A. Bruckner and C. A. Levinson, *Phys. Rev. 97*, 1344 (1955).
25. H. A. Bethe and J. Goldstone, *Proc. Roy. Soc. (London) A238*, 551 (1957); J. Goldstone, *Proc. Roy. Soc. (London) A239,* 267 (1957); N. M. Hugenholtz, *Physica 23*, 481 (1957); H. A. Bethe, B. H. Brandow, and A. G. Petschek, *Phys. Rev. 129*, 225 (1963). The theory is summarized in a series of articles by B. D. Day (p. 719), R. Rajanaman and H. A. Bethe (p. 745), and B. Brandow (p. 771), *Rev. Mod. Phys. 39* (1967).
26. See, for example, R. W. Zwanzig, "Statistical Mechanics of Irreversibility," in *Lectures in Theoretical Physics* (delivered at the Summer Institute for Theoretical Physics, University of Colorado, Boulder, 1960), W. E. Brittin, B. W. Downs, and J. Downs, eds. (Interscience, New York, 1961), Vol. II, p. 106.
27. R. Brout and I. Prigogine, *Physics 22*, 221 (1956); R. Balescu and I. Prigogine, *Physics 25*, 281, 302 (1959).
28. L. Van Hove, *Physics 21*, 517 (1955); *23*, 441 (1957).
29. C. J. Isham, A. Salam, and J. Strathdee, *Phys. Rev. D3*, 867 (1971).
30. T. D. Lee, N. M. Kroll, and B. Zumino, *Phys. Rev. 157*, 1376 (1967).
31. The idea of combining electromagnetism and weak interactions is very old. See, e.g., J. Schwinger, *Ann. Phys. (N.Y.) 2*, 407 (1957); S. L. Glashow, *Nucl. Phys. 10,* 107 (1959); *22*, 529 (1961); A. Salam and J. Ward, *Nuovo Cimento 11*, 568 (1959); *Phys. Rev. Lett. 13*, 168 (1964). The first successful attempt in this direction is due to S. Weinberg, *Phys. Rev. Lett. 19*, 1264 (1967); *27*, 1688 (1971), who has been followed by G.'t Hooft, *Phys. Lett. B33,* 173 (1971); *B35*, 167 (1971); A. Salam and J. Strathdee, preprint IC/71/145 of the International Center for Theoretical Physics (ICPT), Trieste. Similar theories have been proposed also by other authors.
32. To talk about organizing a chapter of physics (or even all present knowledge of the observable world) into a single and unique deductive logical system is not at all at variance with the assertion that the construction of physics is based, at least in part, on inductive processes. These certainly play a very important role in the initial phase of each chapter of physics, dominated by experiments aiming to establish fundamental laws. This phase is followed (logically but not neces-

sarily temporally) by a second phase, in which the mathematical deduction acquires an essential role by allowing the derivation, from the already recognized fundamental laws, of a number of consequences that are tested by means of *ad hoc* experiments. This second phase is often arrived at, giving rise to a third phase, the axiomatic phase, in which the principles suggested by the experiments are re-examined critically, generalized as much as possible, and reduced to the indispensable minimum. From these the whole chapter of physics is derived as a deductive logical system and as such is recorded in consultation treatises. In this form, a chapter of physics usually appears very different from the same chapter in its initial heuristic phase. In practice, the three phases sketched above do not follow one to the other in temporal order. The progress usually takes place through a number of steps, belonging to one or the other of the three phases mentioned above, but which are mixed in a complicated order, determined by frequent returns to previous stages of development.

33. C. F. von Weizsäcker, *La Physique: une unité a construire* (Publication de l'Université de Lausanne, XXIII, 1965).
34. C. F. von Weizsäcker, "The Unity of Physics," in *Quantum Theory and Beyond,* T. Bastin, ed. (Cambridge U.P., New York, 1971). See also the review of Weizsäcker's work published by F. J. Zucker in *Boston Studies Philos. Sci. 5*, 474 (1969).
35. E. P. Wigner, "Relativity and Quantum Theory," *Rev. Mod. Phys. 29*, 255 (1957).
36. H. Bondi, *Assumption and Myth in Physical Theory* (Cambridge U.P., New York, 1967).
37. This point has been discussed by M. Ageno in "Punti di contatto fra fisica e biologia," in *Problemi di Attualità di Scienza e di Cultura*, Quaderno n. 174 (Accademia Nazionale dei Lincei, 1972), who, among others, gives the following example: The theory of biological phenomena is not only based on observations and experiments in the Galilean sense but also on unique and unrepeatable events such as the extinction of dinosaurs or of any other species that has disappeared in the course of evolution.
38. In this connection, L. Van Hove pointed out to me that mathematics itself can be no longer regarded as a closed theory. Indeed, as shown by Gödel's undecidability theorems, there are always propositions that can be rightfully formulated *but* can be *neither* proved *nor* disproved on the basis of the axioms. On such propositions, progress means extension of the system of axioms, which is openness of the theory. This is, for Van Hove, a strong reason to believe that a full mathematization of physics in the sense of one closed theory is unlikely to be achieved and unreasonable to ask for.
39. As other examples of closed theories one can mention classical mechanics, quantum mechanics, and special relativity. The basic equations of all these theories require the introduction from the outside of the masses and the interactions that are necessary for specifying the system to which the theory is applied. These data orginate either directly from experiments on the system under consideration or from another theory that very often belongs to a deeper layer of knowledge.
40. Two examples may be mentioned here. The first one is Dirac's theory of fermions, which has the typical features of a closed theory. But the addition of the Pauli term for explaining the anomalous magnetic moment of the nucleon has shown that in reality it had a certain openness. A second example

is provided by Fermi's theory of beta decay, which also, in its original form, appeared as a closed theory. When, however, shortly after Fermi's original paper, Gamow and Teller proposed a different expression for the weak interaction, the theory acquired some kind of openness, the limit of which was clarified a few years later by the recognition of the existence of only five Lorentz invariant interactions, two of which were those proposed by the authors mentioned above. This kind of openness has been fully exploited after the discovery of the nonconservation of parity by making a number of experiments that have allowed the selection, among the various possibilities, of the (V–A) interaction.

41. I wish to express my thanks to a few friends and colleagues for reading and commenting on my manuscript: M. Ageno, N. Cabibbo, U. Fano, V. Somenzi, and L. Van Hove.

J. TUZO WILSON *is a professor of geophysics at the University of Toronto, Canada, and since 1967 he has been Principal of Erindale College of that university. A Fellow of the Royal Society of London and the Royal Society of Canada, he is a Foreign Associate of the U.S. National Academy of Sciences and past president of the International Union of Geodesy and Geophysics. He has been honored with many medals and decorations both scientific and civil.*

2 *The Physical Study of the Earth and the Scientific Revolution It Has Caused* / J. TUZO WILSON

I. INTRODUCTION

A career in science is a chancy one. Some who are clever and work hard achieve only minor, plodding success. Others, no brighter and perhaps less assiduous, are remembered for major discoveries. Much depends upon when one lives, for the progress of science is uneven.

Some bursts of activity are more important than others, and Kuhn[1] has distinguished a few of the greatest as scientific revolutions. He considers that these have been more than just periods of rapid progress, but rather quantum jumps in scientific thought when one research tradition has been abandoned in favor of a new one. The names of Copernicus, Lavoisier, Darwin, and Einstein recall such episodes because their ideas changed the way in which astronomers, chemists, biologists, and physicists, respectively, viewed their subjects. Each revolution has been followed by a great flowering of ideas; and, indeed, it was Kepler, Galileo, and Newton, the men who followed Copernicus, who set the stage upon which the whole of modern science has been acted out.

Other great advances, for example the elucidation of the structure of DNA, did not alter the philosophical framework or paradigm of any branch of science and hence did not constitute scientific revolutions.

This concept of Kuhn's is widely accepted, although there is debate about which events were true revolutions. In this paper I have adopted the view that the acceptance of the hypothesis of continental drift has been a true scientific revolution, for it has involved a change in view from regarding the continents as fixed and the ocean basins as permanent to thinking of the continents as mobile and ocean basins as ephemeral. It is also interesting to speculate whether a true scientific revolution is not a change from a mere codification of the observations of practical men to a true understanding of nature. If so, is it true that there can only be one revolution in each branch of science? Can anything ever again have so great an effect on planetary astronomy as the change from Ptolemaic to Copernican views? Can any change in philosophy about biology ever again match the importance of the recognition of evolution? Can any period in the progress of physics ever match the first quarter of this century, or are deeper insights and fresh revolutions to be expected? If these ideas are even partly true, this decade may be the most exciting in the history of earth science. Certainly after a century of slow and uneventful progress all geologists and geophysicists, except a few of the most conservative, have recently accepted the new philosophy and embarked upon an active period of revising their ideas.

II. THE INTRACTABLE EARTH

The development of geology began early, along with other aspects of natural history, and started with the classification and mapping of ores, minerals, rocks, and fossils. Geologists were great travelers, and they led some of the first explorations of North America and Siberia. Their findings played an important part in Darwin's recognition of evolution. A century ago, geologists were leaders of the scientific community. Certainly Sir William Logan and Sir William Dawson were the most eminent scientists in Canada. Prominent among the founders of the National Academy of Sciences were geologists, including J. W. Powell, Benjamin Silliman, and W. B. Rogers, who also founded Massachusetts Institute of Technology. Sir Roderick Murchison, Leopold von Buch, and M. A. D'Orbigny were important figures in Europe. Nevertheless, other great discoveries soon aroused fresh interest in organic chemistry, biology, and physics. Geology became a backwater.

The difficulty was that the earth is hard to study, and until the discovery of radioactivity, the development of seismology, and the advent of high-pressure geochemistry no one could understand the nature of the earth's interior. Geologists were like doctors of old who could only examine the external appearance of their patient. Even mapping every wrinkle on land told little of the earth's interior and nothing of that two thirds of the surface covered by oceans. Classical physics dealt at length with geomagnetism, gravity, and tidal observations, but before the development of electronics, measurements were so few that the results were of only minor use. For several reasons, the earth is hard to study. Most natural objects and phenomena studied by scientists are small enough to be treated in a laboratory or active enough for their behavior to be observed. Only the earth, planets, and stars are vastly large and ponderously sluggish. Most objects including stars are sufficiently numerous to be classified, but to us the earth is unique. Not only can we not see the earth's interior, but the indirect observations possible give only a crude representation of its nature. The earth is complex. On all counts it is intractable. Because of these problems, the study of the earth is difficult, and it has been the last science to develop.

III. EARLY STIRRINGS

From earliest times, the earth was axiomatically regarded as a permanent *terra firma.* Before 1800, only a few had vaguely questioned this tenet, among them Leonardo da Vinci and James Hutton, who found in uplifted marine fossils evidence for vertical motions. The first to suggest clearly that continents had moved great distances seems to have been A. Snider in 1858.[2] In 1889, Osmond Fisher without mentioning Snider's ideas emphasized that the heat and probable plasticity of the earth's interior had probably caused some mobility of the surface. Fisher might well be regarded as the first geophysicist. Although he was active before that word was coined, he combined to an unusual degree a knowledge of classical physics, mathematics, and geology. It should be noted that such men have contributed most of the new ideas in earth sciences. Some have been physicists who learned geology, others geologists with an appreciation of quantities and dimensions. Orthodox physicists and members of the geological establishment, including those in the mining and petroleum industries and geological surveys, have for the most part clung as long as possible to old-fashioned ideas about the earth.

Fisher's book *Physics of the Earth's Crust* shows a surprisingly modern point of view, and in it he strenuously attacked his fellow physicists.[3] He disagreed that the theory of contraction ascribed to Newton could be the cause of mountain building. He argued against the opinions of William Thomson (later Lord Kelvin) that the earth must be both rigid and short-lived. He drew attention to W. Hopkins' early work of about 1840 on convection currents in the earth and concluded that they would not have ceased to flow as soon as the crust began to solidify. He did, however, accept G. H. Darwin's origin of the moon from the Pacific Ocean.

Apart from the last, his ideas were clear and modern. Instead of describing the earth as static he emphasized the evidence for mobility—for horizontal movement in the extensive folding of mountains and for vertical movement in the uplift of marine strata, in the rise of sites of former ice sheets, and in the sinking of coral islands (as noted by Charles Darwin). He suggested that a liquid or at least a plastic substratum lies below the crust to allow for changes in elevation and for "a certain amount of lateral shift towards mountain ranges." He thought that convection currents flow in the substratum.

He concluded that, "All these phenomena are perfectly well known to geologists, but appear to have been unaccountably ignored by the distinguished physicists who have discussed the condition of the interior of the earth. They cannot be explained by the supposition that the earth is solid." One cannot fault Kelvin for not knowing of the existence of radioactivity, but it does seem surprising that he did not realize that the great heat of the earth's interior might reduce its rigidity and that he took no account of the earth's great size, which by the theory of dimensions must render it weak.[4]

Fisher held that in the plastic shell beneath the crust "ascending currents are situated beneath the oceanic areas and descending currents beneath the land . . . to produce compression along their common boundary . . . to compress the crust locally at some distance from mid-ocean . . . and that the American theory, that mountain ranges have been formed out of thick deposits might thus receive an explanation . . . . Where the currents ascend beneath the ocean they would give rise to a tensile stress, the correlative of the compression of the land. Fissures would thus be produced, which would open volcanic vents, and, when filled with solidified lava, become dykes of igneous rock in the suboceanic crust . . . . We recognize two principal types of volcanic regions, coastline and oceanic. We believe the former to be connected with the agencies which have raised the continents which they skirt . . . . The oceanic volcanoes on the other hand appear unconnected with compressive action, for the oceanic islands consist almost all of them of volcanic rocks . . . . They occupy a medial position with respect to coast lines, being in the Atlantic widely parallel to opposite shores."[3]

Fisher thus presented a remarkable forecast of many of the ideas that were later to make the hypothesis of continental drift acceptable. It was his misfortune to be a prisoner of his time—a time before the existence of modern seismology, paleomagnetism or marine geology, and geophysics, which were to make a more ingenious theory of drift acceptable.

A curious feature of the study of the earth before that acceptance was the division of its students into three groups, largely isolated from one another. One group, called geologists, investigated such diverse subjects as paleontology, mineralogy, and structural geology, which have little in common except that they all deal with aspects of the earth's surface and require little knowledge of mathematics and physics. Another group, called geophysicists, studied geomagnetism, seismology, and geodesy, subjects that have little except a knowledge of basic physics in common with one another and had even less with geology until modern instruments began to produce abundant data. Geochemists formed a third group.

I should like to suggest that it was the prevailing paradigm of an immobile earth that made any coherent and unified study of the earth impossible and that caused this fragmentation. To take a simple analogy, one can examine whirlpools from many points of view, but it would be impossible to unite the studies or to understand the nature of whirlpools if one refused to admit that the water in whirlpools is moving.

In only one field did the geologists and geophysicists meet. It was in the search for economic deposits of petroleum and ores. There geophysicists successfully invaded a domain previously solely occupied by geologists and drillers.

During World War I, physicists who served as artillery sound rangers gained a knowledge of ground waves, which they applied to develop seismic exploration for petroleum. Later they developed other methods of prospecting. In many universities these men or their students were the first to teach geophysics as an independent discipline, but as long as the false paradigm prevailed any integration of the earth sciences remained impossible. Today this union of the earth sciences is under way, although it may take a new generation to complete the task.

IV. THE FIRST, ABORTIVE ATTEMPT AT REVOLUTION (1910–1930)

Until the last quarter of the nineteenth century, the chief evidence favoring drift was the similarity in the shapes of the opposite shores of the Atlantic, but charts were poor and proposals were not taken seriously until Alpine geologists, notably M. Bertrand, H. Schardt, M. Lugeon, and E. Argand, showed that the structure of the Alps requires a horizontal shortening of hundreds of kilometers across the Alpine–Himalayan mountains. Taylor[5] seems to have been the first to put these two ideas together by proposing that if continents migrated they would leave young oceans behind them and pile up young mountains before them. He supposed "that the mid-Atlantic ridge has remained unmoved, while the two continents on opposite sides of it have crept away in nearly parallel and opposite directions" and also that a change in the oblateness of the earth's figure had produced forces whereby a surface "gently sloping towards lower latitudes and situated beneath the earth's crust just within the zone of rock flowage, would seem to afford a basal slope down which the crustal sheets might move." He did not pursue the subject but was quickly followed by H. D. Baker and Alfred Wegener.

Wegener[6] was the first really persistent champion of continental drift, and by the time of his early death in 1930 he had published much evidence in support of the hypothesis, but he had attracted few disciples. Nearly all geologists[7] and geophysicists[8] had completely rejected his views. Partly, no doubt, this was because of their novelty, but it was also ironical that he, a physicist, had got his physics wrong. He wanted to move continents by the *polfluctkraft*, a real but quite inadequate force acting in the wrong direction. He used imprecise geodetic data that were shown to be unreliable. He held that each continent moved as a raft although there was no evidence of disturbances along many coasts.

V. THE YEARS OF STAGNATION (1930–1955)

In 1930, when Wegener died on the Greenland ice cap, his views had been generally rejected. At about that time I was a student in three universities sucessively in Canada, England, and United States, and at none of the geology departments was drift given the slightest credence.

Indeed Wegener's only important followers were some of the Alpine geologists and a school in the southern hemisphere led by the brilliant South African, du Toit,[9] of whom Daly wrote that he was "the greatest field geologist who ever lived." The argument for the fit of the southern continents was particularly strong, and du Toit went to India, Australia, and South America and published comparisons showing their similarity with Africa.

Two other great geologists, Daly[10] and Holmes,[11] certainly liked the idea of continental drift, but they did not fully support it, I suspect overawed by the dicta of physicists like Jeffreys who dismissed drift as impossible: "It is quantitatively insufficient and qualitatively inapplicable. It is an explanation which explains nothing which we wish to explain."[8]

That was the accepted attitude. That continental drift was ignored in North America and in most other places until after 1954 is plain from the forty-four review papers presented in that year by distinguished geologists and geophysicists at the Bicentennial of Columbia University.[12]

Within the space of a few years, a series of great discoveries completely altered the prevailing opinion. One had already been made but had passed unnoticed. This was the realization that some large faults have offset rocks on their two sides horizontally through scores and even hundreds of kilometers. The first to be recognized were the Great Glen fault in Scotland,[13] the Alpine fault in New Zealand,[14] and the San Andreas fault in California.[15] Wilson[16] lists many more.

It was also found that very large submarine faults called fracture zones offset the crests of midocean ridges by hundreds of kilometers but stop in a puzzling way at continental margins.[17–19]

In 1954, groups from London and Cambridge made the first discovery that produced many converts. This was paleomagnetic evidence that Europe and North America had drifted apart.[20,21]

A third key was the recognition in 1956 by Ewing and Heezen[22] that the seismically active submarine ridges could be connected together to form a continuous system about the earth,[23] which lay along the median axes of all oceans except the Pacific and formed the largest mountain range on earth (Fig. 1).

In 1957, other important developments were the realization that the world's seismological networks were not good enough to distinguish nuclear explosions from earthquakes and the start of construction of new, uniform, and much better stations and arrays. These improved the quality of seismograms and hence the power of seismology, already the most valuable tool for investigating the earth.

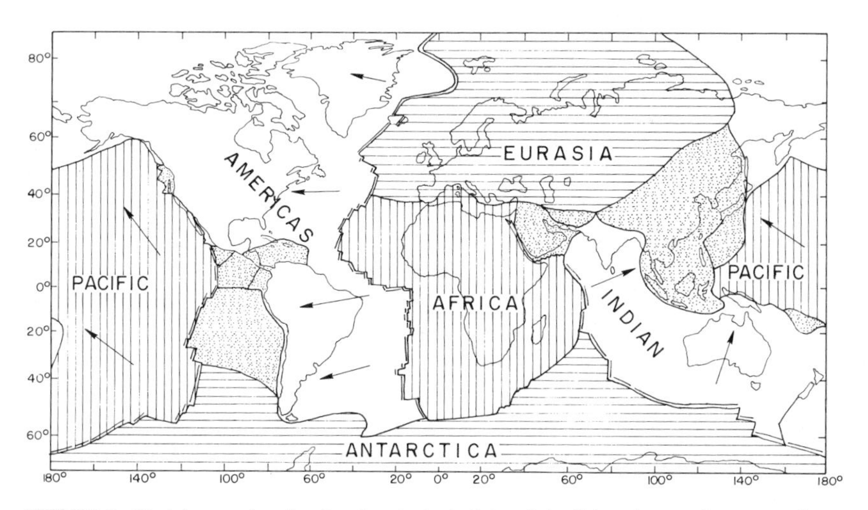

FIGURE 1 Sketch map showing the six principal plates of the lithosphere and some smaller plates. Active midocean ridges are shown as double lines. Arrows indicate direction of motion of plates in relatively rapid motion (i.e., several cm yr^{-1}).

VI. THE REVOLUTION SUCCEEDS. 1960–1972: PLATE TECTONICS ACCEPTED

By 1960, Hess had become converted to a belief in continental drift.[24] Acknowledging the debt he owed to his predecessors, he proposed that, "The mid-ocean ridges could represent the traces of the rising limbs of convection cells while the circum-Pacific belt of deformation and volcanism represents descending limbs. The Mid-Atlantic Ridge is median because the continental areas on each side of it have moved away from it at the same rate—a centimeter a year. This is not exactly the same as continental drift. The continents do not flow through oceanic crust impelled by unknown forces, rather they ride passively on mantle material as it comes to the surface at the crest of the ridge and then moves laterally away from it. On this basis the crest of the ridge should have only recent sediments on it, recent and Tertiary sediments on the flanks and the whole Atlantic Ocean and possibly all the oceans little sediment older than Mesozoic." This statement marked an important turning point in North American geology.

The next discovery arose by chance. Vacquier[25] had initiated a precise magnetic survey off the west coast of North America which unexpectedly revealed that the magnetic field was striped in long, narrow, parallel magnetic anomalies.[26] Later, Pitman and Heirtzler[27] were to find a similar pattern over the Reykjanes Ridge near Iceland, but the origin of these strikingly symmetrical patterns was a mystery.

While Ewing was executing sweeping worldwide oceanographic surveys to collect data and other Americans were doing the same on a smaller scale, Bullard with more limited resources concentrated on investigating a few small areas thoroughly. In one of these, Vine and Matthews[28] found a pattern of linear magnetic anomalies parallel to the crest of an active ridge in the Indian Ocean. They explained the anomalies as the effect of reversals in the earth's magnetic field (known since David in 1904) upon the magnetization of the ocean floor as it had been generated and had spread away from the crest. This discovery, which was to prove so momentous, also passed unnoticed, for a year later only one short reference was made to it during an important symposium on continental drift held in London[29] (Fig. 2).

In 1965, Vine[30] recognized how his discovery could be applied to explain the anomalies found by Raff and Mason,[26] who by chance had mapped the junction of a section of midocean ridge with the end of San Andreas fault as mapped by Benioff.[31]

At that junction, the horizontal shearing and many earthquakes along the fault gave way to a spreading of the ocean floor about a ridge normal

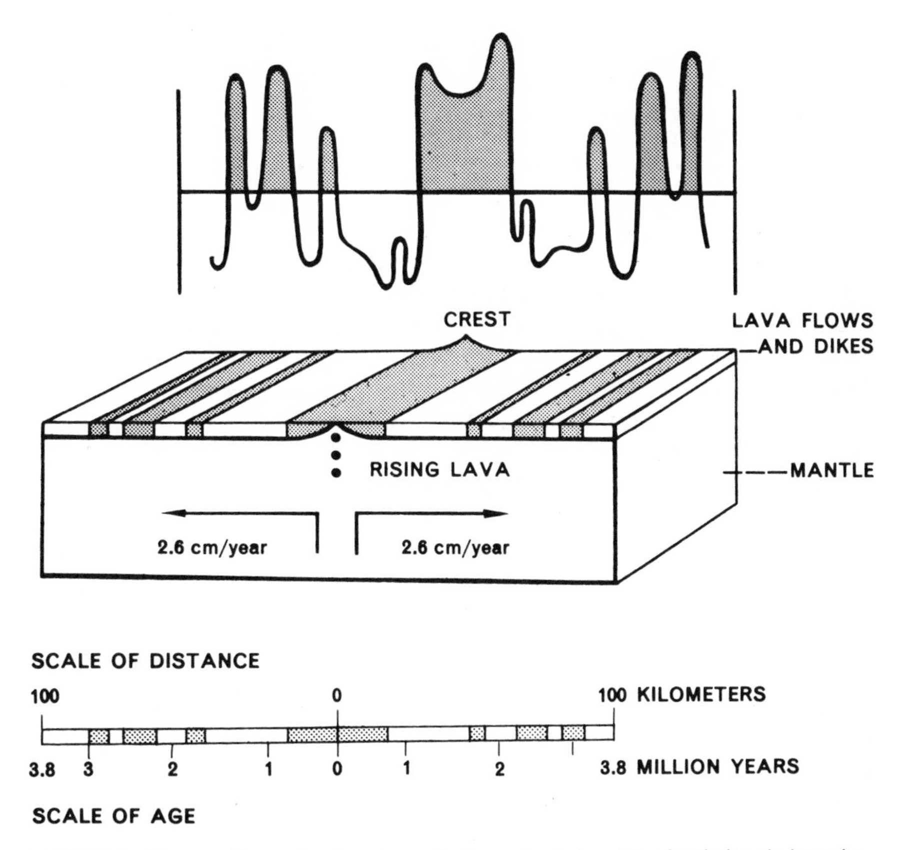

FIGURE 2 Diagrams illustrating from top to bottom a typical profile of variations in intensity of the geomagnetic field across a midocean ridge, the pattern due to field reversals of anomalies in stripes parallel to a spreading ridge, and scales of distance and age.

to the fault. This spreading produced the pattern of magnetic reversals but caused only minor earthquakes. This explained a puzzle pointed out by Benioff that an active fault with a large offset appeared to end abruptly. It became apparent that parts of the crust are moving relative to one another in three different ways. The plates move apart to form midocean ridges; they move together to form active mountains, island arcs, and trenches; and they slide past one another to form horizontal shears called transform faults (Fig. 3). It was also realized that the three different features, midocean ridges, transform faults, and subduction zones (as the compressional forms were christened)[32] were not isolated features but were "connected into a continuous network of mobile belts about the earth which divide the surface into several large rigid plates.

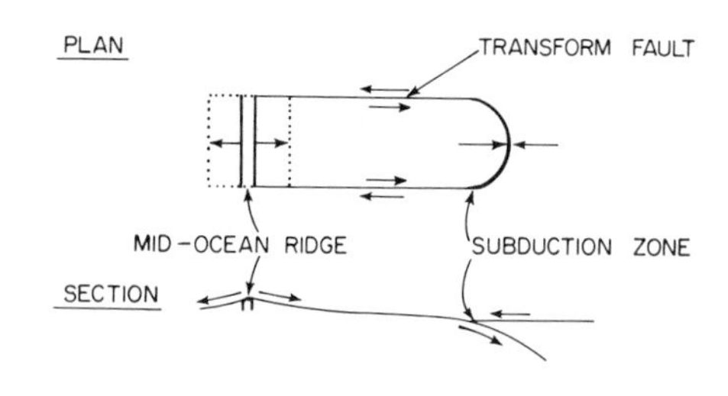

FIGURE 3 Idealized diagram of the motion of a small plate. It is moving away from a midocean ridge by generation of new lithosphere and sea-floor spreading. It is shearing past another plate along transform faults. It is being reabsorbed into the mantle along a subduction zone under an island arc or mountain range.

Any feature at its apparent termination may be transformed into another feature of one of the other two types. For example, a fault may be transformed into a mid-ocean ridge. At the point of transformation the horizontal shear motion along the fault ends abruptly by being changed into an expanding motion across the ridge or rift with a change in seismicity."[33]

Also in 1965, Bullard *et al.*[34] used a computer to fit the two sides of the South Atlantic together by rotation about a pole of which they determined the location. They also fitted together the two sides of the North Atlantic by rotation about another pole, although it seems likely that the actual motion was more complex and involved different poles of rotation at different times (Fig. 4).

At this stage, Hess persuaded Vine to join him in the Department of Geology at Princeton, where he had already enlisted the help of two applied mathematicians, Elsasser and Morgan. Vine also worked with Heirtzler and his colleagues at Columbia University to interpret their vast accumulation of magnetic data.[27,35,36] The results were spectacular and afforded an interpretation of the geology of more than half the world with a precision unique in the history of earth sciences; this soon led to the widespread acceptance of sea-floor spreading. Heirtzler, by a wise choice of the South Atlantic as standard, drew up a chronology of reversals that extended that time scale back to the late Cretaceous. By radiometric dating of many young lava flows Cox *et al.*[37] and McDougall and Chamalaun[38] dated all the reversals of the past four million years, and Heirtzler extrapolated this 20 times. Subsequent drilling and dating of cores from the ocean floor[39] have shown that the extrapolation to 80 million years is approximately correct. This demonstrates how extraordinarily persistant and uniform motions of the lithospheric plates can be.

Meanwhile, Morgan[40] completed a partial statement by Mackenzie and Parker and combined the ideas of sea-floor spreading, rotation, magnetic imprinting, and fracture zones to produce a clear statement of plate tectonics. Le Pichon[41] elaborated this into the present theory of plate tectonics (Fig. 1). Sykes[42] and Isacks *et al.*[43] were able to show

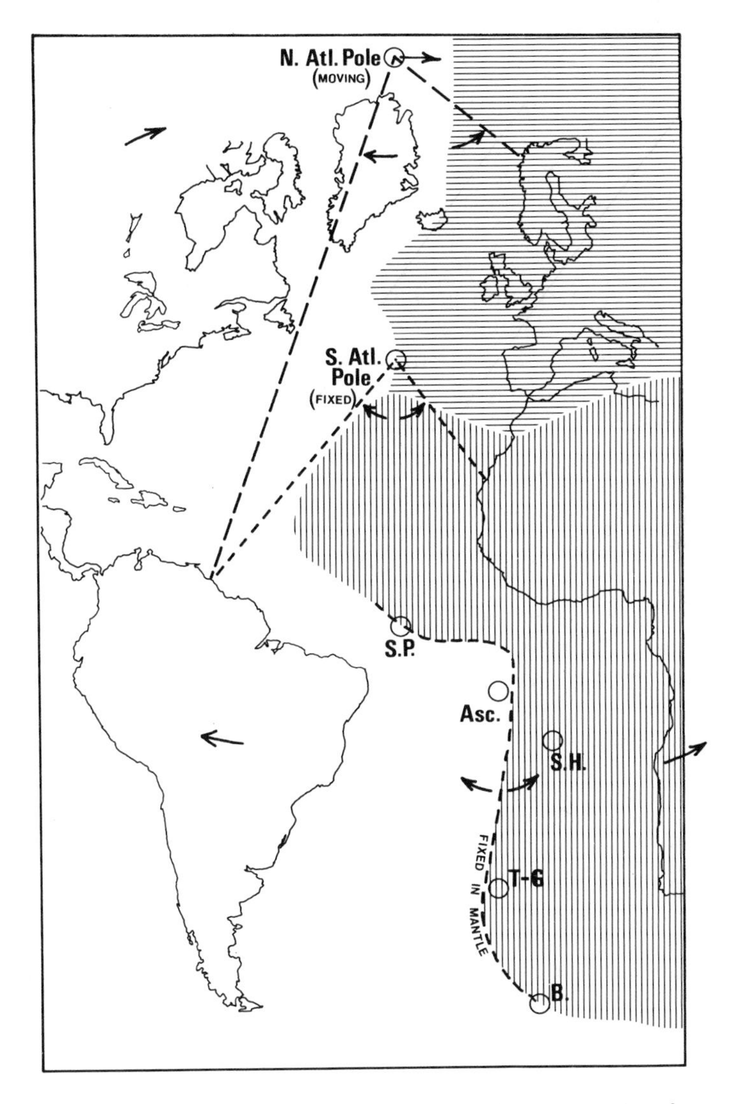

FIGURE 4 Sketch map of opening of the Atlantic Ocean by rotation of plates. If the Mid-South Atlantic Ridge was fixed relative to the mantle from 110 million to 25 million years ago, so was the corresponding pole of rotation. Since the Americas plate cannot rotate about two poles fixed relative to the mantle, the North Atlantic Pole and Ridge must then have been moving over the mantle. Islands indicated are Saint Paul's Rocks (S.P.), Ascension (Asc.), Saint Helena (S.H.), Tristan-Gough (T-G), and Bouvet (B).

that this theory explained the directions of motion and distribution of earthquakes with an elegance that had never been approached before.[44]

Drilling off the shores of the Red Sea and South Atlantic had demonstrated the existence of great deposits of Miocene salt in one and of mid-Cretaceous salt around the borders of the other. This discovery had been a puzzle until it was realized that this could be explained by the evaporation of seawater in narrow seas formed as these basins started to open. This explanation of the origin of evaporite beds and salt domes, with which much offshore oil is associated, did much to convince petroleum geologists of the truth of continental drift.

It remains true that a few distinguished men, including Jeffreys,[45] Beloussov,[46] Meyerhoff and Harding,[47] and Keith[48] remain opposed. In their favor it must be pointed out that if continents are fixed, one map of the world serves for all time; but with drift any number of past motions are possible, and proponents can be in error if they choose an inappropriate pattern of motion. In consequence, many of the points made by opponents have been valid and useful, but most earth scientists feel that their objections can be circumvented.

Since 1968, the chief effort has been to collect data, to clarify the motion of complex regions, and to extend the history backwards into time.[49-52]

The most significant improvement in technique has been the employment of the ship *Glomar Challenger* to drill and core the seafloor.[53] Special attention still needs to be given to the details of plate boundaries and motions in the Arctic Sea, the Mediterranean Sea, the Southwest Pacific Region, the Indian Ocean, and Southeastern Asia for which no complete reconstructions are yet agreed.

VII. A LIFE CYCLE OF OCEAN BASINS

Most early proponents of continental drift were content to deal with drift only since the fragmentation of a single supercontinent that apparently existed about 200 million years ago. If drift began then, one must consider two kinds of geology; but it seems unreasonable to explain the Himalayas by drift and to have to invent another origin for the older Appalachians. This led to the concept that drift had begun much earlier, a view supported by some interpretations of paleontological data and now by much evidence from paleomagnetism.[54]

A century ago, paleontologists noted that, whereas the earliest fossils of Cambrian age over most of Europe belonged to a single recognizable realm and those of the same age in most of North America were different, in some narrow strips of land along the coasts of the Atlantic the realms of fossils appear to have been transposed. Thus Scotland had North American forms, and New England had European ones.

In 1966, it was tentatively suggested that, if an early proto-Atlantic had existed in Cambrian time and that if this had closed and then reopened, nearly but not exactly in the same place, the process could explain the transposition of coastal regions between continents (Fig. 5). Many paleontologists now support this idea and its extension to other coasts.[55-57]

Such transfers have great implications, still to be exploited, but the process leads to the concept of life-cycles for ocean basins.

"If continental drift has been going on for an appreciable part of geological time, at such rapid rates as recent work suggests, it means that a succession of ocean basins may have been born, grown, diminished, and closed again. Since ocean basins are the largest features of the earth's surface and would dominate other features it seems useful to outline the stages in their life-cycle in terms of present examples. This makes it apparent that each stage has its own characteristic rock types and structures."[58] It seems likely that each stage has characteristic ores, physiographic features, and areas and degrees of metamorphism. Thus this life cycle of the largest features of the earth's surface provides a framework about which to rewrite the whole of regional geology, geophysics, and even geomorphology.[59]

VII. CONVECTIVE PLUMES IN THE MANTLE AND THE DRIVING MECHANISM

For well over a century, many have postulated that convection currents have flowed or are still flowing in the mantle. As it has generally been

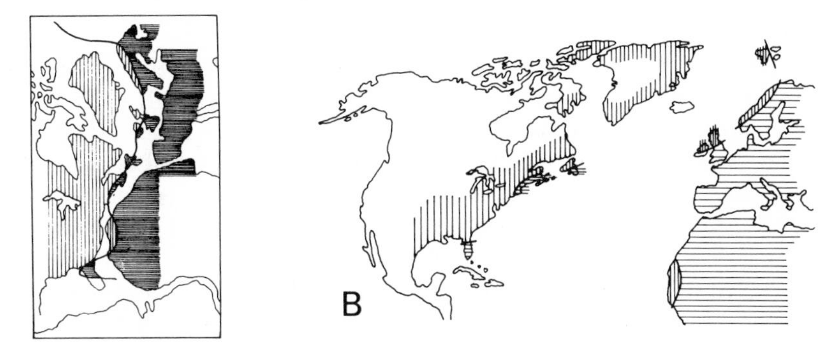

FIGURE 5 A. Sketch map to show how two earlier continents (differentiated by shading) that had been separate closed between Ordovician and Pennsylvanian periods. B. Another map to show how parting along a rift close to, but not identical with, the line of union transposed slices of the pre-Ordovician continents.

considered that the lithosphere is rigid to a depth of at least a few tens of kilometers and that the flow is in a deeper asthenosphere at rates of the order of only a few centimeters a year, no direct observation of these currents is possible. The movement of lithospheric plates may provide indirect evidence, but since these plates are large and rigid and interfere with one another their motion is not a precise guide to currents acting on their lower surfaces. Many theoretical papers have considered convection in a spherical shell.[11, 60-64] Most of these have considered whole mantle convection in order to produce cells of comparable dimensions to continents and oceans. Unfortunately, seismic evidence suggests that the lower mantle is extremely rigid and flow more likely in a thin and shallow shell—the asthenosphere. Furthermore, the patterns suggested by theory do not closely resemble those of the earth's surface features. Some have tried laboratory experiments, but for reasons mentioned by Elsasser[65] modeling of the earth's interior is difficult. Ramberg[66, 67] has achieved the most success as will be described later.

Any system of currents must meet stringent conditions. The midocean ridge system of spreading zones and the young mountain system of subduction zones have both existed for at least several tens of millions of years. Spreading from the South Atlantic Ridge and the East Pacific Rise appears to have been at a very steady rate. Some midocean ridges must move relative to one another, and hence some, at least, must migrate over the mantle. (The ridges form nearly complete rings around both Africa and Antarctica, and there are no subduction zones within these rings, so that if the southern continents were once together the ridges must then have been against the coasts of Africa and Antarctica and hence the ridges have migrated outward.)

Some physicists, notably Orowan[68] and Elsasser[65] have felt that these conditions cannot be met by the kind of convection cells usually visualized.

Morgan[69, 70] has proposed a new theory. It is based upon the proposal made in 1963 for the origin of the Hawaiian Islands.[71] It has long been known that they lie along a ridge of volcanic rocks that get progressively older north-westwards from the active volcanic island of Hawaii, and it was suggested that Hawaii might owe its active volcanism to the rise of a long-lived and stationary plume of hot material in the deep mantle beneath it, and that as the Pacific floor had moved over this plume the chain of extinct volcanoes had been formed, and that they indicated the locus of past movements. Thus, as Walter Sullivan suggested, the Hawaiian chain might be likened to a trail of smoke drifting with the wind away from a chimney[72] (Fig. 6).

It had also been proposed that if an active plume lay on a spreading ridge at the junction of two plates it would give rise to two lava trails or

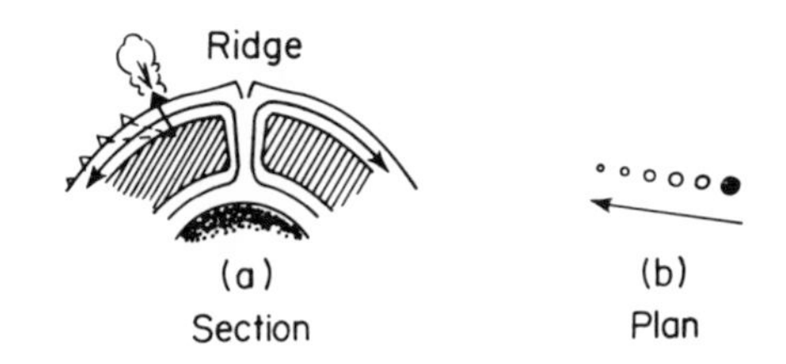

FIGURE 6 Sketch to indicate how near-surface movement over a deep-seated plume could give rise to a chain of islands like the Hawaiian Islands, progressively older from one end to the other.

inactive ridges, one on each plate, such as those linking Iceland or Tristan da Cunha to both coasts of the Atlantic (Figs. 7 and 8).

Morgan extended this concept by proposing that the plumes not only mark the locus of absolute motion of plates relative to the deep mantle but also provide the mechanism that drives the plates. He recognized this to be thermal convection of a special type in which the currents flow upward in a few dozen pipes, each a hundred or so kilometers in diameter, at rates of possibly more than a meter a year, and that all the rest of the mantle forms the downward part of the circulating system by settling at rates of less than a millimeter a year (Fig. 9).

The concept of convective plumes can be stated in quite precise terms. Several dozen volcanically active hot spots have been identified that may mark the site of plumes. According to this interpretation each has had a life of some tens of millions or as much as a hundred million years. They have diagnostic characteristics that are clearest for hot spots located in ocean basins.

1. Each is an uplift, marked by elevated basement rocks on land and shoal water at sea, including two islands in which sea-floor or

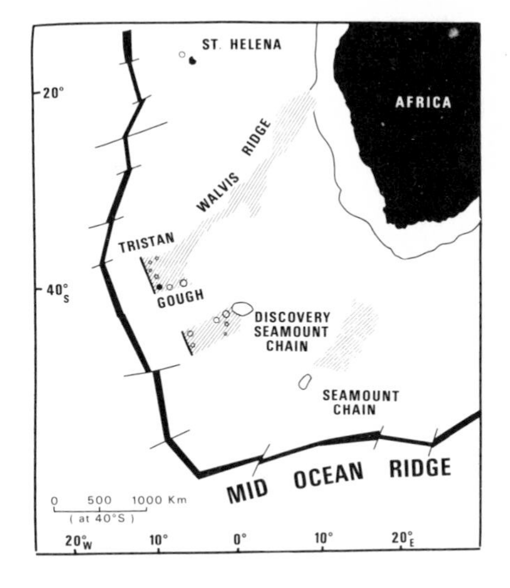

FIGURE 7 Map of part of the African plate showing volcanic centers over supposed plumes. Until the plate became stationary 25 million years ago, the plumes near the Mid-Atlantic Ridge formed trails leading to the continent, but since then the ridge has moved west from the plumes.

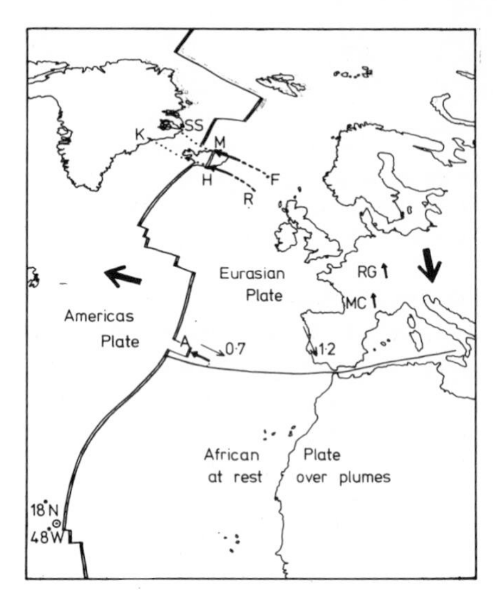

FIGURE 8 Map showing by broad arrows the direction of motion of the Americas and Eurasian plates since Africa became stationary 25 million years ago and by small arrows some resulting volcanic plume traces due to rotation of the Eurasian plate about a pole at 18°N and 48°W. Note how the midocean ridge has moved west off the plumes beneath Azores (A), Hecla (H), and Myvaten (M).

FIGURE 9 Sketch map of the world showing the location of supposed plumes rising from deep in the mantle. Notice that those reaching the surface are concentrated either beneath plates believed to be fixed (Africa, Nazca, and Southeast Asia) or near midocean ridges, which are plate boundaries.

mantle rocks have been raised above sea level (Macquarie Island and St. Paul's Rocks).

2. The uplifts are capped by active volcanoes producing alkaline basalts, andesites, and rhyolite, as well as the tholeiite basalts common on the sea floor. These lavas have distinctive isotopic ratios and geochemical patterns.
3. Negative anomalies of gravity probably accompany them.[73]
4. In the oceans one or sometimes two lateral seismic ridges stretch away from them.
5. They are areas of high heat flow.
6. The islands over hot spots are geochemically and isotopically distinct from the sea floor.[74]

No physical theory has yet been given, but Morgan[69] believes that one may be possible. Similar diapiric uplifts are common in nature, for example, salt domes, igneous, and shale diapirs; volcanic pipes; and thunderheads. Kanasewich *et al.*[75] have produced evidence that directly below the Hawaiian hot spot the core–mantle interface is anomalous. The theory meets Lliboutry's requirement from considerations of energy requirements that more of the mantle than the thin asthenosphere be involved[76] but avoids the difficulty imposed on other convection current hypotheses by the very high viscosity of the greater part of the lower mantle.[77]

This pattern of convection is also supported by Ramberg's model experiments in a centrifuge.[67] His findings suggest that if convection is to open features as wide as oceans then the currents must rise from deep in the mantle and that such currents would not rise in uniform ridges but in a series of domes. He therefore supports the idea of convective plumes (personal communication at International Geological Congress, 1972).

If we assume that these ideas are correct, then certain conclusions follow almost automatically from the geometry of a sphere.

IX. TWO TYPES OF OCEAN WITH STATIONARY AND MOVING SPREADING RIDGES

Inherent in the original concept of rising plumes is the notion that they remained fixed in location in the deeper mantle, as volcanic pipes, salt domes, and other rising diapirs do in the crust. This is supported by the knowledge that several chains of islands in the Pacific are parallel (or concentric) with the Hawaiian chain and that each has an active volcanic island or seamont at its southeastern end. The recent demonstration that the three chains can each be extended into an L-shaped pattern with the second arm of each chain also concentric about a second pole has strengthened the case. It suggests that the Pacific plate changed its direction of motion at a date marked by the elbows that were passed about 25 million years ago.[72]

If the plumes are fixed or nearly so, Burke and Wilson[78] have suggested that there should be two types of ocean basin. In one type, which may be regarded as dominant, the basin is expanding symmetrically about a stationary ridge that connects a series of fixed plumes. In the other case, geometry and the interference of plates forces the expanding ridge off the plumes that created the ridge. The result is the periodic creation of a new ridge subparallel to the old.

Consider the South Atlantic. From Upper Cretaceous to Miocene time (80 million to 25 million years ago) this basin was expanding equally in either direction from a midocean ridge that connected the plumes marked by St. Paul's Rocks, St. Helena, Tristan-Gough Islands, and Bouvet Island. This expansion took the form of rotation of the American and African plates equally in opposite directions about a pole of rotation lying near the Azores in the North Atlantic (Fig. 4). This pole, like the median ridge, stayed fixed relative to the mantle at a position that Bullard *et al.*[34] determined to be 44°N and 30°W.

Now consider the North Atlantic and the Arctic Sea. This section opened by rotation of the Eurasian and American plates about another pole that lay in the Arctic Sea at 88°N and 28°W. According to all interpretations of plate tectonics there is only one American plate embracing both North and South America and the Western Atlantic floor. Since we have just pointed out that the pole of rotation of the American plate fixed relative to the mantle is near the Azores, the pole of rotation relative to Eurasia in the Arctic Sea cannot also be fixed relative to the mantle. If the pole of rotation between Eurasia and the American plate is moving over the mantle, the ridge in the North Atlantic of which the segments are radial to that pole must also be moving.

This simple geometric argument is supported by evidence that the ridges in the north form a complex pattern. Vogt *et al.*[79] have interpreted this as due to a series of three successive spreading ridges that moved their location from Baffin Bay and the Alaskan side of the Arctic Sea progressively eastward to their present location next to Norway and Siberia (Fig. 10).

The evidence of discrete abandoned ridges suggests that ridges have a life of their own and can exist for a period after they have been pushed off their founding plumes before a new subparallel ridge forms through the plumes[80] (Fig. 10).

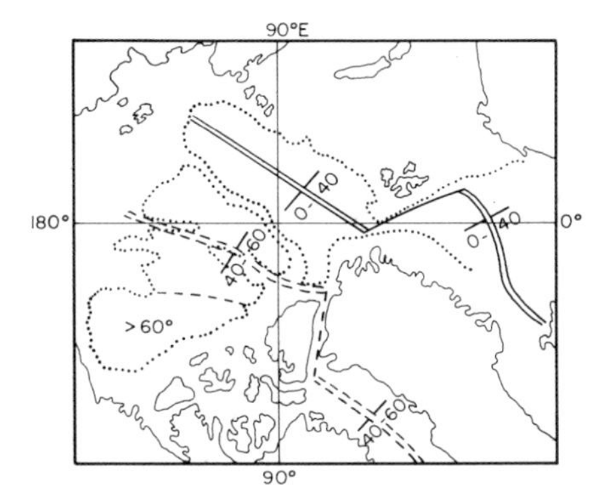

FIGURE 10 Map of the arctic regions showing by double lines the active mid-ocean ridge that has been spreading for the past 40 million years, by dotted lines the position of a ridge spreading between 60 million and 40 million years ago, and a third part older than 60 million years (after P. R. Vogt *et al.*).

By an extension of these ideas one can suggest that the East Pacific Rise has also been a dominant or fixed spreading ridge, for the ring of subduction zones around the Pacific has enabled its expansion to proceed independently of the behavior of the South Atlantic. On the other hand, the Indian Ocean and Tasman Sea like the Arctic are not independent. It is the movement of their poles of rotation and of their ridges that have produced more complex patterns on their floors.

X. TWO TYPES OF MOUNTAIN WITH THE LEADING EDGE ON CONTINENTAL AND OCEANIC PLATES

The idea that one can determine motion of plates relative to the mantle as well as relative to one another enables one to distinguish between continents that are stationary over the mantle and those that are moving.

We have seen that the southern part of American plate is rotating westward over the mantle. Other arguments, less strong, but still I believe tenable, suggest that the Nasca plate lying between the Peru–Chile trench and the East Pacific Rise is nearly stationary on the mantle.[81]

If this is the case, the Andes have formed along the leading edge of an advancing continent and the Peru–Chile trench has passively formed at the edge of the continent where it is constantly being overridden and pushed ahead over the mantle (Figs. 11 and 12). This type of Andean mountain has several characteristics. Such mountains can only form on one coast (the leading edge) of a continent at one time. Because the trench and subduction zone are close to the coast, the Benioff zone, which marks the path of the Pacific floor down under the mountains, is close to the coast. This generates andesite volcanoes above it that are close to the coast also. Their heat and lavas rising among the deltaic sediments along the coast mixed together volcanic (eugeosynclinal) rocks and sedimentary (miogeosynclinal) rocks and metamorphosed the whole into granitic batholiths.[82-84]

The Sierra Nevada and the Coast Ranges of British Columbia are similar.

Other arguments too lengthy to repeat here suggest that Southeastern Asia may be an example of a continental plate stationary relative to the mantle that produces another type of mountain system. Island arcs from Burma, through Indonesia to the Philippines and Marianas have formed on three sides, due, according to plate tectonic theory, to the advance westward of the Pacific floor toward Asia and the advance northward

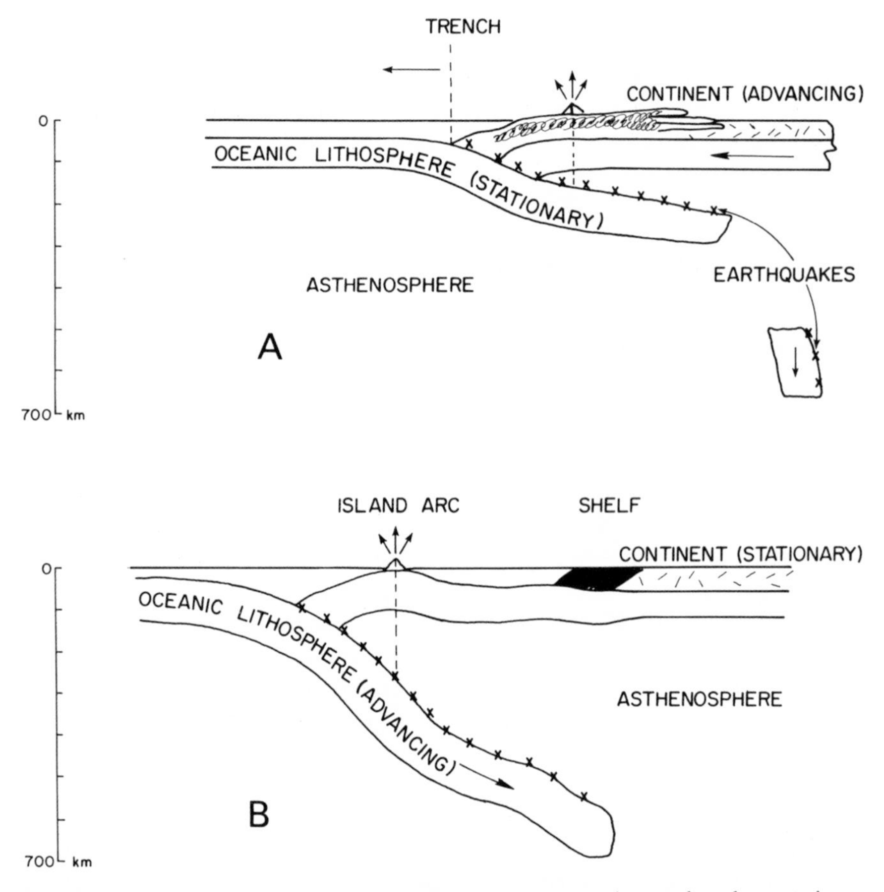

FIGURE 11 A. Section showing a continent advancing relative to the mantle and to a stationary plate of oceanic lithosphere. In this case, the trench and volcanoes form along the coast. B. Section showing an oceanic plate advancing relative to a continent and the mantle. In this case island arcs have moved the trench and volcanoes out from the coast line.

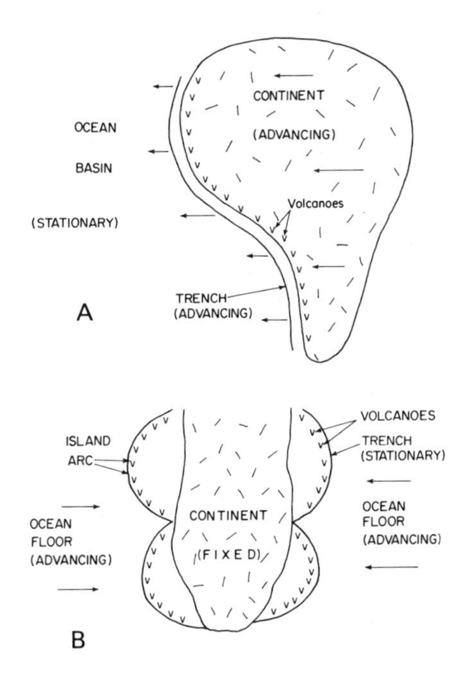

FIGURE 12 Maps corresponding to the sections in Fig. 11 showing that volcanoes and a trench form only on the leading edge of an advancing continent but can be formed all around a stationary one.

of the Indian Ocean floor toward Asia. In the case of a stationary continent and moving ocean floor, mountains can thus form on any or all sides of the continent simultaneously (Figs. 1, 11, and 12).

Furthermore, the subduction zones take the form of island arcs not marginal trenches. This is held to be due to the freedom of movement that activity imparts to the oceanic plates. Given this freedom (which is lacking where the oeeanic plates are stationary on the mantle) they assume circular patterns that involve no distortion and minimum energy. This is analogous to the fact that if a tennis ball is pinched the depressions are circular, as Frank[85] pointed out. Presumably the subduction zones started along the coast and moved offshore into arcuate patterns opening young marginal seas behind them.[86] Since the arcs are offshore, the rising volcanic products in the island arcs are well separated from the shelf sediments along the coasts, and major batholiths have no chance to form.

XI. TWO TYPES OF CONTINENT: STATIONARY OR MOVING OVER THE MANTLE

The fact that during the Paleozoic Era mountains formed at overlapping periods of time on all sides of North America (in the Appalachian,

Ouachita, Antler, and Innuitian disturbances) suggests that North America was then stationary or nearly so and that these mountains had the form of island arcs. This is supported by the good separation between the sequences of sedimentary and volcanic rocks in these mountains (e.g., the Valley and Ridge and Piedmont provinces, respectively).

Furthermore, for reasons dealt with by Burke *et al.*[80] it appears that Africa may have become stationary about 25 million years ago, forcing the Mid-Atlantic Ridge to move west off its plumes, and that since Africa has been stationary it has assumed a structure of basins, swells, and vertical uplifts different from other continents as Krenkel[87] and Cloos[88] noted. This may reflect differential heating between different parts of the mantle beneath it. In moving continents, such effects are smeared out and do not appear.

If so, it is of interest to note that during the Paleozoic Era North America developed a succession of swells and basins, for example, the Michigan, Willison, and Peace Rover Basins, whose origin has long puzzled geologists but which this hypothesis might explain.

XII. CONCLUSION

This paper is a review of an immense amount of work, and I only regret that it is impossible to know, remember, or record it all. Inevitably much of equal importance has been omitted, but I hope that enough has been said to show that acceptance of mobility of the earth's surface opens a new era of better understanding of processes and history that amount to a true scientific revolution. If so, past examples suggest that the excitement of important discoveries that have marked the last few years may continue for a little longer. Already they have made possible a unified physical approach to problems of the solid earth, which is solving many geological problems by modes of thought acceptable to physicists.

At the present time, new ideas are being published every month, and I can only acknowledge a general gratitude to all upon whose work I have drawn, with particular thanks to my colleagues K. Burke and W. F. S. Kidd, and express regrets for many omissions, particularly easy to make in so new and broad a subject.

REFERENCES

1. T. S. Kuhn, *The Structure of Scientific Revolutions*, 2nd ed. (Chicago U.P., Chicago, Ill., 1970).
2. A. V. Carozzi, *Geol. Soc. Am. Bull. 81*, 283 (1970).
3. O. Fisher, *Physics of the Earth's Crust*, 2nd ed. (Macmillan, London, 1889).
4. M. K. Hubbert, *Geol. Soc. Am. Bull. 62*, 355 (1951).

5. F. B. Taylor, *Geol. Soc. Am. Bull. 21,* 179 (1910).
6. A. Wegener, *The Origin of Continents and Oceans,* 4th ed., 1929 transl. (Dover, New York, 1966).
7. W. A. J. M. van Waterschoot van der Gracht, *Theory of Continental Drift* (Am. Assoc. Petrol. Geol. Publ., 1928).
8. H. Jeffreys, *The Earth: Its Origin, History and Physical Constitution,* 5th ed. (Cambridge U.P., Cambridge, England, 1970).
9. A. L. du Toit, *Carnegie Inst. Wash. Publ. 381,* 1 (1927).
10. R. A. Daly, *Our Mobile Earth* (Charles Scribner's Sons, New York, 1926).
11. A. Holmes, *Trans. Geol. Soc. Glasgow 1928–1929 18,* 559 (1931).
12. A. Poldervaart, *Geol. Soc. Am. Spec. Paper 62,* 119 (1955).
13. W. Q. Kennedy, *Q. J. Geol. Soc. London 102,* 41 (1946).
14. H. W. Wellman, *New Zealand J. Sci. Tech. 33,* 409 (1952).
15. M. L. Hill and T. W. Bibblee, Jr., *Geol. Soc. Am. Bull. 64,* 443 (1953).
16. J. T. Wilson, in *International Upper Mantle Project Report 6*, L. Knopoff, ed. (Physics Dept., U.C.L.A., Los Angeles, Calif., 1968), p. 32.
17. H. W. Menard and R. S. Dietz, *J. Geol. 60,* 266 (1952).
18. B. C. Heezen, E. T. Bunce, J. B. Hersey, and M. Tharp, *Deep-Sea Res. 11,* 11 (1964).
19. H. W. Menard, *Marine Geology of the Pacific* (McGraw-Hill, New York, 1964).
20. S. W. Carey, *Continental Drift: A Symposium* (Univ. Tasmania, Dept. Geol., Hobart, Tasmania, 1958).
21. E. Irving, *Palaeomagnetism* (Wiley, New York, 1964).
22. M. Ewing and B. C. Heezen, *Science 131,* 1677 (1960).
23. B. Gutenberg and C. F. Richter, *Seismicity of the Earth and Associated Phenomena* (Princeton U.P., Princeton, N. J., 1954).
24. H. H. Hess, in *Petrological Studies–a Volume in Honor of A. F. Buddington* (Geol. Soc. Am., New York, 1962), p. 599.
25. V. Vacquier, *Nature 183,* 452 (1959).
26. A. D. Raff and R. G. Mason, *Geol. Soc. Am. Bull. 72,* 1267 (1961).
27. W. C. Pitman, III, and J. R. Heirtzler, *Science 154,* 1164 (1966).
28. F. J. Vine and D. H. Mathews, *Nature 199,* 947 (1963).
29. P. M. S. Blackett, E. C. Bullard, and S. K. Runcorn, *Phil. Trans. Roy. Soc. (London) A258,* 1 (1965).
30. F. J. Vine and J. T. Wilson, *Science 150,* 485 (1965).
31. H. Benioff, *Continental Drift* (Academic Press, New York, 1962), Chap. 4.
32. D. A. White, D. H. Roeder, T. H. Melson, and J. C. Crowell, *Geol. Soc. Am. Bull. 81,* 3431 (1970).
33. J. T. Wilson, *Nature 207,* 343 (1965).
34. E. C. Bullard, J. E. Everett, and A. G. Smith, *Phil. Trans. Roy. Soc. (London) A258,* 41 (1965).
35. F. J. Vine, *Science 154,* 1405 (1966).
36. J. R. Heirtzler, G. O. Dickson, E. M. Herron, W. C. Pitman, III, and X. Le Pichon, *J. Geophys. Res. 73,* 2119 (1968).
37. A. Cox, R. R. Doell, and G. B. Dalrymple, *Science 144*, 1537 (1964).
38. I. McDougall and F. H. Chamalaun, *Nature 212,* 1415 (1966).
39. A. E. Maxwell, R. P. Von Herzen, K. Inghwa Hsu, J. E. Andrews, T. Saito, S. F. Percival, Jr., E. D. Milow, and R. E. Boyce, *Science 168*, 1047 (1970).
40. W. J. Morgan, *J. Geophys. Res. 73,* 1959 (1968).
41. X. Le Pichon, *J. Geophys. Res. 73,* 3661 (1968); *75,* 2793 (1970).

42. L. R. Sykes, *J. Geophys. Res. 72,* 2131 (1967).
43. B. Isacks, J. Oliver, and L. R. Sykes, *J. Geophys. Res. 73,* 5855 (1968).
44. M. Barazangi and J. Dorman, *Bull. Seismol. Soc. Am. 60,* 1741 (1970).
45. H. Jeffreys, *Nature 222,* 706 (1969).
46. V. V. Beloussov, *Tectonophys. 13*, 95 (1972).
47. A. A. Meyerhoff and J. L. Harding, *Tectonophys. 12*, 235 (1971).
48. M. L. Keith, *J. Geol. 80,* 249 (1972).
49. J. D. Phillips and D. Forsyth, *Geol. Soc. Am. Bull. 83,* 1579 (1972).
50. D. P. McKenzie and J. G. Sclater, *Geophys. J. 25,* 437 (1971).
51. A. G. Smith and A. Hallam, *Nature 225,* 139 (1970).
52. W. C. Pitman, III, and M. Talwani, *Geol. Soc. Am. Bull. 83,* 619 (1972).
53. M. Ewing, J. L. Worzel, A. U. Beall, W. A. Berygren, D. Bukry, C. A. Burk, A. G. Fischer, and E. A. Pessagno, Jr., *Initial Reports of the Deep Sea Drilling Project* (U.S. Government Printing Office, Washington, D.C.) and subsequent vols.
54. A. Brock, in *Continental Drift, Sea Floor Spreading and Plate Tectonics,* D. H. Tarling and S. K. Runcorn, eds. (Academic Press, London, 1973), Chap. 1.3.
55. J. R. Ross and J. K. Ingham, *Geol. Soc. Am. Bull. 81,* 393 (1970).
56. J. W. H. Monger and C. A. Ross, *Can. J. Earth Sci. 8,* 259 (1971).
57. H. B. Whittington and C. P. Hughes, *Phil. Trans. Roy. Soc. (London) B263,* 235 (1972).
58. J. T. Wilson, *Am. Phil. Soc. Proc. 112,* 309 (1968).
59. P. R. Vogt, *Nature 226,* 743 (1970).
60. C. L. Pekeris, *Mon. Not. Roy. Astron. Soc. (Geophys. Suppl.) 3,* 343 (1935).
61. F. A. Vening Meinesz, *The Earth's Crust and Mantle* (Elsevier, Amsterdam, 1964).
62. S. K. Runcorn, *Phil. Trans. Roy. Soc. (London) A258,* 228 (1965).
63. D. C. Tozer, in *The Earth's Mantle,* T. F. Gaskell, ed. (Academic Press, New York, 1967), Chap. XI.
64. E. R. Oxburgh and D. L. Turcotte, *J. Geophys. Res. 76,* 1315 (1971).
65. W. M. Elsasser, *Am. Phil. Soc. Proc. 112,* 344 (1968).
66. H. Ramberg, *Gravity, Deformation and the Earth's Crust* (Academic Press, London, 1967).
67. H. Ramberg, *Phys. Earth Planet. Inter. 5*, 45 (1972).
68. E. Orowan, *Sci. Am. 221,* 102 (1969).
69. W. J. Morgan, *Nature 230,* 42 (1971).
70. W. J. Morgan, *Geol. Soc. Am. Mem.* (H. H. Hess volume) (in press).
71. J. T. Wilson, *Can. J. Phys. 41,* 863 (1963).
72. E. D. Jackson, E. A. Silver, and G. B. Dalrymple, *Geol. Soc. Am. Bull. 83,* 601 (1972).
73. W. M. Kaula, in *The Nature of the Solid Earth*, E. C. Robertson, ed. (McGraw-Hill, New York, 1972), Chap. 15.
74. V. M. Oversby, *Geochim. Cosmochim. Acta 36,* 1167 (1972).
75. E. R. Kanasewich, R. M. Ellis, C. M. Chapman, and P. R. Gutowski, *Nature, Phys. Sci. 239,* 99 (1972).
76. L. Lliboutry, *J. Geophys. Res. 77,* 3759 (1972).
77. F. Press, in *The Nature of the Solid Earth,* E. C. Robertson, ed. (McGraw-Hill, New York, 1972), Chap. 7.
78. K. Burke and J. T. Wilson, *Nature 239,* 387 (1972).
79. P. R. Vogt, N. A. Ostenso, and G. L. Johnson, *J. Geophys. Res. 75,* 903 (1970).
80. K. Burke, W. S. F. Kidd, and J. T. Wilson, *Nature* (in press).

81. J. T. Wilson and K. Burke, *Nature 239,* 448 (1972).
82. D. E. James, *Geol. Soc. Am. Bull. 82,* 3263 (1972).
83. P. C. Bateman and J. P. Eaton, *Science 158,* 1407 (1967).
84. J. Gilluly, *Geol. Soc. Am. Bull. 82,* 2382 (1971).
85. F. C. Frank, *Nature 220,* 363 (1968).
86. D. E. Karig, *J. Geophys. Res. 76,* 2542 (1971).
87. E. Krenkel, *Geologie und Bodenschatze Africas,* 2nd ed. (Akad. Verlag, Leipzig, 1957).
88. H. Cloos, *Geol. Rundschau 30,* 401, 637 (1939).

SIR FRED HOYLE *is one of the world's leading astronomers. For many years he was the Plumian Professor of Astronomy and Experimental Philosophy at Cambridge. At present he holds appointments both at the California Institute of Technology in Pasadena and at the University of Manchester in England. He is a foreign fellow of the American Academy of Arts and Sciences and the United States National Academy of Sciences. He has served as Vice President of the Royal Society and as President of the Royal Astronomical Society.*

3 *The Crisis in Astronomy*/

FRED HOYLE

All astronomical objects, whether they are distant or close-by, are projected together on the sky. The astronomer must devise methods for deciding which are far away and which are close. Indeed it is a principal task for the astronomer to determine actual distances from the earth to the many kinds of object that populate the universe.

For the smallest distances, to the sun and planets, distances within our own solar system, a method like that of stereoscopic vision is used, the two "eyes" being telescopes sited at fairly widely separated places on the earth. The same stereoscopic method is used for distances to the nearest stars, except that instead of using different places on the earth itself, different points on the earth's annual orbit around the sun are used. But how of the very great distances, distances going far outside the confines of our galaxy? Here astronomers have been in the habit of using the so-called method of red shifts, which I will now explain.

Imagine a piano systematically out of tune, with the property that every note is lowered in the same proportion from a normal piano. We could say that our imaginary piano was red shifted compared with a normal piano, and we could write $1 + z$ for the factor by which the imaginary piano was lowered in its pitch compared with a normal piano. A pianist could perfectly well play all his usual pieces. He would simply say the absolute pitch of the music had been changed. In actual fact, musicians for about two centuries have been gradually raising their absolute pitch, so the modern pianist does not play classical sonatas at the pitch intended by their composers.

Except that astronomers are concerned with the frequency of light waves instead of with the frequency of sound waves, exactly the same thing applies to distant galaxies as it did to our imaginary piano. Corresponding to the notes of the piano we have particular emissions for various kinds of atom, "spectrum lines," as they are called. The pitches of the spectrum lines are lowered systematically compared with lines from similar atoms in the terrestrial laboratory, and the amount by which they are lowered can again be expressed by the factor $1 + z$. The number z is usually referred to as the red shift. When z is small, near zero, the lowering is small–the frequencies are nearly the same as in the laboratory. When z is large, the frequencies are all significantly less than in the laboratory. The largest value of z so far reported for any astronomical object is close to 3, so $1 + z$ is close to 4. This means that all "pitches" from this particular object are systematically lowered by nearly two "octaves."

A crude simple way to estimate the distance of a galaxy is from how bright it seems to be–its apparent brightness or magnitude m as it is usually called. A standard lamp will appear fainter the farther away it happens to be, and the same property will apply to galaxies. But only approximately, because galaxies are only approximately the same as each other–they are not strict standard lamps. However, a faint-looking galaxy will usually be considerably more distant than a bright galaxy.

Hubble found the remarkable relation between z and m shown in Fig. 1. [The pg subscript on m (m_{pg}) means apparent magnitude determined photographically.] The convention used in astronomy is for m to become larger the fainter the object happens to be. Thus a first-magnitude star, $m = 1$, is brighter than a second-magnitude star, $m = 2$, and so on. The galaxies plotted at $m = 16$ in Fig. 1 are 15 magnitudes fainter than a first-magnitude star. So the more distant a "standard lamp," the larger is the associated value of m. This means that distance increases as one goes from left to right in Fig. 1. Hence Fig. 1 tells us that the more distant a galaxy, the larger the red shift z, generally speaking.

Figure 1 refers to a mixed bag of galaxies. Next, Hubble chose galaxies by a uniform criterion. He found several clusters of galaxies, apparently very similar to each other, and he chose the fifth brightest member of each such cluster. This gave the relation between z and m shown in Fig. 2, a relation with much less "scatter" than shown in Fig. 1.

It will be useful in relation to later discussion to replot Fig. 2 in the form shown in Fig. 3, where the red shift z is given directly on the left-hand scale instead of the logarithm of z. The curve shown in Fig. 3 corresponds to the straight line of Fig. 2. [The bottom scale of Fig. 3 has

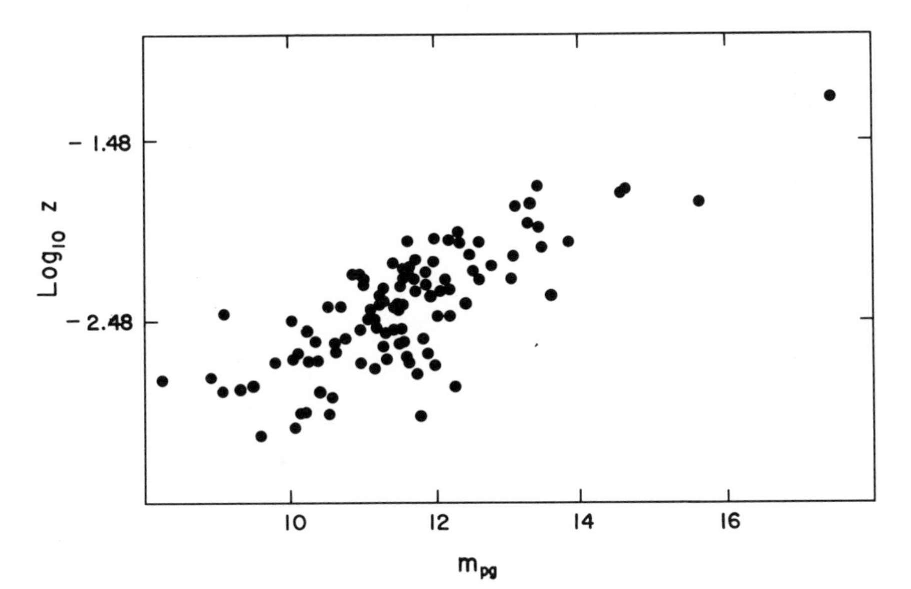

FIGURE 1 Plot of m_{pg} against log z for field galaxies, taken from Hubble. The curious numerical values of the ordinate divisions arise because Hubble plotted $\log_{10} cz$, with c the velocity of light in kilometers per second.

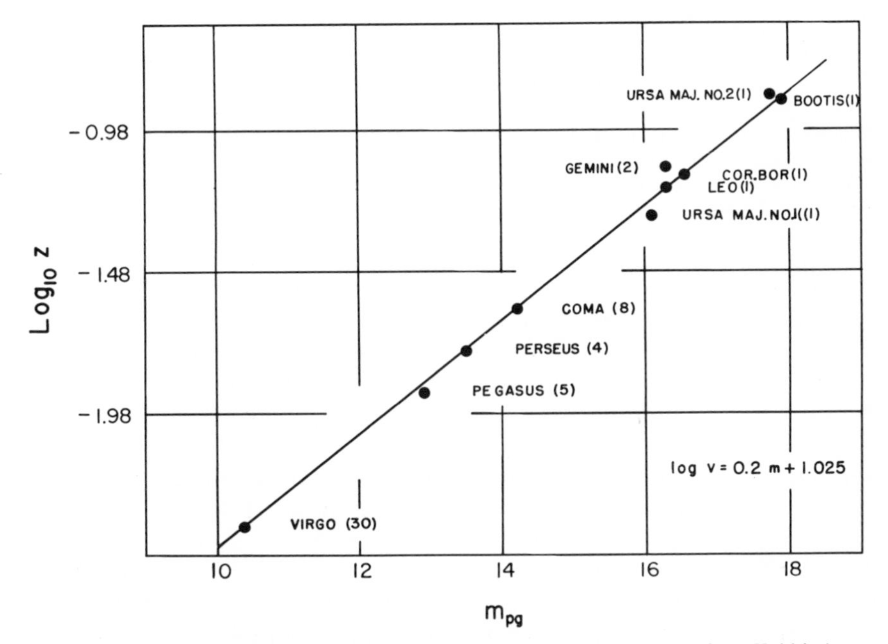

FIGURE 2 Plot of m_{pg} against log z for fifth-ranking cluster galaxies taken from Hubble (see Fig. 1 caption for remark on ordinate values).

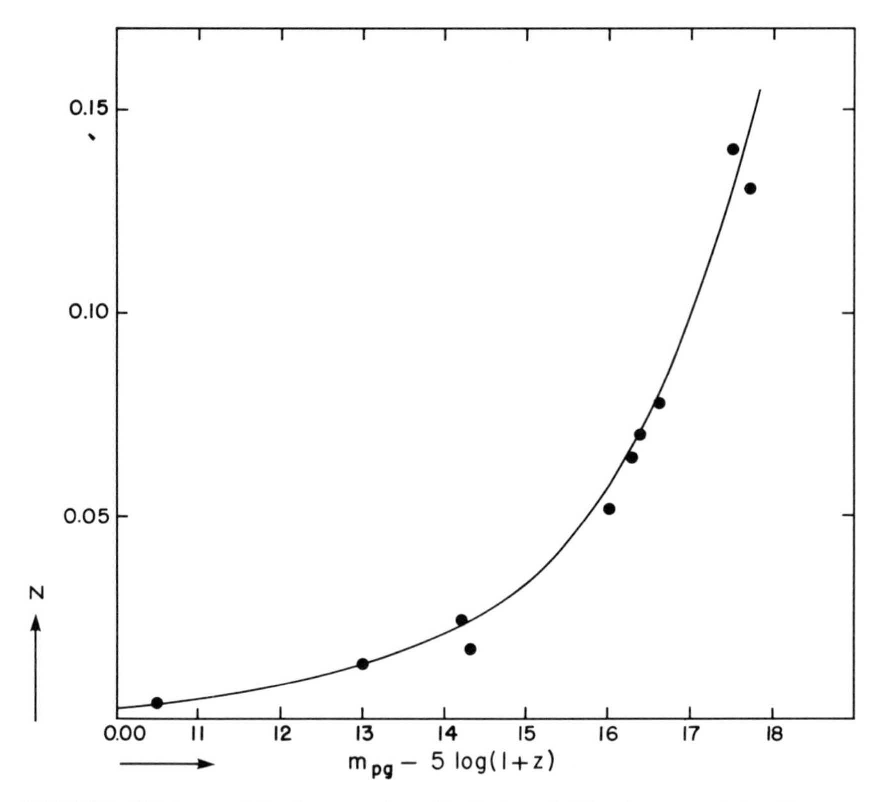

FIGURE 3 This is essentially the same plot as Fig. 2, the red shift values z now being displayed directly instead of the logarithm of z. The bottom magnitude scale has been changed slightly to take account of energy loss from the red shift effect. This figure should be compared with Fig. 7, which is the corresponding plot for quasars.

also been slightly changed to $m_{\text{pg}} - 5 \log (1 + z)$ instead of m_{pg}. By taking account of loss of energy due to the red shifting effect, the values on the bottom scale are then more strictly comparable with each other.]

These results of observation seem to imply that red shift is related to distance, in the sense that large distance means large red shift. Since red shifts are measurable for galaxies and for other distant objects, often without undue difficulty, we have here a convenient method for determining extremely large distances. The method has first to be calibrated, however. That is to say, distances must be determined otherwise, by a different method, for a few sample cases. These cases then serve to relate the bottom magnitude scale of Fig. 2 to actual distances. With this done, red shift z determines distance. This is the method of red shifts used by astronomers for establishing the scale of the universe.

Figure 4 shows a compact chain of five galaxies known by the catalogue designation VV172. The measured value of z for the galaxy second from the top is $z = 0.12$, the measured values for the other four are all close to $z = 0.053$. If the second galaxy is truly a member of the chain, we have here a gross discrepancy from a unique relation of red shift to distance, since all five galaxies would then be at essentially the same

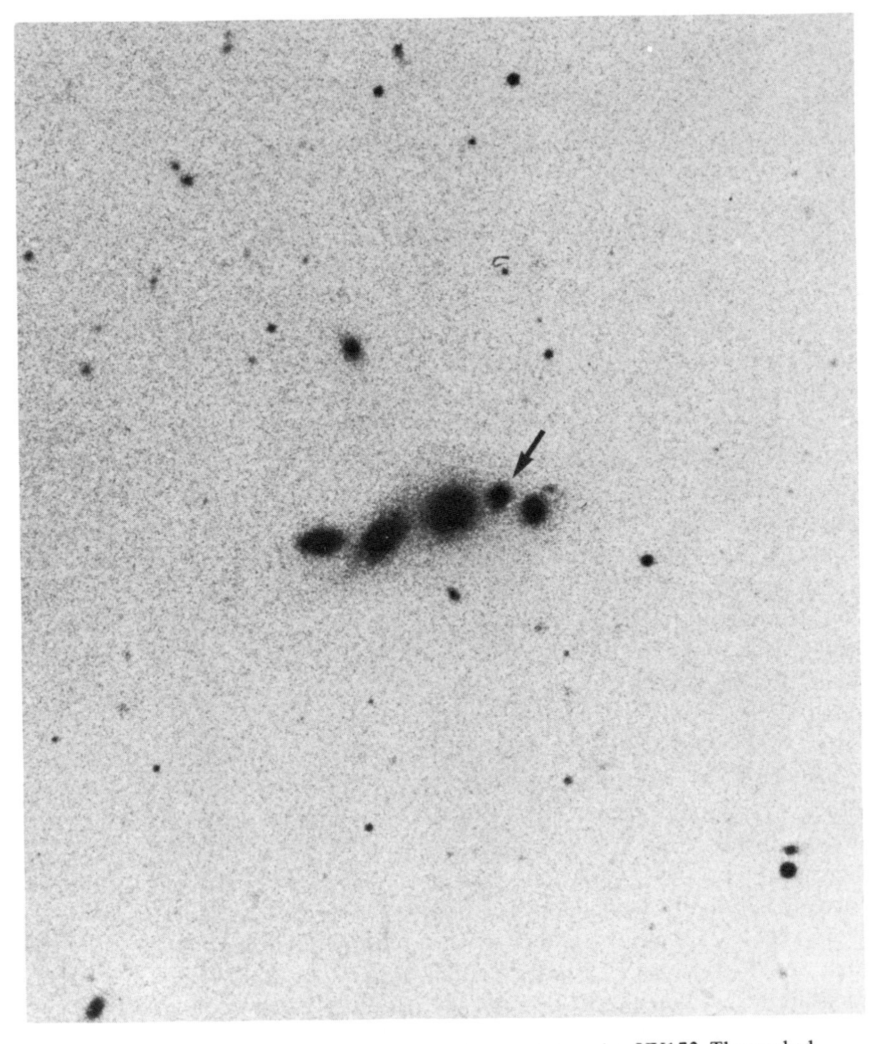

FIGURE 4 This chain of five galaxies has the catalogue designation VV172. The marked galaxy has been found by W. L. W. Sargent to have $z = 0.12$, whereas the other four have $z = 0.053$. If the marked galaxy is truly a member of the chain we have a serious discrepancy here from the usual distance–red shift relationship. (Courtesy: Hale Observatories)

distance away from us. But the second galaxy could be much more distant than the other four. Its apparent membership of the chain could be an illusion caused by projection on the sky. While such a situation would be unlikely, improbable occurrences happen all the time. Certainly on the strength of VV172 alone it would be unwise to reject the general rule suggested by Fig. 3.

Figure 5 shows a compact cluster of six galaxies known as Seyfert's sextet. The small spiderlike spiral in the middle of the group has $z = 0.066$, while the others all have red shift values close to $z = 0.015$. Another coincidence? Possibly, but it is curious to find marked discrepancies occurring in two out of half a dozen cases of compact groups of galaxies—not more than about half a dozen such cases have so far been examined from this point of view.

Figure 6 shows the galaxy with catalogue designation NGC 7603 connected to an appendage by a clearly marked arm and also by a faintly marked luminous crescent that sweeps through the appendage. It goes beyond reasonable scepticism to deny a real connection in this case. But if we accept that the galaxy and the appendage are connected, and hence are at essentially the same distance away from us, we have to face up to the measured red shifts, $z = 0.027$ for the galaxy, $z = 0.053$ for the appendage. And if the strict red shift–distance relationship is breaking down for Fig. 6, it is not unreasonable to suppose that the apparent discrepancies of Figs. 4 and 5 are also real.

What many astronomers fear is that any weakening of a strict relationship between red shift and distance will place in jeopardy the whole line of reasoning upon which the subject of cosmology rests, indeed that a well-ordered system of logic will become replaced by a kind of scientific anarchy. I suppose this has always been the point of view whenever anything unusual and far-reaching has happened in science. Actually I, myself, doubt whether the usual line of reasoning will in fact be greatly changed, at any rate as far as it affects most galaxies. The excellent correlation of Fig. 2 indicates that the kind of new effect that seems to be present in Fig. 6 is likely to prove the exception rather than the rule. This is for galaxies. The situation may indeed be drastically changed for the quasi-stellar objects (QSO's), the so-called quasars.

The QSO's are a new kind of object, discovered less than ten years ago. Many of them show very large red shifts, much larger than has ever been measured for a galaxy. Values of z greater than 1.0 are often found, whereas almost all galaxies have red shift values of z less than 0.25. On the usual view relating red shift to distance, such QSO's would be very far away, more distant than any known galaxy. But if, for QSO's, the red shift is not related to distance, then, of course, we have no reason to

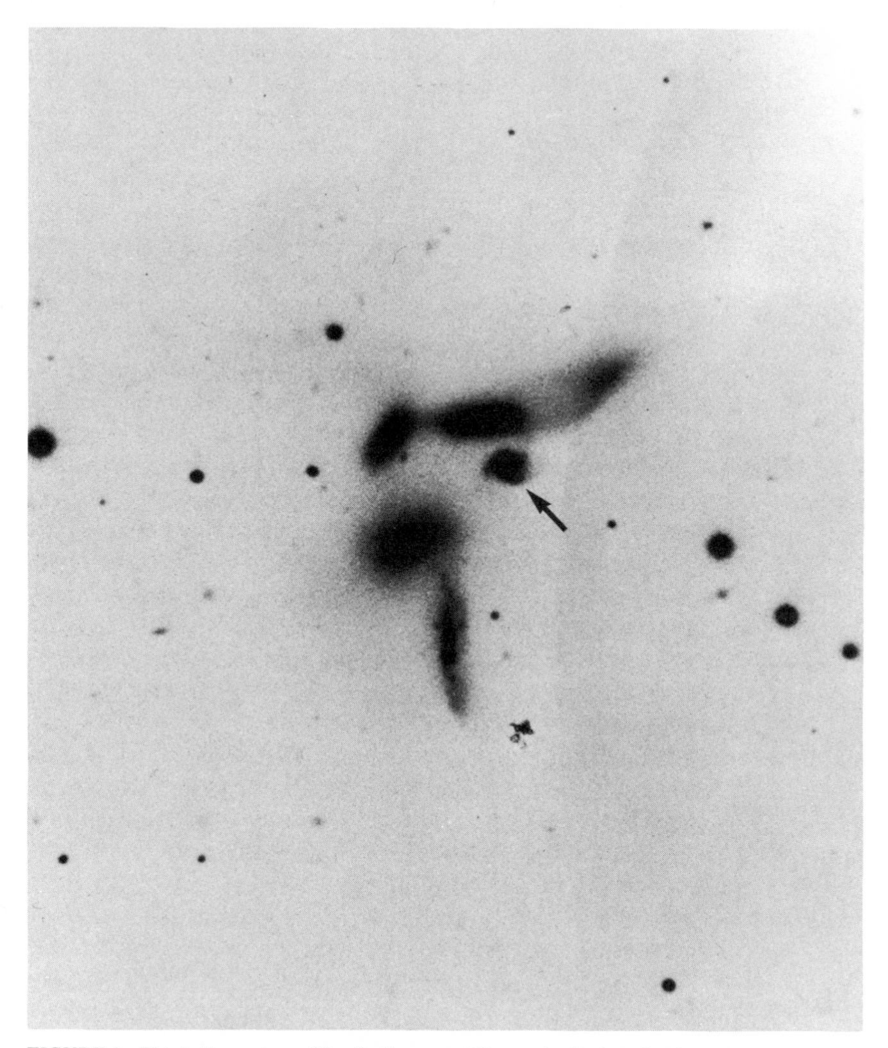

FIGURE 5 This is the system of Seyfert's sextet. The marked galaxy has been found by W. L. W. Sargent to have $z = 0.066$, whereas the other members have $z = 0.015$. If the marked galaxy is truly a member of the group, we have another serious discrepancy from the usual distance–red shift relationship. (Courtesy: Hale Observatories)

regard their distances as being very large. The first indication that red shifts might not be related to distances came when a plot similar to Fig. 3 was constructed for QSO's. The situation is shown in Fig. 7, which is immediately seen to be grotesquely different from Fig. 3. There is no indication at all of the QSO's all falling close to the line appropriate to the usual red shift–distance relationship. Either the QSO's

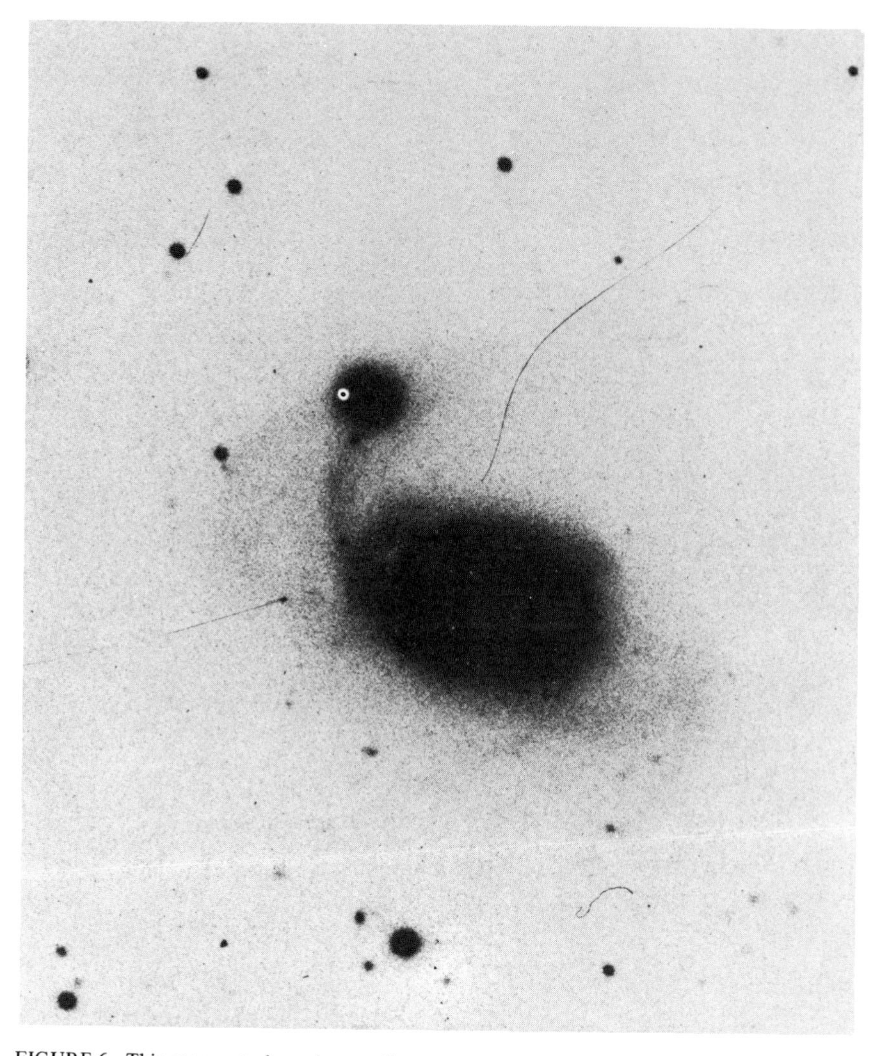

FIGURE 6 This appears to be a clearcut discrepancy from the usual distance–red shift relationship. The main galaxy with catalogue designation NGC 7603 has $z = 0.027$, whereas H. C. Arp finds $z = 0.053$ for the appendage. (Courtesy: Hale Observatories)

do not satisfy the usual red shift–distance relationship, or they must differ enormously one to another in their intrinsic properties. To save the usual point of view, it is necessary to suppose that, unlike the galaxies of Fig. 3, it is not possible to define a uniform class of QSO's.

For five years or more, astronomers have mostly taken refuge in this last possibility. Recently, however, a new line of evidence has emerged that is making it increasingly difficult to maintain such a point of view.

Figure 8 shows the galaxy with catalogue designation NGC 4651. Some ten years ago this galaxy was thought to be a source of radio waves, the source known as 3C 275.1. This association of the radio source with the galaxy was discovered to be in error, however. Instead, the radio waves were found to be coming from the starlike object marked in Fig. 8. This

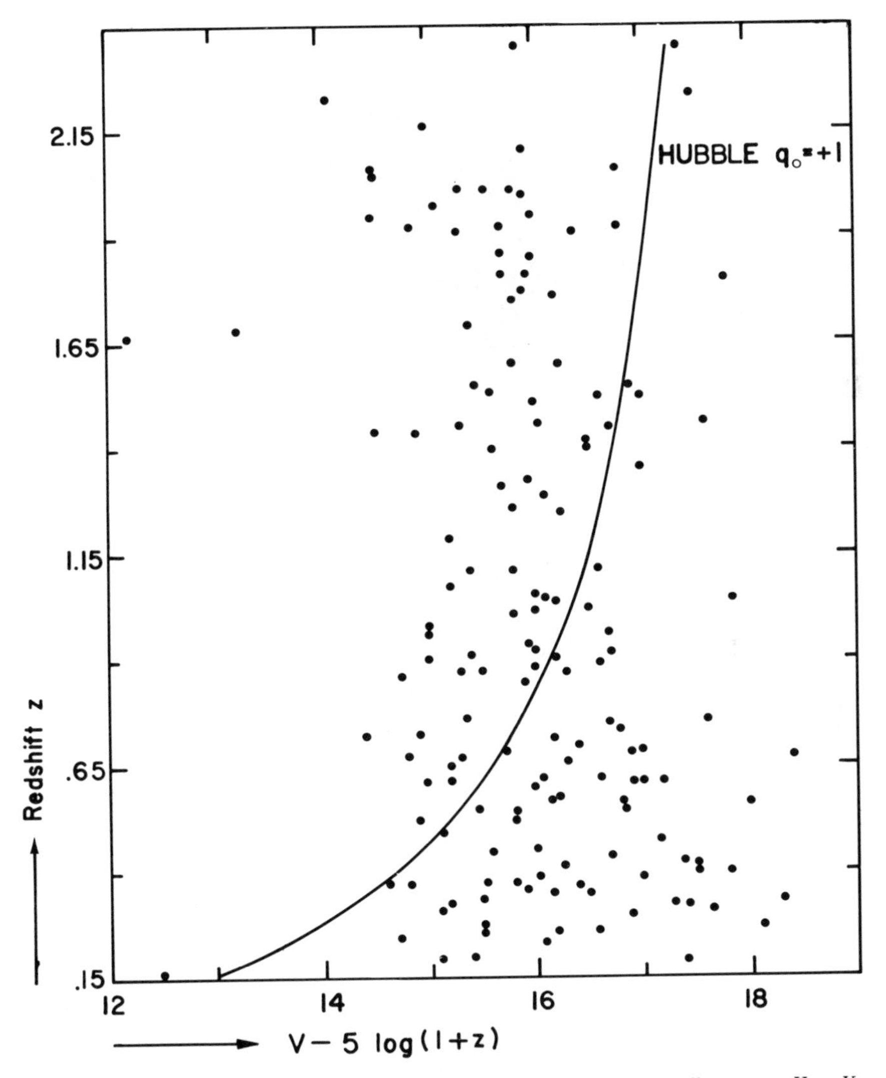

FIGURE 7 Plot of red shifts z against $V - 5 \log_{10}(1 + z)$ for quasi-stellar radio sources. Here V is the visual magnitude (instead of the photographic magnitude, as in Fig. 2). If z is unrelated to distance, this should be a scatter diagram, which it is.

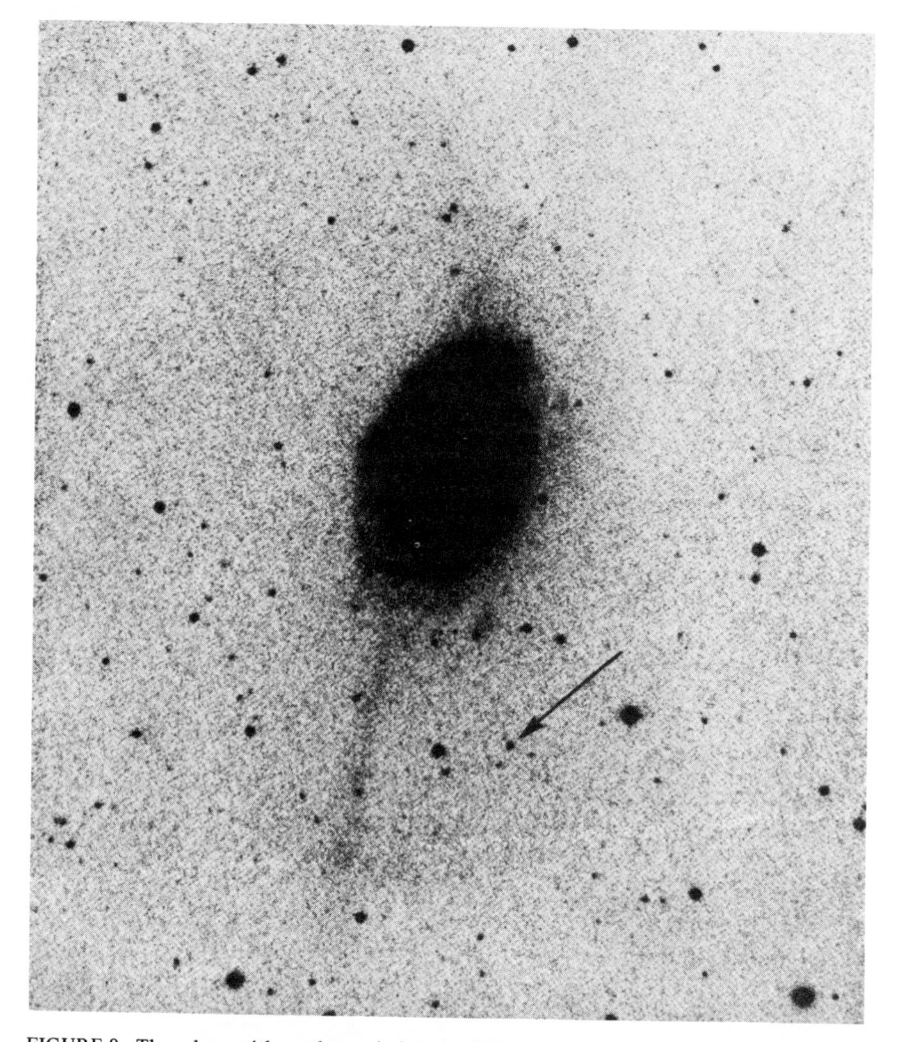

FIGURE 8 The galaxy with catalogue designation NGC 4651 was at first associated with a radio source having catalogue designation 3C 275.1. Then the radio source was found by more accurate measurements to lie on the marked starlike object. This turned out to be a quasar with $z = 0.557$. Since a number of other quasars of large red shifts have now been found lying apparently close to bright galaxies, the question must be seriously asked: Could the quasar really be associated with the galaxy? The probability of all these cases being due to chance projections on the sky is very small. (Courtesy: Hale Observatories)

starlike object is a QSO with red shift $z = 0.557$, whereas the galaxy has a red shift of only $z = 0.0025$. On the usual point of view the QSO would therefore be much farther away from us than the galaxy, and the apparent association of the galaxy and the QSO close together in the sky would be a coincidence. What is the chance of such a coincidence? If we restrict ourselves to galaxies in the New General Catalogue (NGC) and to radio sources of the 3C survey (Third Cambridge Survey) the chance is about 1/10. The situation seen in Fig. 8 would be unusual, but not wildly so.

The case of Fig. 8 has been known since 1965. What is recently new is that four similar cases have now come to light, all involving radio sources of the 3C catalogue and all involving galaxies in the NGC catalogue. The chance of all five cases being accidental is as little as 1 in 100,000. This possibility is so small that one is almost compelled to abandon the usual point of view. However, the following argument has to be considered before we accept this conclusion.

The QSO's of the 3C catalogue comprise about one fifth of all QSO's with measured red shifts. Suppose we widen the category of QSO's by including all of them—all those with known red shifts. This weakens the statistical weight of the five cases mentioned above. The possibility of five such cases occurring at random in the total category of QSO's is about 1 in 30, still small but not nearly so small as before. But then we must also consider new possibilities. Actually there are no new significant cases for galaxies of the NGC catalogue, but if we also widen the category of galaxies very slightly, by including those of the Index Catalogue (IC), the case shown in Fig. 9 appears, and the probability of everything being due to chance is about 1 in 10,000. This is once again too small to be accepted lightly.

One can pursue the argument further by significantly widening the category of galaxy, from the 10,000 or so that appear in the NGC and IC catalogues. Suppose we consider as many as 100,000 galaxies, chosen as the 100,000 with the greatest apparent brightness (or perhaps better, the greatest apparent size). This again weakens the statistical weight of the cases considered so far; but a further case, shown in Fig. 10, now appears. Adding Fig. 10 to all the previous cases, the probability of the whole set being accidental remains at about 1 in 10,000. The fact that the probability remains very low for these different ways of treating the data is a strong indication that the effect in question, namely, the close association of QSO's with nearby galaxies, is not accidental. In all cases, the red shift value for the QSO is much greater than that for the galaxy. For example, in the case of Fig. 10 the galaxy has $z = 0.023$ and the QSO has $z = 1.11$.

Why, one may ask, should all this trouble be taken to avoid the conclusion that red shift values are not uniquely related to distance? Why are astronomers so reluctant to accept this conclusion? Because it has not proved possible to find a straightforward explanation for the discrepancies. We are left therefore only with possibilities that are strange and as yet inscrutable. A straightforward explanation, which at first seems

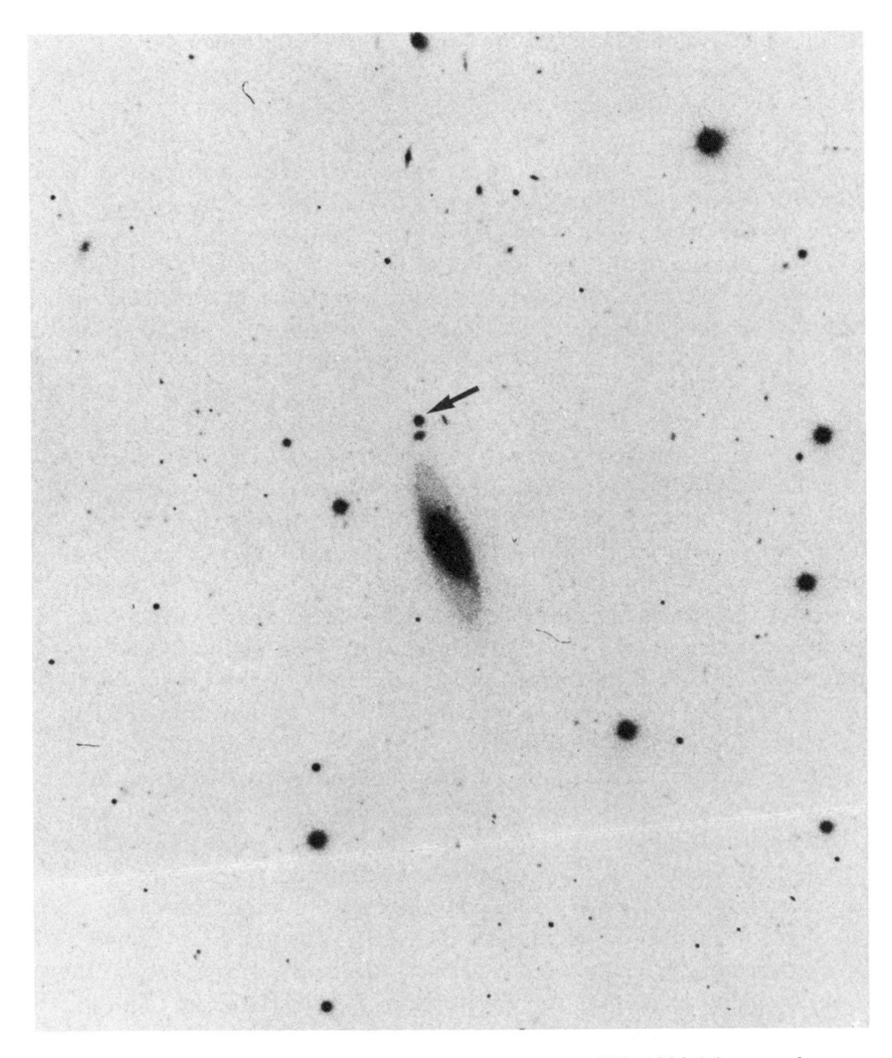

FIGURE 9 A quasi-stellar object (not this time a radio source), PHL 1226, lying near the galaxy IC 1746. The separation in this case is about half a minute of arc. (Courtesy: Hale Observatories)

worth considering, is to suppose that QSO's are shot out of galaxies at high speeds comparable to light. Such a rapidly moving object would show an appreciable excess red shift if it happened to be moving away from us, due to the well-known Doppler effect. But why should the QSO's found close to galaxies all be moving away from us? One would expect some of them to be shot out in directions toward us. Such cases would show a reversed effect: the frequencies of their emitted spectrum lines would be raised higher than similar lines from the parent galaxies. This reversed effect is often described as a blue shift. While one might conceivably argue that because the sample of QSO's lying apparently near galaxies is rather small, it is simply a matter of chance that no blue shift case has yet been found, there is a further serious difficulty. Not one of the 200 or so QSO's plotted in Fig. 7 shows a blue shift when its spectrum lines are compared with those of atoms in a terrestrial laboratory. This seems to dispose of the Doppler shift idea, unless all QSO's arose as debris from a gigantic explosion that took place locally, perhaps

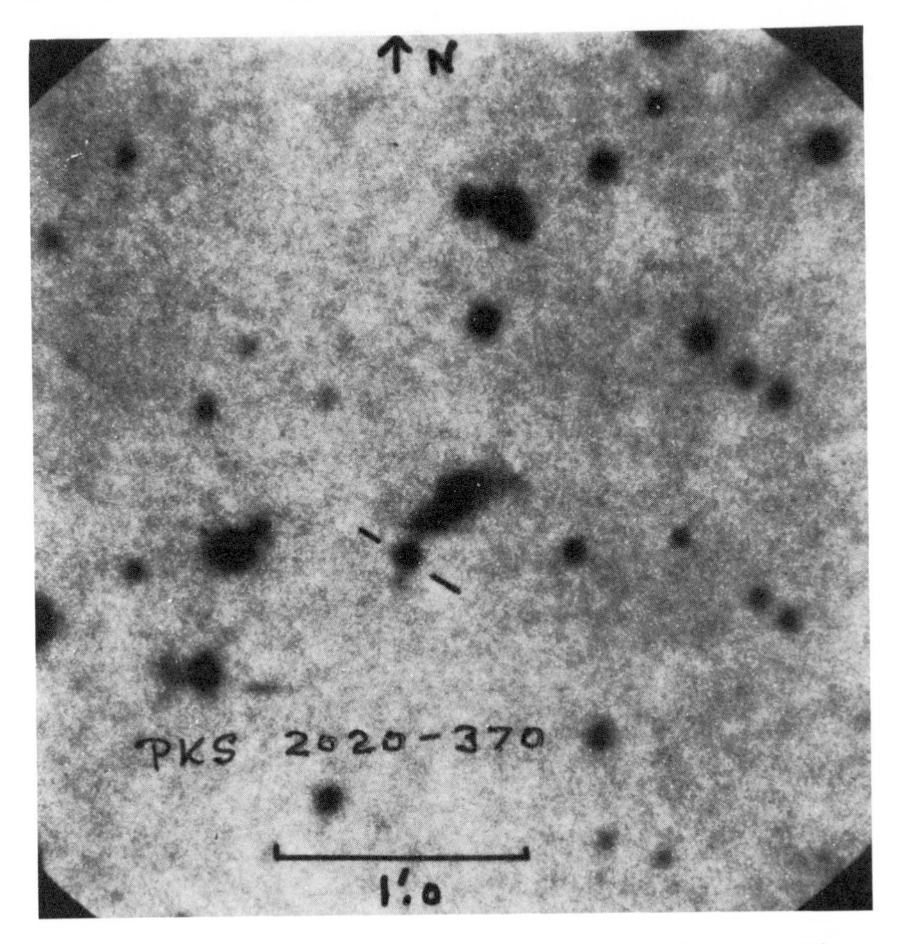

FIGURE 10 The remarkable case of the radio source PKS 2020-370, a QSO with red shift $z = 1.11$ lying within about a quarter of a minute of arc of the galaxy. (Courtesy: Peterson and Bolton)

in a nearby cluster of glaxies. Even so, we would still be left with the difficulty of explaining discrepant cases like that of NGC 7603 and its appendage (Fig. 6).

To make progress, we must ask the general question: what is the phenomenon of red shift due to anyway? The usual answer to this question is a somewhat complicated one. The usual answer lies in terms of geometry. According to Einstein's general theory of relativity, the universe in the large has a geometry different from Euclid's. The more complicated form of Riemannian geometry is used. When large distances are involved, the departure from Euclidean geometry becomes important and shows itself as a red shift. The greater the distance, the greater the difference between the geometries of Euclid and Riemann—and the greater the red shift.

It is possible to generalize Einstein's theory in such a way that this complication of geometry is removed, indeed in such a way that we can continue to use simple Euclidean geometry even as far as the universe in the large is concerned. How are we to understand red shifts according to this alternative point of view? In a more physical way. The frequencies of the spectrum lines emitted by a particular atom depend in part on the atom in question and in part on the mass of the electron. Change the mass of the electron, and the frequencies of the spectrum lines are changed. Suppose now that on a cosmological scale the mass of the electron increases slowly with time. Then it follows naturally that the light from distant galaxies will show the red shift effect and that the greater the distance, the greater the red shift. This is simply because our observation of a galaxy refers to the state of atoms, not as they are now in the terrestrial laboratory but to the atoms as they were at the time the light started on its journey from the galaxy to the earth. This introduces a *time delay*—the greater the distance, the greater the time delay; the smaller the electron mass, the lower the frequencies of spectrum lines and so the greater the red shift. The relation between red shift and distance for galaxies, shown in Fig. 2, can be understood in a completely satisfactory and natural way along these lines.

If the mass of the electron could also vary spatially—i.e., from place to place—as well as with respect to time, we would have the beginning of an explanation for the curious observations discussed above. But as soon as we try to formulate this idea in a precise way, we hit massive difficulties. This is not to say that the idea is wrong but that deep-rooted changes in our physical understanding are needed if we are to give satisfactory expression to it. We are forced to quit the safe ground of well-attended physical theory for the icy waters of uncertainty, and this is something that nobody lightly accepts. Many physicists and astronomers will hope that somehow a way around the curious observational situation will be found. For myself, until a year or so ago I thought this might just be possible, but the accelerating pace of unexpected new discoveries now seems to make such a comforting possibility unlikely.

I will end this paper with a suggestion as to the direction in which physical ideas might well go as a consequence of this developing astronomical crisis. Throughout the history of science there have been two opposing views of the basic nature of physics. Both are associated with famous names: the direct particle-interaction theory associated with Newton and Gauss and the field theory associated with Maxwell and Einstein. In the early days of physics, the direct interaction theory held sway, but throughout most of the present century the field theory has been preferred, since the field theory has seemed to lead to more fruitful developments. But the direct interaction theory has always shown itself capable of yielding the same successes as the field theory, although some problems have certainly proved much harder to solve in terms of the direct theory. The direct interaction theory is never awkward or unaesthetic in its mathematical structure, however. The difficulties have proved conceptual rather than mathematical.

It will be as well to consider an example. The everyday world of experience shows causality. The field theory is made causal at the outset, thereby permitting local problems in the laboratory to be solved without making any reference at all to the universe in the large. The direct interaction theory, on the other hand, has no built-in causality. Indeed, quite the reverse. It requires all purely local problems to lack causality, to be acausal. How then are we to understand everyday experience? By recognizing that our everyday world is not truly local. We are not in fact separated from the universe in the large. Causality arises not locally but from an interaction with the whole universe. Indeed it is the universe in the large that impresses causality on our everyday world. This is the situation in the direct theory. Putting it very generally, in the field theory the universe obeys and is controlled by locally determined laws. In the direct interaction theory this situation is essentially reversed—the local situation is controlled in part by the physical laws (acausal laws) and by the structure of the universe in the large.

The crisis in astronomy today is quite simply this: do the observational data force us to the view that the red shift–distance relationship is discrepant in some cases? If it does, then new physical ideas are evidently needed. These are likely to be far-reaching. In particular, it seems that the direct particle interpretation of physics will be greatly helped in its long controversy with the field theory interpretation. So critical are these consequences that the utmost care must be taken in assessing the data. For me, personally, the exact state of the data at any given moment is less important than the trend of the data. There seems no doubt that the trend is toward forcing us, whether we like it or not, across a bridge into wholly unfamiliar territory. Either the bridge must be crossed or we must judge the astronomical data that have come along in the past few years to be extremely freakish.

Supported in part by the National Science Foundation under grants GP-28027 and GP-27304.

GIULIANO TORALDO DI FRANCIA *is a professor of physics at the University of Florence and director of the Research Institute for Electromagnetic Waves of the National Research Council of Italy. He is president of the Italian Physical Society and editor of* Il Nuovo Cimento. *From 1966 to 1969, he was president of the Commission for Optics of the International Union of Pure and Applied Physics.*

4 *Optics* / G. TORALDO DI FRANCIA

Before the starting of almost any human activity, whether of ordinary everyday life or of scientific investigation, there is a preliminary stage when the subject collects information about the outer world. This can be achieved in several ways, by taking advantage of different physical processes. However, in comparatively recent times physicists have come to realize that the most efficient processes of this kind are those that go under the general heading of *scattering*. One aims a suitable beam of particles at the *target* or *object* to be investigated and observes the recoil particles or, more generally, the end products of the process (Fig. 1). A classical attitude is to start from the results of this observation and subsequently to derive by theory and computation some properties of the target that are assumed to be more fundamental than the mere scattering data. Alternatively, one might take a very cautious attitude and assume that the scattering matrix, *without further elaboration*, gives complete information and fully describes the object as far as that kind of primary particle is concerned.

In order for us to be able to collect information from a scattering process, two conditions must be fulfilled:

(A) A beam of primary particles must be available with sufficient intensity for the end products to be distinguishable from the background *as well as from the particles spontaneously emitted by the target.*

(B) The observer must be provided with a detector of suitable *sensitivity* and *resolving power.*

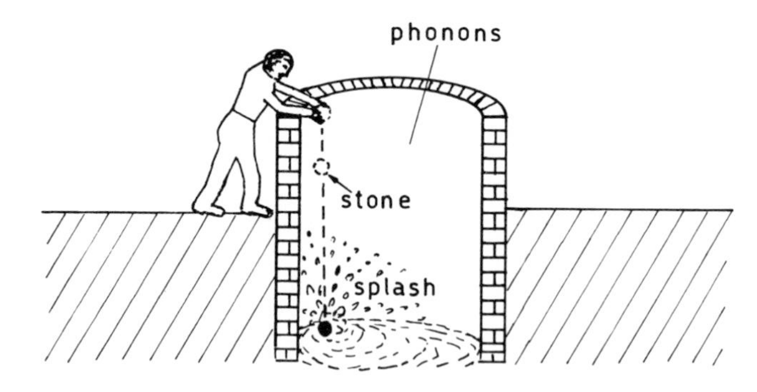

FIGURE 1 The primary particle can be a stone and the end product phonons. The man dropping the stone derives from the phonons many properties of the target. He can measure how deep is the well and knowledge about there being water, mud, or nothing at the bottom.

Thermodynamically, condition (A) implies a situation of nonequilibrium and is related to the entropy balance. At thermodynamic equilibrium no information can be acquired about the details of the surroundings. In order to collect information we must have a source of negative entropy or *negentropy*.

The negentropy principle of information was developed many years ago by L. Brillouin. The negentropy of an isolated system in a given state may be defined as the difference between the maximum admissible (or equilibrium) entropy of the system and the entropy in that state. Negentropy can be either dissipated through irreversible processes or converted into information. Information, in turn, can be transformed back into negentropy. The second principle of thermodynamics must be written in the form

$$\Delta (N + I) \leqslant 0,$$

where N stands for negentropy and I for information *measured in the same units*. One *bit* of information is equivalent to $k \ln 2$ thermodynamical units, k being the Boltzmann constant. In other words, one bit of information must be paid for with at least $k \ln 2$ units of negentropy.

Now, the only important source of negentropy for us is the sun. In other words, the sun–earth system has a lot of negentropy to be spent. Negentropy from the sun is carried by radiation of three main types: (1) neutrinos, (2) electromagnetic radiation, and (3) solar wind.

Neutrinos reach the ground freely but cannot be used for acquiring information about terrestrial objects. First, they are virtually never scattered by material objects, and, second, condition (B) cannot be fulfilled, at least at present.

As far as the other two types of radiation are concerned, we find ourselves in a very peculiar condition. *We are living inside a black box*, whose walls are represented by the atmosphere. The atmosphere is virtually impenetrable to electromagnetic and particle radiation.

Luckily enough, the builder of the box has made a small oversight. He has left inadvertently a tiny *crack* in the wall, which lets in visible light, plus a bigger *hole* for microwaves (Fig. 2). Nature has been terrific in exploiting the possibilities offered by the radiation coming through the tiny crack. An overwhelming majority of living creatures is provided with detectors for visible light, and very often such devices are extremely refined and sophisticated.

This wonderful display of efficiency leads us forcibly to ask the following question: Why did nature disregard microwave radiation coming through the bigger hole?

There are indeed very good reasons. An obvious reason is that a microwave detecting device endowed with reasonable resolving power, as required by condition (B), would have to be too big, as compared with the normal size of an animal (Fig. 3).

But there is something else. As is well known, the average number of photons per degree of freedom contained in electromagnetic radiation at temperature T is represented by

$$n = 1/[\exp(h\nu/kT) - 1].$$

Now in the case of visible light there results $n \simeq 0.05$ for solar radiation and $n \simeq 10^{-26}$ for terrestrial radiation. There follows that the scattered solar photons can be perfectly distinguished from the virtually

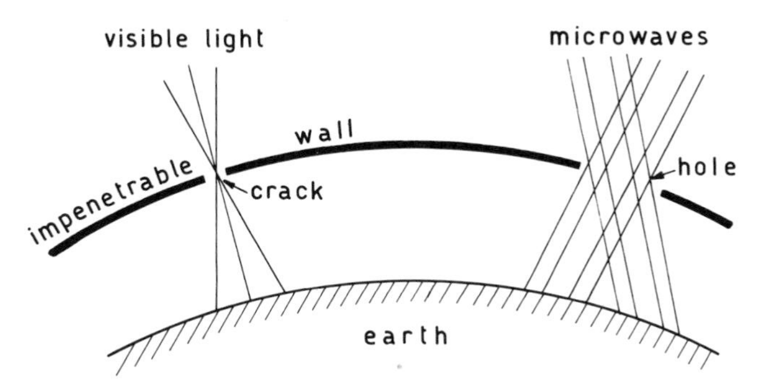

FIGURE 2 We are living inside a black box.

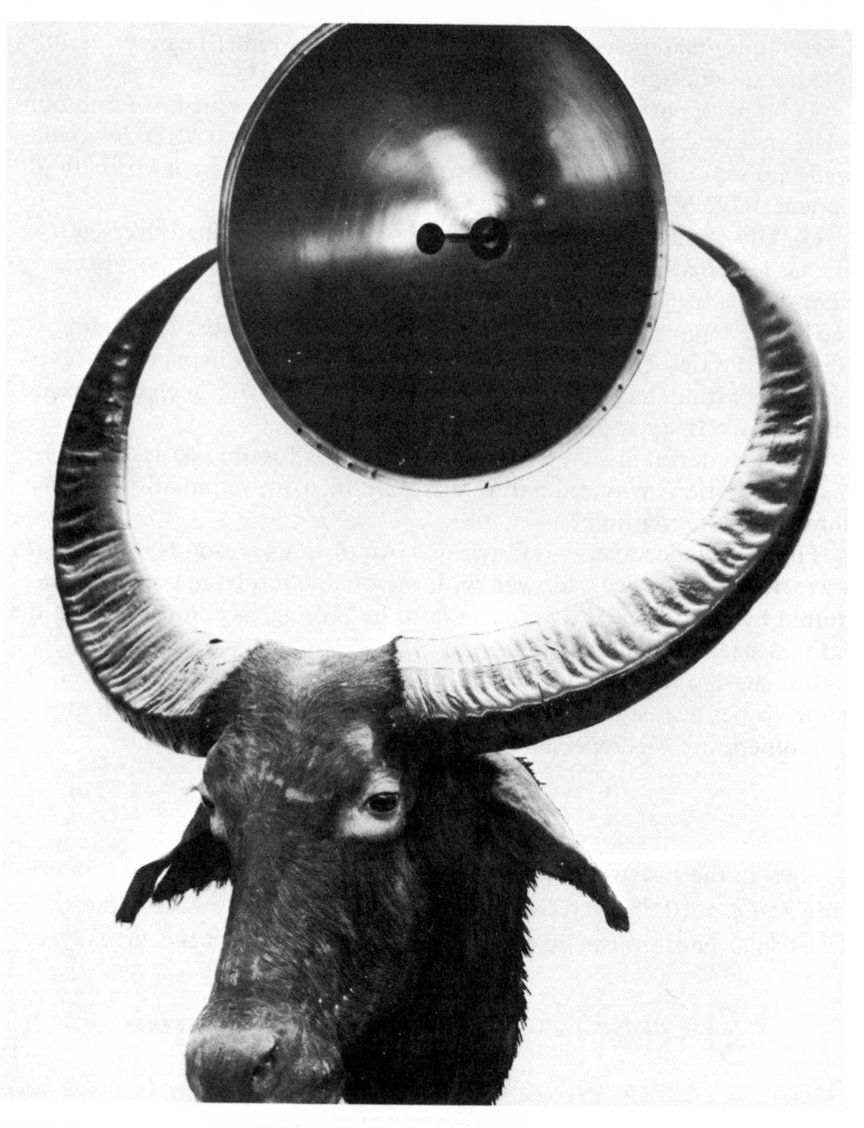

FIGURE 3 A mythical species that became extinct because the animal could not run sufficiently fast or when there was a strong wind, it had to run too fast!

nonexisting photons emitted by the target, even when solar light does not arrive directly to the target but is previously scattered one or more times by other objects (atmosphere, moon, walls of a room). Incidentally, the absence of a background noise renders it useful to employ a detector of the highest sensitivity, of the order of one photon. The human eye has such a sensitivity.

The situation is different for microwaves. In this case we have $h\nu << kT$ both for solar and for terrestrial radiation, so that $n \simeq kT/h\nu >> 1$. Since the sun surface is about 20 times as hot as the earth surface, the value of n for solar radiation is only 20 times greater than the value for terrestrial radiation. Even in the case of direct illumination, and very small absorption, a target that scatters within an angle of more than 2.5° sends out more noise than scattered radiation (Fig. 4). Nature has decided not to go about all the trouble necessary to provide animals with such an inefficient system for collecting information.

This is probably the reason why of all possible scattering processes that can be used to obtain information about the physical world, light scattering has virtually been for so many thousand years the only one known to human beings. Consequently, *optics*, as is revealed by its Greek etymology, has been thought to be necessarily related to the eye and vision.

Now we can ask ourselves: Is there any sensible reason to go on and preserve as a separate science the science that deals with the collection of information by means of light scattering?

The limitation to a very small band of electromagnetic radiation seems to be due rather to an accidental condition of the sun–earth system than to fundamental or conceptual reasons.

Of course, the hardware used to deal with visible light is different from that used for other radiations. Consequently, the *technology* of optical instruments may still represent a separate body of knowledge. And we must not forget that very peculiar and intriguing optical instrument the human eye. This system is still the object of profound investigations.

However, the emphasis of present-day research is placed rather on a number of methods and devices that do not belong specifically or necessarily in the visible range of electromagnetic radiation. The main problem and *leitmotiv* of the research is to derive information about an object from the radiation scattered by it or sent out spontaneously. On this account one may even be tempted to say that optics encompasses a great part of modern physics.

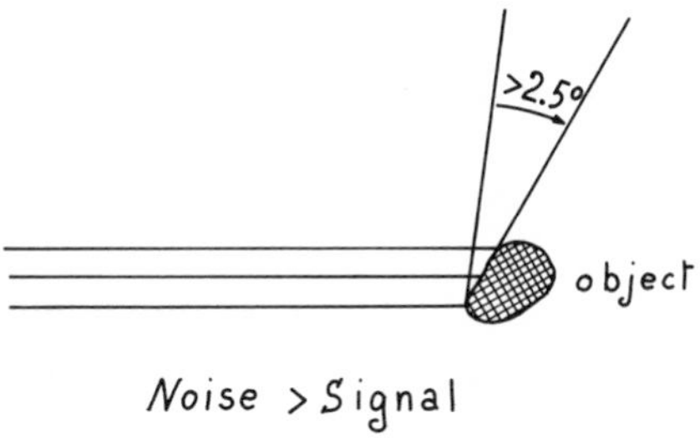

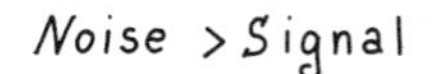

FIGURE 4 Whenever an object scatters light or other radiation at an angle greater than 2.5° the background noise is of the same level as the incoming signal.

Man realized very early that when solar radiation was not available (by night or inside a cave), he could easily produce more or less the same radiation by means of fire. From that discovery he proceeded to the construction of a host of artificial sources like oil and gas lamps, candles, and torches.

He did not know that some animals, like bats, were provided by nature with sources and detectors of a different type of radiation, namely *ultrasonic radiation*, and that the system could be very efficient for gathering information, especially for measuring distances.

No wonder that when an extremely serviceable physical agent like electricity was discovered, it was first applied to the production of conventional light.

Meanwhile the art of gathering information from light had developed to a high degree of refinement, on the basis of geometrical optics. Conventional optical instruments were virtually brought to perfection. Wave optics was considered only as a sort of unfortunate disturbance that set a limit to *resolving power*, due to the appearance of the *Airy disk* of diffraction.

The first important step toward a conceptual revolution was due to Rutherford and Laue. At about the same time, they showed that two kinds of radiations, different from visible light, could be applied to gather essential information about atoms and crystals, respectively. The *inverse scattering problem* was making its decisive appearance in physics.

Since then, the method has been applied with all conceivable sorts of radiation and with enormous success.

It took some time for optics researchers to realize that what they had been doing for a long time was nothing but a particular case of what a greater and greater number of physicists were doing with different radiations. The eye, like the other optical instruments of classical type, is an analog computer that elaborates the information carried by scattered light and presents it in a convenient form to the mind of the observer. However:

1. Information can be displayed also in different and sometimes more convenient ways.
2. A lot of information carried by the scattered light is missing in the image formed by a conventional instrument.

Two main factors are responsible for missing information:

2(a). Some *real* and all *evanescent* waves scattered by the object are not collected by the instrument, consequently their information is lost.

2(b). Classical instruments do not measure the *phase* of scattered light, and the corresponding information is also lost.

As is well known, the latter shortcoming was removed between the wars by F. Zernike with the invention of the phase-contrast method and in general with the method of the coherent background. The idea is to introduce a proper phase shift in the light scattered by a small object and to make it interfere with the light of a coherent background. Mere phase differences are thus converted into amplitude differences.

Today, this idea appears so simple and natural that one may miss its historical significance and wonder why it had such an influence on all later developments. For the first time after the pioneering efforts of E. Abbe, *coherent light* entered the scene as an important tool for investigating the structure of the visual world. It turned out to be so useful that soon after the Second World War many refined methods and devices were developed to take full advantage of the possibilities it offered. It is of interest to note that this happened long before coherent light could be produced with substantial efficiency by means of lasers. Zernike's phase contrast is very efficient but only applicable to particular microscopic objects. The large-scale utilization of interference with a coherent background came with the introduction of holography, whose first idea occurred to D. Gabor in 1948. Gabor had in mind electron microscopy. He was concerned with the difficulty of correcting the spherical aberration of electron lenses and thought it possible to compensate spherical aberration by wavefront reconstruction. However, his method has never found a practical application in electron microscopy and has instead come into prominence for the use with laser light.

Let S represent the signal, namely, the complex amplitude of a beam of coherent light scattered by the object, and R the coherent background, or reference beam (Fig. 5). When they impinge simultaneously on a photographic plate, the exposure E is proportional to the square modulus of their sum or to

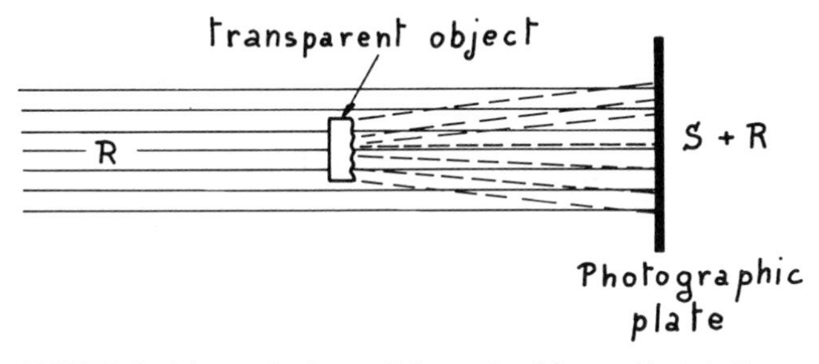

FIGURE 5 A beam of coherent light scattered by an object; in-line method.

$$E = SS^* + RR^* + RS^* + R^*S. \tag{1}$$

If a positive is made with a resultant $\gamma = -2$, the amplitude transmission will be proportional to E. Therefore, if we illuminate with the reference beam, we will get for the transmitted amplitude

$$A = (SS^* + RR^*)R + R^2S^* + RR^*S. \tag{2}$$

Under the usual conditions of holography, the first term represents the reference beam almost ummodified, the last term represents the signal multiplied by the constant (or almost constant) factor RR^*, while the second term represents the complex conjugate of the signal, or "twin image."

In his original work, Gabor had to use the in-line method, where the signal and reference beams were approximately in one line. This was because of the limited coherence of the light used (high-pressure mercury lamp), which did not allow interference at large angles. As a result, the image, the twin image, and the reference beam were very inconveniently superimposed. This difficulty was overcome by Leith and Upatnieks in 1962, when lasers were available, by taking a skew reference beam. Thus in the reconstruction the three beams were angularly separated (Fig. 6). The assumption that $\gamma = -2$ is useful for an elementary discussion but can easily be dropped. If $\gamma \neq -2$ one obtains besides the ordinary and twin images a series of "harmonic terms" or of angularly separated images. Black-and-white holograms, bleached holograms, and phase holograms have become possible. Black-and-white holograms

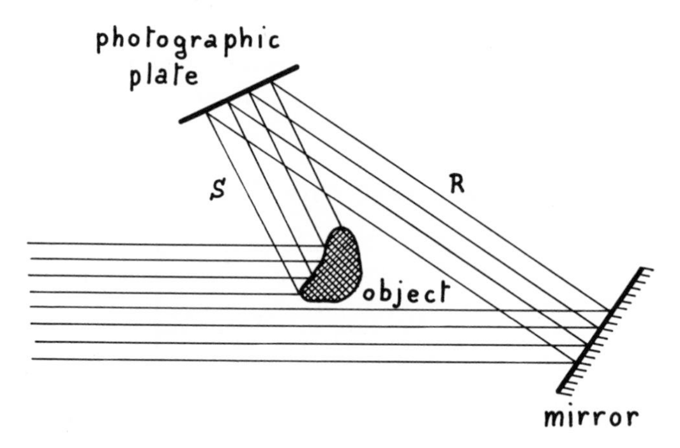

FIGURE 6 A beam of coherent light scattered by an object; skew-beam method.

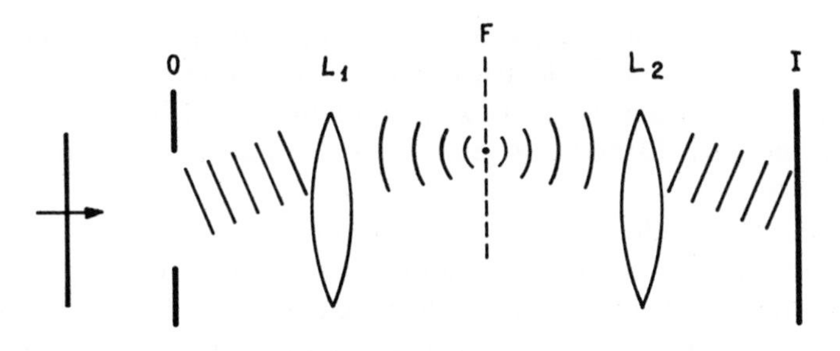

FIGURE 7 Image formation with coherent light through a transparent object.

can be computed and synthesized. By taking advantage of the thickness of the emulsion one can combine the principle of Lippman's plates with holography, as suggested by Denisyuk, and get color holograms. One of the good points of holography, which makes it very useful for application, is that information about a given wavefront remains "frozen in" in the emulsion and can be linearly added to the information of a different wavefront. This property is applied in holographic interferometry. Two holograms of a given object, taken at different times, are recorded on the same plate. The two wavefronts obtained in the reconstruction can interfere, and from the interference pattern one can tell whether the object has moved or has been deformed between both recordings. This property has important industrial applications. Gabor has said in 1971, when looking back to his intended electron-optics application: "Why should one bother in light optics with such a complicated two-stage process, when coherent light was so weak and uncomfortable to use, when we had such perfect lenses, even achromatic ones? Little did I think at that time that after 24 years the application of holography in electron microscopy would still be in a primitive stage, while the simple optical experiments, which I considered only as model or feasibility experiments, would give rise to a new branch of optics, with some 2000 papers and a dozen books!"

Image formation with coherent light can be described in a typical case of a transparent object (Fig. 7). A coherent plane wave illuminates the plane of the object O and is scattered by "inverse interference" into many plane waves with different directions, plus a set of evanescent waves. Each plane wave is brought to a focus by lens L_1 at a point of the focal plane F conventionally termed the "pupil plane." Lens L_2 transforms the wave back into a plane wave. All the output plane waves are brought together to direct interference at the plane of the image I.

Each one of the scattered plane waves represents a *spatial frequency* or a component of the Fourier spectrum of the object. This analysis shows that there is an analogy between an optical signal as a function of space coordinates and an electrical signal as a function of time. The ordinary techniques of frequency manipulation and filtering can be applied. The only difference is that in optics there appear two dimensions instead of one.

Let us first consider the question of resolving power.

If the instrument could collect all the scattered waves and bring them back to interference onto plane I, the image would be similar to the object. However, due to the finite aperture of the instrument some real waves and all the evanescent waves do not enter the system and are missing in the image. There is a cutoff in the spatial frequencies of the image, and, consequently, the finest details of the object are lost. This is tantamount to saying that an infinite number of different objects should have one and the same image. The image is *ambiguous*.

The question of how many and what objects have one and the same image is closely related to the problem of finding the number of degrees of freedom of the image. This question dates back at least to Laue's discussion of the degrees of freedom of electromagnetic radiation. In more recent times it has been argued that, due to the limitation of the spatial frequency band of the image, one can apply the sampling theorem, with the result that the number S of degrees of freedom is proportional to the object area times the entrance solid angle of the instrument, divided by the wavelength squared. This has been called the "Shannon number."

This conclusion has been repeatedly refuted for the reason that when the object has finite extension, the frequency distribution over the pupil plane is an analytic function and as such can be completely known when we know its behavior even over the limited domain of the pupil. This remark is true, but it does not have any practical value. The question has been completely clarified by means of the prolate spheroidal functions, analyzed by Slepian and Pollack. An object distribution represented by a (properly scaled) spheroidal function ψ_n has an image similar to the object, i.e., an image represented by the same distribution multiplied by a factor λ_n. Now the ψ_n form a complete set of orthogonal functions for the object, which can therefore be expanded in series of the ψ_n. Each coefficient of the series corresponds to a degree of freedom of the object. Now it turns out that $\lambda_n \simeq 1$ up to $n \simeq S$, while for $n > S$ the factor λ_n drops practically to zero. The corresponding degree of freedom is lost for the image, even when we use any physically conceivable type of detector. Therefore S represents the "physical" number of degrees of freedom.

Of course, the evaluation of S represents only a preliminary stage for the application of information theory to optics. A complete theory can be developed only by taking into account noise. Noise in optics can arise from different sources. But one source that can never be eliminated is represented by the photon nature of light.

The following rough argument, due to Gabor, is very instructive. A photographic plate collects energy $RR^* = E$ from the reference beam, $SS^* = e$ from the signal, and $RS^* + R^*S = 2\sqrt{Ee}\cos\Phi$ from their interference. The last term, which represents the interference fringes, carries the information and can be made as large as we want by increasing E, however small the signal energy e. In the limit one could transmit information without energy. Of course, this is absurd. The signal cannot be recognized as soon as it is drowned in the fluctuations of E. Now the mean-square amplitude of the fringes is $2Ee$. We postulate that the signal becomes unrecognizable at some minimum energy ϵ at which the mean-square fluctuation of E exceeds this by a factor k. Thus,

$$\langle \Delta E^2 \rangle = 2kE\epsilon,$$

or, putting $E/\epsilon = n$,

$$\langle \Delta n^2 \rangle = 2kn.$$

By putting $k = \frac{1}{2}$ we get Poisson's law:

$$\langle \Delta N^2 \rangle = n,$$

which indicates that the fluctuation of the energy is of the nature of shot noise. One will recognize that the result is the correct one as long as the photons are distributed over a very great number of cells of phase space. Otherwise there is a correction term, due to the fact that photons are not classical particles. Of course, the problem is similar to that of the fluctuations of blackbody radiation in an enclosure, discussed by Einstein.

Whatever the source of noise, one can do something to code and to process the information to reduce its effects. For instance, one can filter out unwanted frequencies from the plane of the pupil, a procedure pioneered by Maréchal and Croce. We just mention a few interesting cases.

The object can be periodic. It gives rise to a series of dots on the image plane. By blocking these frequencies by means of black dots (Fig. 8), one removes the image of the "perfect" object and will reveal only its faults. Conversely, if one lets through only the periodic frequencies by means of a diaphragm with holes, one can remove the faults and restore a perfect image of the object.

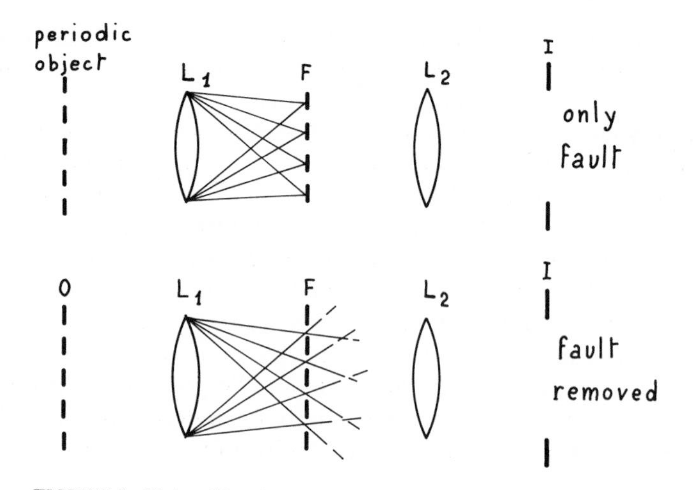

FIGURE 8 If the object is periodic, the scattered waves are discreet and we can stop them at the focal plane, removing everything that relates to the object and only the fault appears in the image.

In general, the restoration of an image damaged by blurring or other causes requires the setting up of a filter for both amplitude and phase in the plane of the pupil. This was a formidable task, until holography was applied to it by van der Lugt and Stroke. Of course, holography allows a plate to be built that gives to the impinging wavefront any wanted modulation of amplitude and phase.

One of the first things that become possible by this filtering technique is the restoration of images blurred by defocusing, movement, or other causes. Another application of great interest is in pattern recognition. If the pattern to be recognized is p, its Fourier spectrum P is formed in the focal plane F and is holographically recorded. The hologram will contain both P and P^*. The latter represents the filter to be used. When P^* is multiplied with the Fourier spectrum of the object, one obtains in the image plane the correlation function of the object and the pattern. Bright spots appear in the places where the object contains the pattern to be searched. It is obvious that this procedure suggests the possibility of building a reading machine. However, no practical device appears so far to have been built on this principle. One of the serious problems one meets when applying this technique is that the pattern may have different size and orientation from that used in the recording. Several proposals have been made to overcome this difficulty. Some partial good results have been obtained, but the problem still appears to be serious.

All these developments appear to have been made possible by coherent light. What is coherent light? For a long time people were contented with the vague notion of coherent light being highly monochromatic light coming from a very small source. A quantitative analysis was pioneered by Zernike and developed extensively in the last few decades. Let us denote by $V(t)$ the complex wavefunction (or *analytic signal*) whose square modulus $V(t)\,V^*(t) = I(t)$ represents light intensity.

For two points P_1, P_2 we define the *mutual coherence function*

$$\Gamma_{12}(\tau) = \langle V_1(t+\tau)\, V_2^*(t) \rangle$$

as a time average taken over a sufficiently long period of time. The normalized function

$$\gamma_{12}(\tau) = \Gamma_{12}(\tau)/(\langle I_1 \rangle \langle I_2 \rangle)^{1/2}$$

is called the *complex degree of coherence.* It can be shown that the maximum value of $|\gamma_{12}|$ represents the visibility of the fringes (Fig. 9) formed by light coming from P_1 and P_2 in a Young interferometer.

There is a beautiful theorem (see Fig. 10), due to van Cittert and Zernike, on the value of γ_{12} for two points illuminated by a plane quasi-monochromatic source S: γ_{12} equals the normalized complex amplitude that would be produced at P_2 by a spherical wave centered at P_1 and diffracted through an aperture equal to S. This easily explains why Michelson could measure star diameters by means of his stellar interferometer (Fig. 11). The shortcomings of such an interferometer are well known. Scintillation and the difficulty of maintaining the arm length constant to a quarter-wavelength make it impossible to exceed the distance of a few meters.

However, one can show that if light can be considered as a random Gaussian process, the fluctuations of *intensity* obey the equation

$$\langle \Delta I_1(t+\tau)\, \Delta I_2(t) \rangle = I_1 I_2 |\gamma_{12}(\tau)|^2.$$

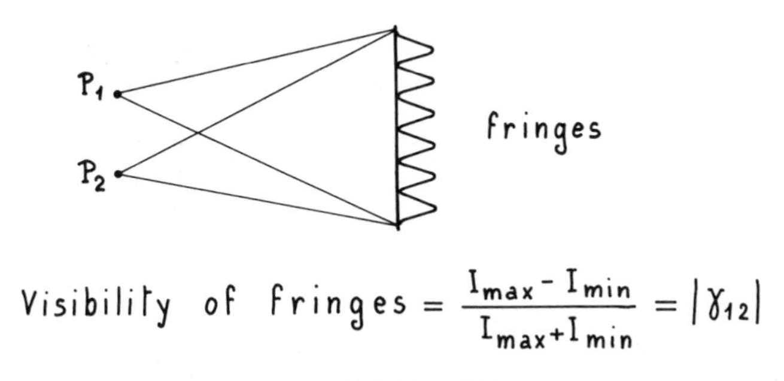

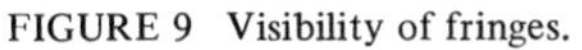

FIGURE 9 Visibility of fringes.

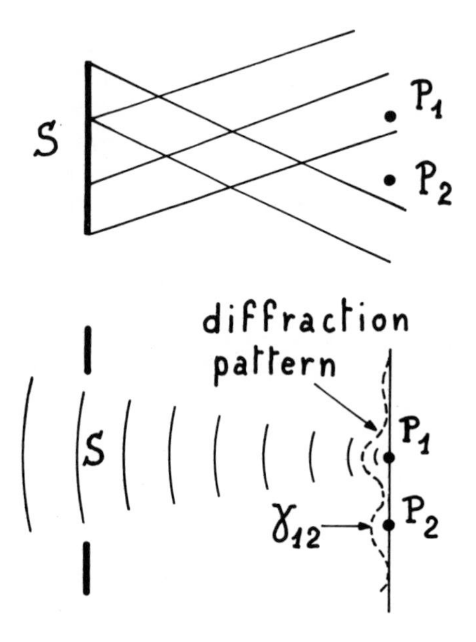

FIGURE 10 Two points illuminated by a plane quasi-monochromatic source.

Now intensity fluctuations are much slower than amplitude oscillations, and the measurement is much easier. This is the basis of the intensity interferometers that Hanbury-Brown and Twiss have constructed both for visible light and microwaves. Radiation is collected by two mirrors at P_1 and P_2 and sent to two phototubes that reveal intensity. Signals from both phototubes are sent to a multiplier and correlated. In this way one can measure $|\gamma_{12}|$ for two points even hundreds of meters apart.

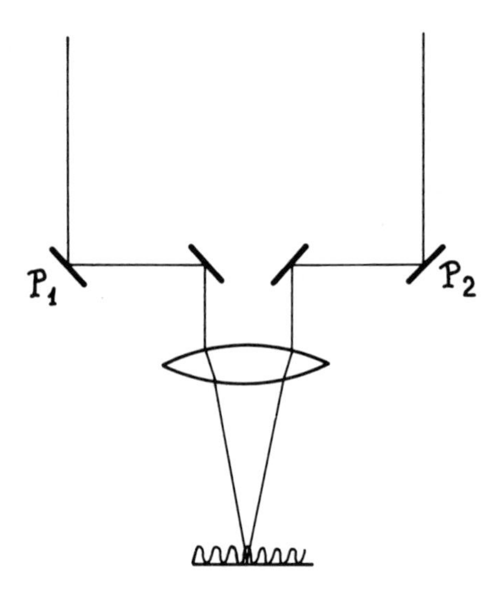

FIGURE 11 Michelson's stellar interferometer.

By recording the signals, one can also compare fluctuations at places thousands of kilometers apart.

Phototubes can be sensitive to individual photons, and one may wonder if classical wave theory is adequate to treat these phenomena. However, a semiclassical theory shows that, due to the boson nature of photons, fluctuations of photocounts obey substantially the same laws as fluctuations of intensity.

A more important question is whether and when the assumption that light be a Gaussian random process is justified. It is certainly justified when light is generated by many independently excited and spontaneously emitting atoms. But what about lasers?

The question can be clarified only by an appropriate quantum treatment. Pioneer work in this direction has been done by R. Glauber.

As is well known, the electric field **E** can be represented by an expansion

$$E = \Sigma \left[a_k u_k(\mathbf{r}) e^{-i\omega kt} + a_k^+ u_k^*(n) e^{i\omega kt}\right],$$

where a_k, a_k^+ represent the annihilation and creation operators. The annihilation and creation parts of **E** will be represented by $\mathbf{E}_+$ and $\mathbf{E}_-$, respectively.

If by $x_1 \ldots x_{2N}$ we denote $2N$ points of space-time and by ρ the density matrix of the states of the field, we can introduce the correlation function of Nth degree by

$$G^{(N)}(x_1 \ldots x_{2N}) = tr \left[\rho \mathbf{E}_-(x_1) \ldots E_-(x_N) E_+(x_{N+1}) \ldots E_x(x_{2N})\right].$$

It turns out that $G^{(1)}(x_1 x_1)$ is proportional to the probability P_{x1} of having a photoelectron produced at x_1, i.e., to the intensity at x_1. Similarly, $G^{(N)}(x_1 \ldots x_N, x_1 \ldots x_N)$ is proportional to the probability $P_{x1} \ldots x_N$ of revealing one photoelectron at each point $x_1 \ldots x_N$ (N-fold coincidence). Let us introduce a normalized correlation function:

$$g^{(1)}(x_1 x_2) = G^{(1)}(x_1, x_2) / \left[G^{(1)}(x_1, x_1) G^{(1)}(x_2, x_2)\right]^{1/2}$$

and similar expressions $g^{(N)}(x_1 \ldots x_{2N})$ for the higher-order correlation functions. A field will be said to be coherent in the first order if $g^{(1)} = 1$, in the Nth order if $g^{(N)} = 1$ everywhere.

If $g^{(N)} = 1$, the probability of $P_{x_1 x_2} \ldots x_N$ of an N-fold coincidence turns out to be equal to the product

$$P_{x_1 x_2} \ldots {}_{x_N} = P_{x_1} P_{x_2} \ldots P_{x_N}$$

of the probabilities of revealing one photon at each point $x_1 \ldots x_N$. Therefore, such probabilities are independent, and there is no correlation in the photocounts. Coherent light obtained with ordinary sources (i.e., filtered Gaussian light) is coherent only to the first order and therefore shows twofold correlations, while the field radiated by a classical antenna is coherent to any order. Laser light lies in between. It is coherent to a high, but not infinite, order.

The invention of the laser is just one more example of something that has occurred since the beginning of mankind. At first, man has played only a role analogous to that of Maxwell's demon, selecting from the random noise of all natural objects and phenomena those improbable cases that were advantageous for his survival. Then, little by little, he has learned to build the improbable and useful things right away, without having recourse to noise. Thus coherent light from a natural or conventional source is only filtered noise. Making a laser is something much more clever.

However, noise can be made use of in a very subtle way. We are accustomed to thinking that random fluctuations mar the signal sent out by an object and diminish obtainable information. Nevertheless, random fluctuations carry some valuable and quantitative information about the object. This was first shown by Einstein with his theory of Brownian motion. Another good example is Rayleigh scattering of light by the sky. A more recent and striking case is represented by correlation interferometry. Correlation of light fluctuations gives us precise information about the source. Noise can be useful.

Departing from filtered noise and natural phenomena occurring around us very often brings us to discovering nonlinearities.

When a young student first encounters the mathematical expressions of elementary physical laws, he may be tempted to conclude that *nature is linear*. Is this conclusion right?

A more correct statement seems today to be that *nature is analytic*, which in turn is a modern rephrasing of the old statement: *natura non facit saltus*, or nature does not jump. Any curve representing a physical phenomenon has a continuous tangent and *within a limited interval* can be replaced by a straight line (Fig. 12). The important philosophical question to answer is why the interval of interest is in most cases limited to the straight portion of the curve. Why is the usual departure of the parameters from their equilibrium values so small? Is this an essential feature of the human world? We live in a degenerate world, very close to absolute zero—the world of molecular forces. But even if we include the sun in our system, we remain in the realm of modest energies as compared with the *high energies* that we know to be physically pos-

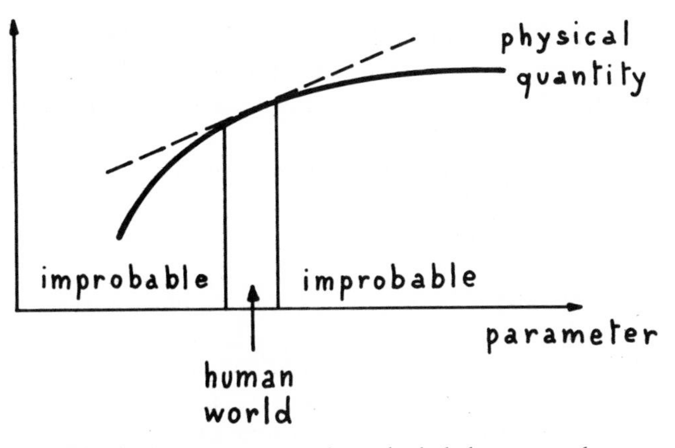

FIGURE 12 Any curve representing a physical phenomenon has a continuous tangent and within a limited interval can be replaced by a straight line.

sible. Near equilibrium only the lower states tend to be filled, and the more we want to depart from equilibrium, the more ingenuity and effort we must spend in order to make a very improbable case to occur.

Laser light can easily be made a million times hotter than the sun's surface. No wonder, therefore, that the straight-line limits are exceeded and optics may become nonlinear.

Classical optics is based on the circumstance that in an ordinary material the electric polarization is proportional to the electric field

$$P = \chi E.$$

In this situation, the waves scattered by the atoms have the same frequency as the incident wave. However, when E becomes sufficiently large, we discover that P is a nonlinear function of E:

$$P = a_1 E + a_2 E^2 + a_3 E^3 + \ldots.$$

The second term of the expansion gives rise to scattered waves with doubled frequency, the third term gives rise to a third harmonic, and so on. In this way, one can multiply the frequency of optical radiation, as was first shown by P. A. Franken. Of course, since two or more photons of the incident radiation give rise to one photon of the harmonic, there is a condition of momentum conservation like $\hbar K_1 + \hbar K_1 = \hbar K_2$. Since $K = 2\pi/\lambda$ is proportional to the refractive index, this is equivalent to an index matching condition $n_1 = n_2$. This condition can be met in an anisotropic medium in a particular direction.

Nonlinear optics has opened a huge field of prospective applications, where many well-known radio-frequency techniques like mixing, heterodyning, and modulating can be transferred to the domain of optical frequencies.

Light, the first gift of God to Man, one of the first physical phenomena to be investigated, still continues to supply a wealth of problems for our mind, a wealth of applications for our welfare. This is a gratifying realization.

However, no talk about optics and its latest applications would be complete if it tried to ignore the fact that with all our light, we are not yet able to illuminate human brains and to defeat human stupidity. Many people working with lasers were simply horrified when they realized that the most effective application of lasers is today represented by the guidance of deadly missiles. Once more, what could have been a monument to man's intelligence has been turned into a shame for all mankind. *Mehr Licht!* More light! Like Goethe on his deathbed, we desperately need more light. More light for the human mind.

GERHARD HERZBERG *holds the position of Distinguished Research Scientist in the Division of Physics of the National Research Council in Ottawa, Canada. He was president of the Royal Society of Canada in 1966 and 1967. His many international honors include fellowship in the Royal Society of London, the Indian Academy of Sciences, the Hungarian Academy of Sciences, the American Academy of Arts and Sciences, the Pontifical Academy of Sciences, and the United States National Academy of Sciences. In 1971, Dr. Herzberg was awarded the Nobel Prize in Chemistry.*

5 *Spectroscopy and Molecular Structure* / GERHARD HERZBERG

I. INTRODUCTION

The serious study of molecular structure on the basis of the spectra of molecules is only a few years older than the International Union of Pure and Applied Physics. It is therefore appropriate that at the Fiftieth Anniversary of the Union the subject of molecular spectroscopy should be included in the discussions.

In 1885, the same year in which Balmer discovered the Balmer formula for the line spectrum of the hydrogen atom, Deslandres first recognized (and published) some of the empirical regularities in band spectra, which are strikingly different from those encountered in atomic spectra. But it was only in 1916 that K. Schwarzschild (a month before his death) first applied the Bohr theory to a diatomic molecular model (the rotator),[1] and it was only in 1919, i.e., three years before the foundation of our Union, that Heurlinger[2] and Lenz[3] first applied Schwarzschild's theory to the interpretation of the empirical regularities in band spectra. Their work was followed in quick succession by the theoretical work of Kratzer,[4] still based on the old quantum theory, and the experimental work of Hulthén, Mecke, Mulliken, and many others.

Naturally, the development of wave mechanics (soon after the foundation of the Union) greatly accelerated the spectroscopic studies of molecular structure. Many empirical features of band spectra could be understood only on the basis of wave mechanics; and, conversely, in several

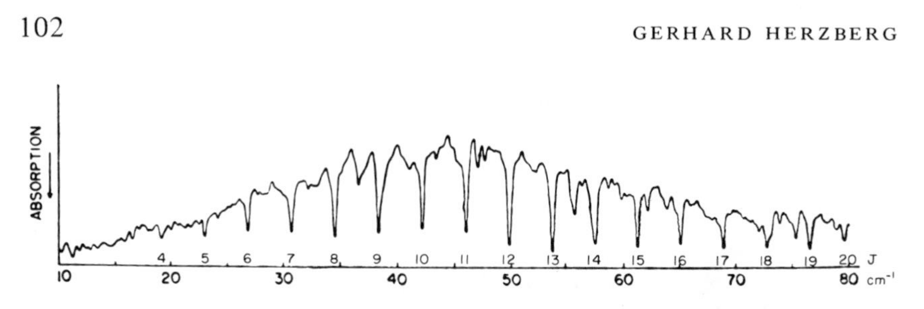

FIGURE 1 Far infrared spectrum of CO after Loewenstein.[10] The numbers attached to the absorption lines are the rotational quantum numbers J of the lower state. The first two lines ($J = 0$ and 1) are outside the range covered by this spectrogram.

instances molecular spectroscopy supplied most welcome confirmations of the predictions of wave mechanics and thus helped to establish its validity for all atomic phenomena.

It is not possible to present in a single paper a systematic development of the subject of molecular spectroscopy.* My aim will be to give some idea of the steps that are needed to derive from the observed spectra information about the structure of molecules. At the same time I shall mention some of the applications in astronomy.

II. FAR-INFRARED, MICROWAVE, AND RAMAN SPECTRA

The simplest spectra occur in the far infrared and in the microwave region. Even though the real exploration of these spectra started only after World War II there had been in the early days of spectroscopy a few important studies,[9] which confirmed in an essential way the predictions of wave mechanics. In Fig. 1 the far-infrared absorption spectrum of CO is shown.[10] It consists of a series (*branch*) of very nearly equidistant lines. The spacing at the short-wavelength end deviates from that at the long-wavelength end by only 0.8%. The longest wavelength member not shown in Fig. 1 lies in the microwave (millimeter) region at $\lambda = 2.60$ mm (corresponding to 3.84 cm^{-1}). It is shown in Fig. 2(a). The wavenumber of this line is very close to the spacing of successive lines in Fig. 1. The same line has recently been observed in emission from interstellar space by Wilson *et al.*,[12] as shown in Fig. 2(b). On the basis of this observation, CO has been found to be by far the most abundant molecule in the interstellar medium.

Early in the development of molecular spectroscopy it was recognized that the far infrared spectra are pure rotation spectra. According to the

*A very detailed presentation of the field may be found in Refs. 5–7, a shorter introduction in Ref. 8

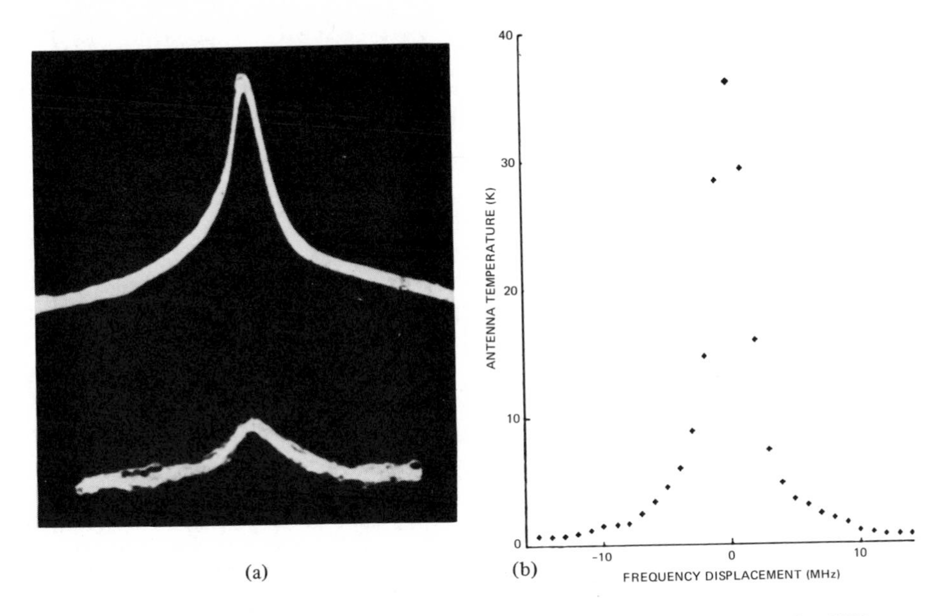

FIGURE 2 Microwave absorption line of CO at λ = 2.60 mm (a) in the laboratory after Gilliam *et al.*[11] and (b) in emission from the Orion nebula after Wilson *et al.*[12] The two oscilloscope traces in (a) are taken at −190°C and −80°C. The diagram (b) gives the antenna temperature of the radio telescopes used. The peak of the line in both cases occurs at 115271.20 MHz = 3.84503_3 cm^{-1}.

old quantum theory,[1] the energy of a model consisting of two mass points (nuclei) at a fixed distance from one another comes out to be

$$E_r = \frac{h^2}{8\pi^2 I_0} J^2, \quad J = 0,1,2,\ldots . \tag{1}$$

Here I_0 is the moment of inertia and J the rotational quantum number, which gives the angular momentum in units $h/2\pi$.

According to wave mechanics, the energy is

$$E_r = \frac{h^2}{8\pi^2 I_0} J(J+1). \tag{2}$$

Figure 3 shows the energy levels as well as the transitions that are possible according to the selection rule $\Delta J = \pm 1$. In spectroscopy, usually instead of the energy we use the corresponding term values:

$$F_0(J) = E_r/hc = B_0 J^2, \tag{1a}$$

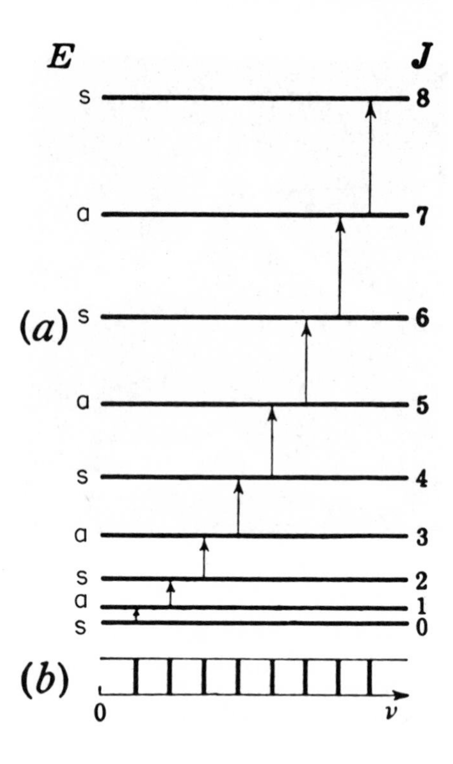

FIGURE 3 (a) Energy levels and (b) schematic spectrum of the rigid rotator. In (a) the transitions possible according to the selection rule $\Delta J = 1$ are indicated.

$$F_0(J) = E_r/hc = B_0 J(J + 1), \tag{2a}$$

where

$$B_0 = h/8\pi^2 cI_0 \tag{3}$$

is the rotational constant, which, except for a constant factor, is the reciprocal moment of inertia.

Using Eq. (1) or (1a) for the energy we obtain for the wavenumbers of the absorption lines

$$\nu = \frac{E'}{hc} - \frac{E''}{hc} = F_0(J + 1) - F_0(J) = 2B_0 J + B_0, \tag{4}$$

while, using Eq. (2), we find

$$\nu = 2B_0 J + 2B_0. \tag{5}$$

In other words, the first line ($J = 0$) in the old theory would occur at B_0, i.e., at a wavenumber that is one half the spacing of successive lines;

in the new theory, at $2B_0$, i.e., at a wavenumber equal to the spacing. It was an important confirmation of wave mechanics that, as was shown by Czerny for HCl in 1925, the first line occurs at a wavenumber equal to the spacing (compare also Figs. 1 and 2).

The determination of the rotational constant B_0 from the far-infrared spectrum allows for diatomic molecules a precise evaluation of the (average) internuclear distance r_0 in the ground state, since

$$I_0 = \mu r_0^2.$$

In this evaluation it is necessary to take account of the deviation of the molecule from the model of the rigid rotator, that is, of the effect of centrifugal stretching, which produces the very slight change of spacing referred to earlier (Fig. 1). The solution of the wave equation for the *nonrigid rotator* leads to the energy formula

$$F_0(J) = E_r/hc = B_0 J(J + 1) - D_0 J^2 (J + 1)^2 \quad \ldots, \tag{6}$$

where D_0 is usually very small compared to B_0. As an example, consider the following values for CO according to the most recent evaluations[13]:

$$B_0 = 1.922529 \text{ cm}^{-1}, \quad D_0 = 6.134 \times 10^{-6} \text{ cm}^{-1}, \quad r_0 = 1.13089 \text{ Å}.$$

The formulas for *linear polyatomic molecules* are the same as those for diatomic molecules except that the moment of inertia depends now on two or more internuclear distances, which can be evaluated from the spectrum only if B_0 values for two or more isotopic molecules have been determined. For example, from

$$B_0(\text{HCN}) = 1.478222 \text{ cm}^{-1} \text{ and } B_0(\text{DCN}) = 1.207749 \text{ cm}^{-1}$$

one obtains

$$r_0(\text{CH}) = 1.064 \text{ Å}, \quad r_0(\text{CN}) = 1.156 \text{ Å}.$$

For *nonlinear molecules*, the rotational energies are more complex. If the molecule is a symmetric top, that is, if two of the three principal moments of inertia are equal, in addition to the total angular momentum **J**, its component **K** in the direction of the figure axis is a constant and the rotational energy is

$$F_0(J,K) = E_r/hc = B_0 J(J + 1) + (A_0 - B_0)K^2, \tag{7}$$

where the rotational quantum number K takes the values $0,1,2, \ldots J$ and where

$$B_0 = h/8\pi^2 cI_0b , \quad A_0 = h/8\pi^2 cI_0a \tag{8}$$

are, except for a constant, the reciprocals of the moments of inertia about the axis of symmetry (a-axis) and about an axis perpendicular to it (b-axis). The energy levels according to Eq. (7) are plotted in Fig. 4. For a molecule that is a symmetric top on account of symmetry we have the selection rule $\Delta K = 0$, and, therefore, each line of the rotation spectrum now consists of $J + 1$ components, which coincide as long as the symmetric top is rigid but are very slightly separated when centrifugal distortion is present. Whether or not these components are resolved, only the rotational constant B_0, and not A_0, can be determined from the far-infrared spectrum.

When all three principal moments of inertia are different, as in H_2O, we have an *asymmetric top.* The energy-level diagram of such a system, illustrated in Fig. 5, is much more complicated than that of the symmetric top, and the same applies to the resulting infrared spectrum. However, once the many resulting spectral lines are assigned, it is possible to determine all three moments of inertia. The rotation spectra of a large number of asymmetric top molecules have been studied in the far infrared and in the microwave region, and their structures have been determined. Figure 6 shows as an example the structure of H_2O derived in this way.

The intensity of the far-infrared (or microwave) absorption is determined by the magnitude of the (electric) dipole moment of the molecule considered. Conversely, the dipole moment can be determined by a measurement of the intensity of absorption. Homonuclear molecules such as N_2, O_2, . . . C_2H_2, C_2H_4, . . . have no dipole moment and therefore no pure rotation spectrum in the infrared.

Pure rotational transitions of molecules can also be produced by light scattering, that is, can be observed in the *Raman effect.* But for Raman scattering the selection rules are different: in the simplest case of a diatomic or linear polyatomic molecule with zero electronic angular momentum we have the rule

$$\Delta J = \pm 2. \tag{9}$$

Figure 7 shows the transitions allowed by this selection rule in an energy-level diagram, as well as the expected Raman spectrum. It is readily seen by introducing the expression (2) for the rotational energy that the

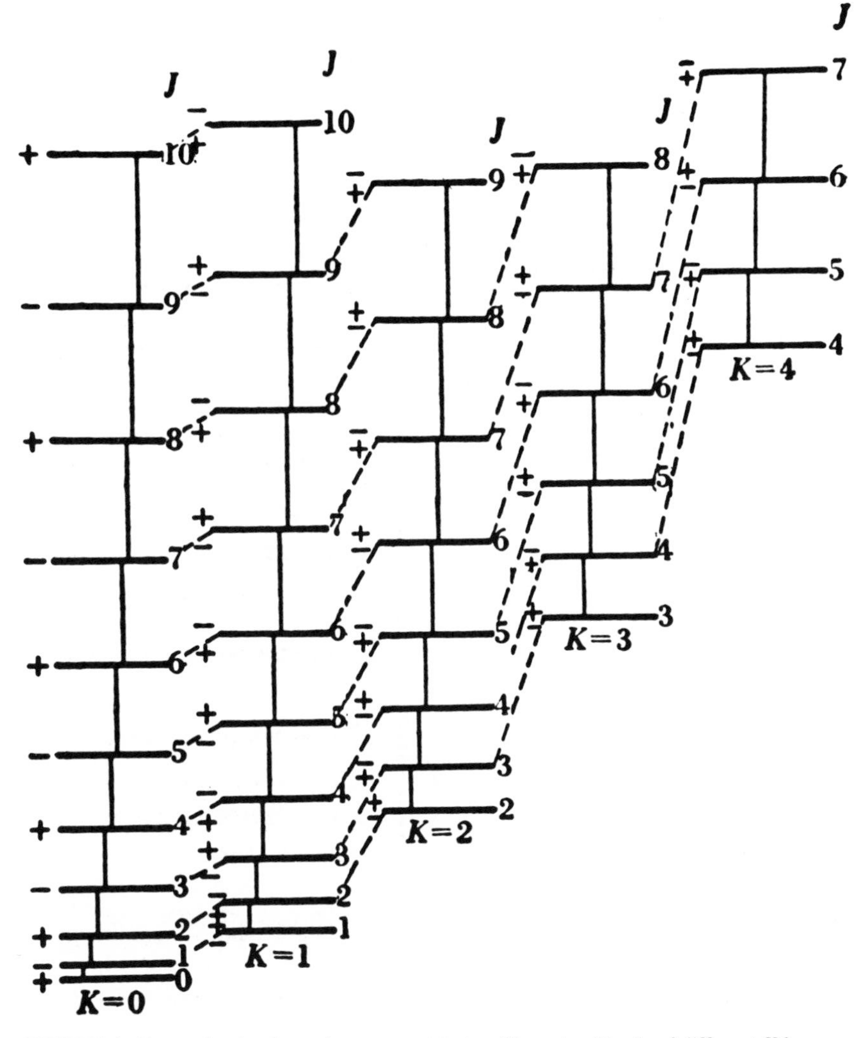

FIGURE 4 Energy levels of a prolate symmetric top. The sets of levels of different K have been plotted side by side. The vertical lines indicate the allowed transitions according to the selection rule $\Delta K = 0$, $\Delta J = +1$. The +, – signs indicate the symmetry of the rotational wavefunction with respect to inversion for a planar molecule. Levels with $K \neq 0$ are doubly degenerate.

Raman shifts, both to longer and shorter wavelengths (Stokes and anti-Stokes lines) are given by

$$\Delta\nu = \pm \frac{1}{hc} (E_r' - E_r'') = \pm [F_0(J + 2) - F_0(J)] = \pm 4B(J + \frac{3}{2}), \quad (10)$$

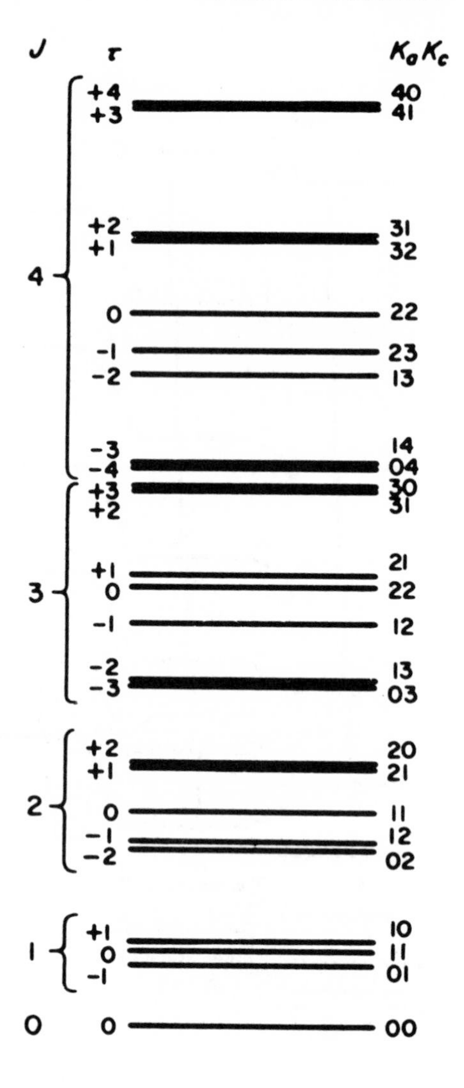

FIGURE 5 Energy levels of an asymmetric top. The energy levels depend strongly on the value of the asymmetry parameter $\kappa = (2B - A - C)/(A - C)$, which here has been assumed to have the value −0.2. At left is J, the quantum number of the total angular momentum, and τ, the parameter that numbers successive levels of a given J. At right the values of K_a and K_c are given, that is, the component of J in the a and c axis, respectively.

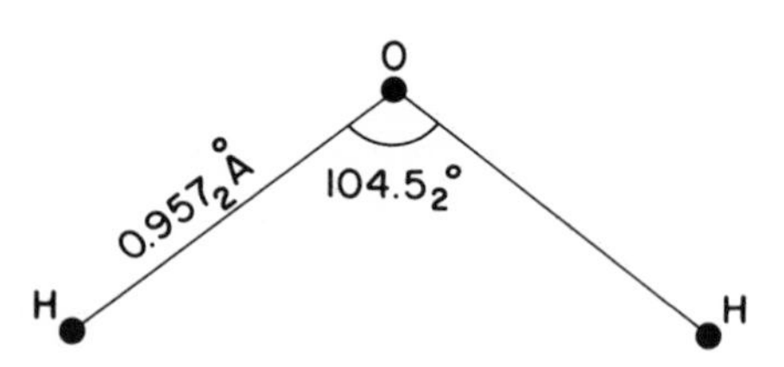

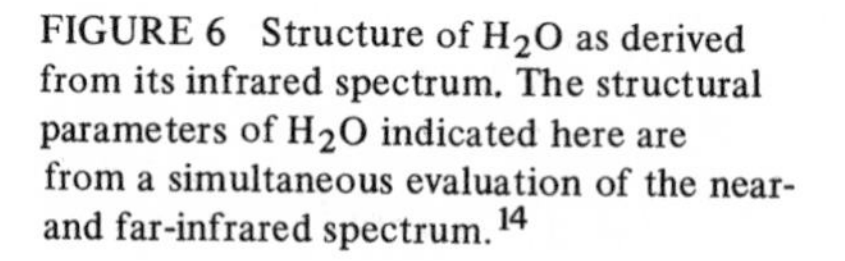

FIGURE 6 Structure of H_2O as derived from its infrared spectrum. The structural parameters of H_2O indicated here are from a simultaneous evaluation of the near- and far-infrared spectrum.[14]

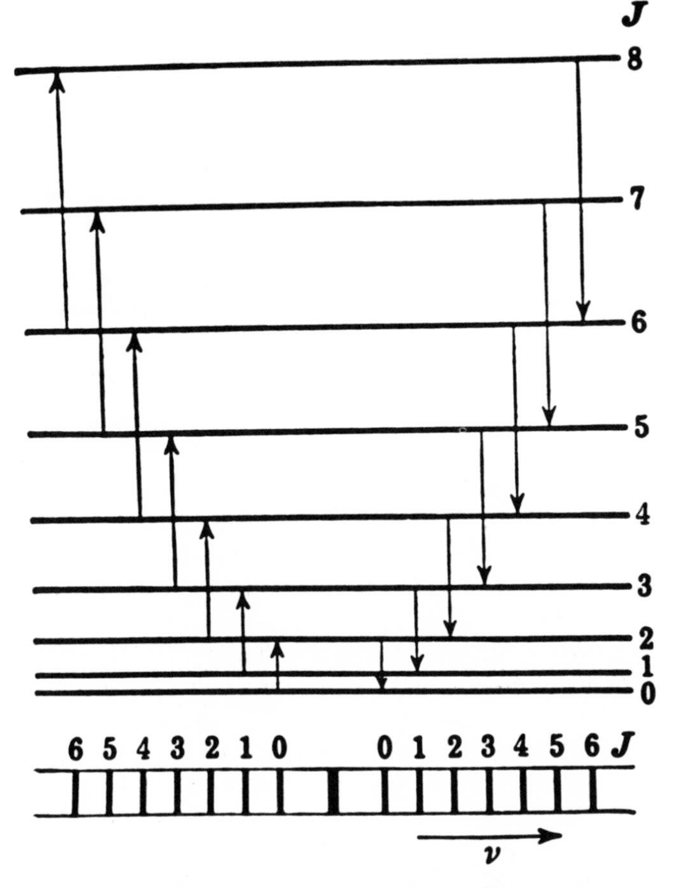

FIGURE 7 Energy level diagram for the rotational Raman spectrum. The vertical arrows indicate the Raman transitions ($\Delta J = 2$), to the left the Stokes lines, to the right the anti-Stokes lines. In the schematic spectrum below, the heavy line in the middle represents the exciting line.

that is, there is a series of equidistant lines on either side of the exciting line, similar to the far infrared spectrum. However, the expected spacing is now $4B$ (instead of $2B$), and the first line has a Raman shift of $6B$. Such spectra have been observed for many molecules. Figure 8 shows a recent Raman spectrum of CO.[15] The spacing of the Raman lines in this spectrum is twice that of the infrared lines of Fig. 1.

The study of the rotational Raman spectrum is particularly important for homonuclear diatomic molecules and symmetric polyatomic molecules, which do not exhibit a far-infrared spectrum. Figure 9 shows

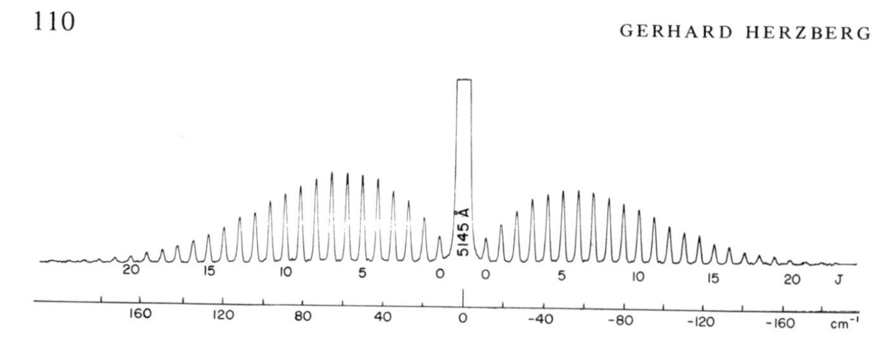

FIGURE 8 Rotational Raman spectrum of CO after Kiefer and Bernstein.[15] This spectrum was produced in CO of 1-atm pressure by means of the 5145 Å line of an argon laser by photoelectric recording. (I am deeply indebted to Drs. Kiefer and Bernstein for taking this spectrum specially for this paper.) At left are the Stokes lines, at right the anti-Stokes lines. Note that in accordance with Eq. (10) the first line is separated from the excited line by $\frac{3}{2}$ times the spacing of successive lines.

early Raman spectra of N_2 and O_2,[16] while Figs. 10 and 11 show Raman spectra of N_2O, CO_2, C_6H_6, and C_6D_6.[17,18] In Table 1 the rotational constants B_0 and internuclear distances r_0 of some of the more important diatomic molecules are listed as determined from the Raman or far-infrared rotation spectra.

A characteristic feature of the Raman spectrum of N_2 in Fig. 9(a) is the *intensity alternation*: the lines are alternately strong and weak. A similar alternation has been observed for H_2, F_2, and others. On the other hand, for O_2 and CO_2 [Figs. 9(b) and 10(b)] there is apparently no intensity alternation, but on closer examination it is found that alternate lines are entirely missing (the separation between the first Stokes and anti-Stokes line is not $2 \times 6B = 3 \times$ separation of successive lines; also the spacing is roughly twice that in N_2 and N_2O, respectively, which would be expected to have B values similar to O_2 and CO_2). This behavior results from the identity of the nuclei, as was first realized by Heisenberg[19] and Hund.[20] The rotational wavefunctions are alternately symmetric and antisymmetric with respect to an exchange of the identical nuclei, i.e., remain unchanged or change sign for such a symmetry operation when J is even or odd, respectively (see the s and a in Fig. 3).

For identical nuclei of zero spin, only the symmetric levels (i.e., the even levels for symmetric electronic and vibrational wavefunctions) occur and we have alternate missing lines. If the identical nuclei have nonzero spin, both even and odd rotational levels occur but with different statistical weights, in the ratio $I + 1:I$. The even (symmetric) levels are the stronger ones when the nuclei follow Bose-Einstein statistics, the odd (antisymmetric) levels are the stronger ones when they follow Fermi-Dirac statistics. When $I = 1/2$ the intensity alternation is 3:1 as for the H_2

molecule. Indeed the spin of the proton was first discovered and determined in this way as was that of the deuteron ($I = 1$) and many other nuclei. Equally important, the statistics that these nuclei obey were first established on the basis of studies of intensity alternation.

The only direct and unambiguous way to establish whether the spin of a nucleus X is zero is by ascertaining whether alternate lines are missing in the spectrum of the molecule X_2. This has been found, e.g.,

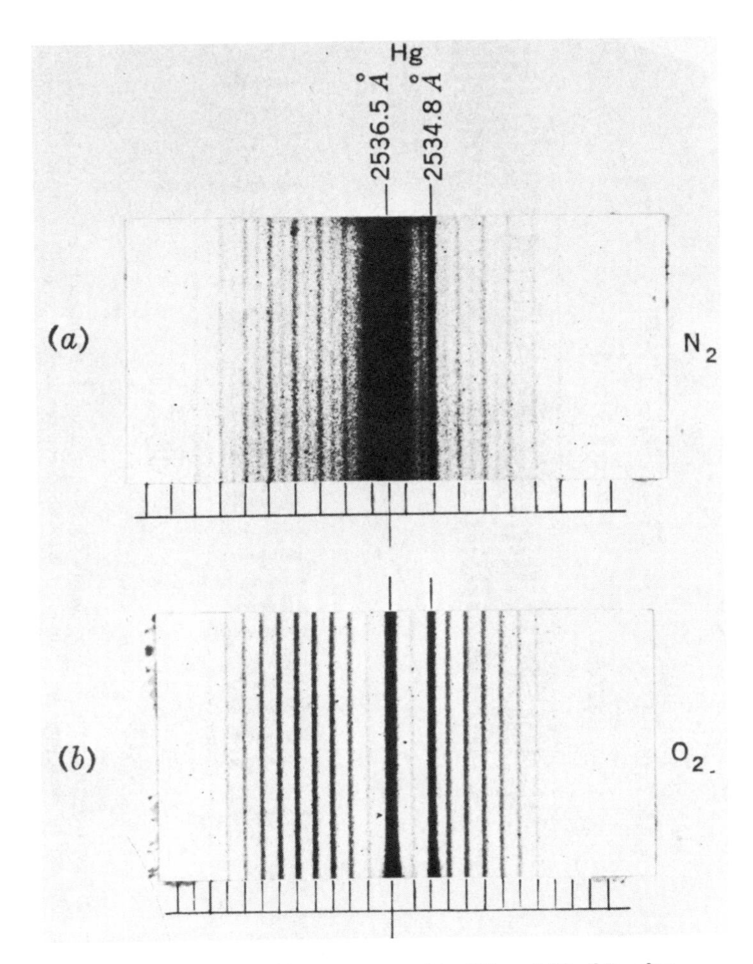

FIGURE 9 Rotational Raman spectra (a) of N_2 and (b) of O_2 after Rasetti.[16] Both spectra are excited by the Hg line at 2536.5 Å. [This line is weaker in (b) than in (a) because of absorption by Hg vapor put in front of the spectrograph.] In (a) only the even (strong) lines are marked, in (b) only the odd lines; the even lines are absent in (b) since $I = 0$ for the O nucleus.

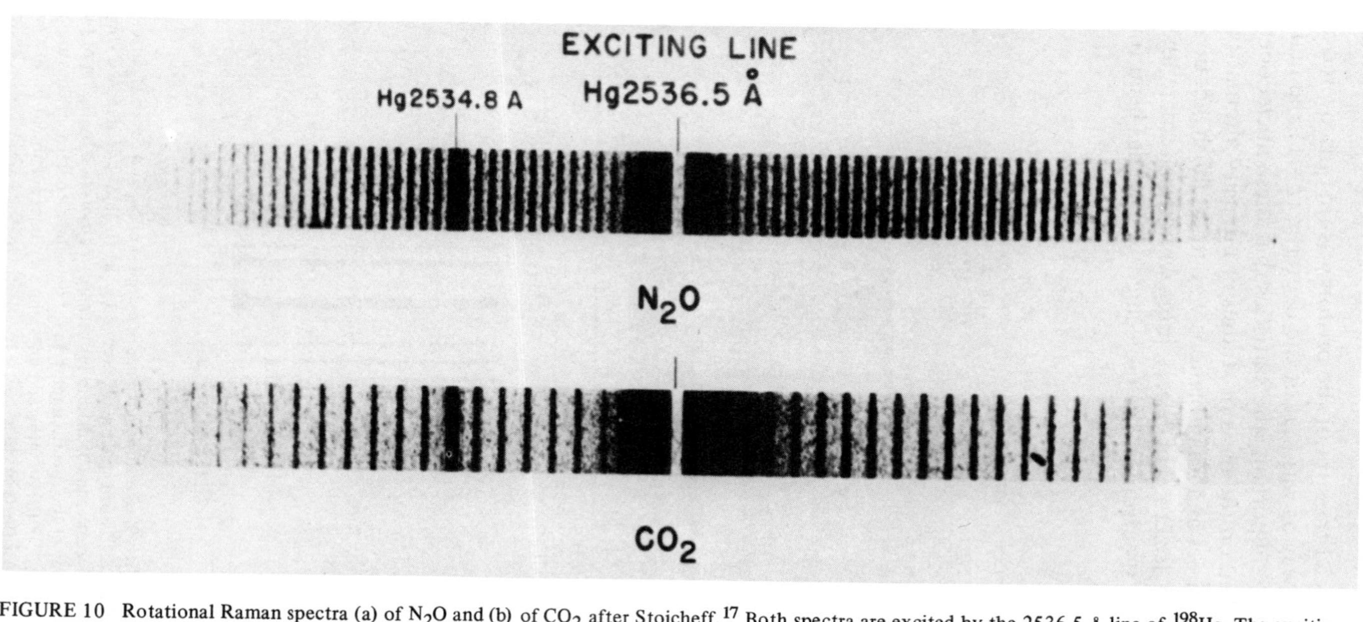

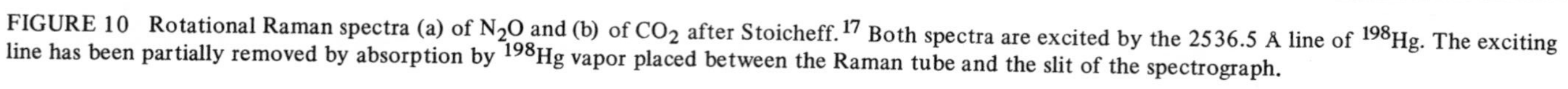
FIGURE 10 Rotational Raman spectra (a) of N_2O and (b) of CO_2 after Stoicheff.[17] Both spectra are excited by the 2536.5 Å line of ^{198}Hg. The exciting line has been partially removed by absorption by ^{198}Hg vapor placed between the Raman tube and the slit of the spectrograph.

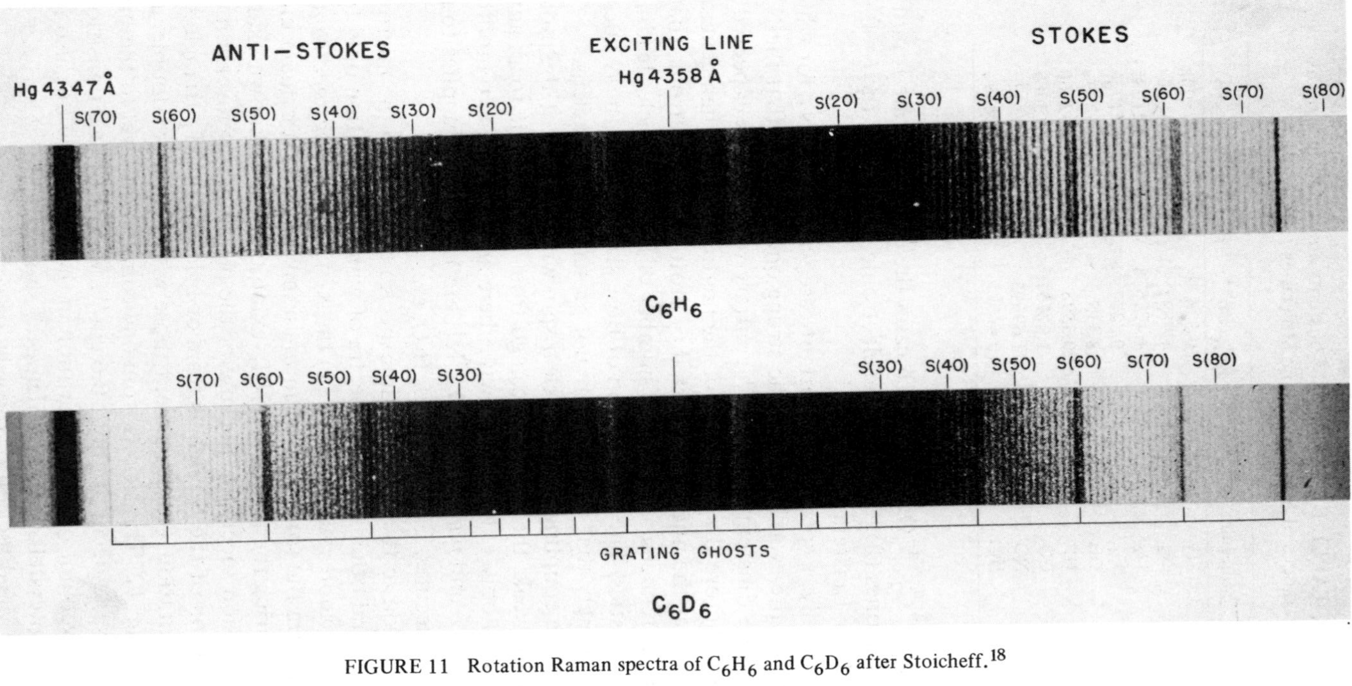

FIGURE 11 Rotation Raman spectra of C_6H_6 and C_6D_6 after Stoicheff.[18]

TABLE 1 Rotational Constants and Internuclear Distances of Some Diatomic Molecules

Molecule	B_0 (cm^{-1})	r_0 (Å)
H_2	59.336	0.7508_6
N_2	1.9891	1.1002
O_2	1.437678	1.21085_1
F_2	0.8827	1.417_9
OH	18.515	0.9799_7
HF	20.5596	0.92560
CO	1.92251	1.13089
NO	1.69605	1.1538
HCl	10.44025	1.2839

for $^4He_2$, $^{12}C_2$, $^{16}O_2$. In all such cases the even levels are found to be the strong ones (for a symmetric electronic state), i.e., the $I = 0$ nuclei follow Bose-Einstein statistics.

Historically, a specially important case was that of N_2 for which Fig. 9 shows that the even levels are the strong ones, that is, that the N nuclei follow Bose-Einstein statistics,[21] contrary to what was expected at the time, i.e., before the discovery of the neutron; it was then generally believed that the nucleus consisted of protons and electrons, that is, for the nitrogen nucleus, of an odd number of Fermi particles leading to Fermi statistics. The observation of Bose statistics for the N nuclei suggested strongly that electrons are not present in the nucleus.

Since the coupling of the nuclear spin with the rest of the molecule is extremely weak, the molecule can go from a symmetric to an antisymmetric level only extremely slowly: there are in effect two modifications, called *ortho*- and *para*-H_2 (or D_2, N_2, etc.). The same applies to polyatomic molecules, C_2H_2, H_2CO, H_2O, etc.

In addition to the pure rotation spectra there are in the microwave region several other types of spectra of which we mention only two: (1) the *inversion spectrum* and (2) the *K-type doubling spectrum.* For these spectra the rotational quantum numbers J and K remain unchanged during the transition. NH_3 exhibits the best-known example of an inversion spectrum. Because of the possibility of inversion (i.e., the transition of the N nucleus from one side of the H_3 triangle to the other) all rotational levels, except those with $K = 0$, are double, as illustrated in Fig. 12. Transitions from one component level to the other are very strong and occur at about 0.6 cm^{-1}. Many of these transitions have been observed not only in absorption in the laboratory but also in emission from interstellar clouds. Figure 13 shows the line $J = 1$, $K = 1$ as observed in the direction of the galactic center.

The *K*-type doubling spectrum arises in nearly symmetric top molecules in which the double degeneracy of levels with $K \neq 0$ is split on account of the slight asymmetry. Figure 14 shows the lowest rotational levels of formaldehyde (H_2CO). Transitions between the two $J = 1$ or $J = 2, 3, \ldots$ levels with $K = 1$ are allowed and have been observed in absorption both in the laboratory and in interstellar clouds. Figure 15 shows, as an example, the $J = 1$ line as observed in the nebula NGC 2024 by Zuckerman *et al.*[23]

It is perhaps of interest to emphasize that pure rotational spectra have been observed for atomic nuclei (see A. Bohr's paper in this volume). Since nuclei have no electric dipole moment, such spectra can only occur on account of the electric quadrupole moment (or magnetic dipole moment if the nucleus has nonzero spin). The selection rule is then $\Delta I = \pm 2$, similar to the Raman selection rule of the molecular rotation spectrum. In the nuclear case, the centrifugal distortion effect is relatively much larger.

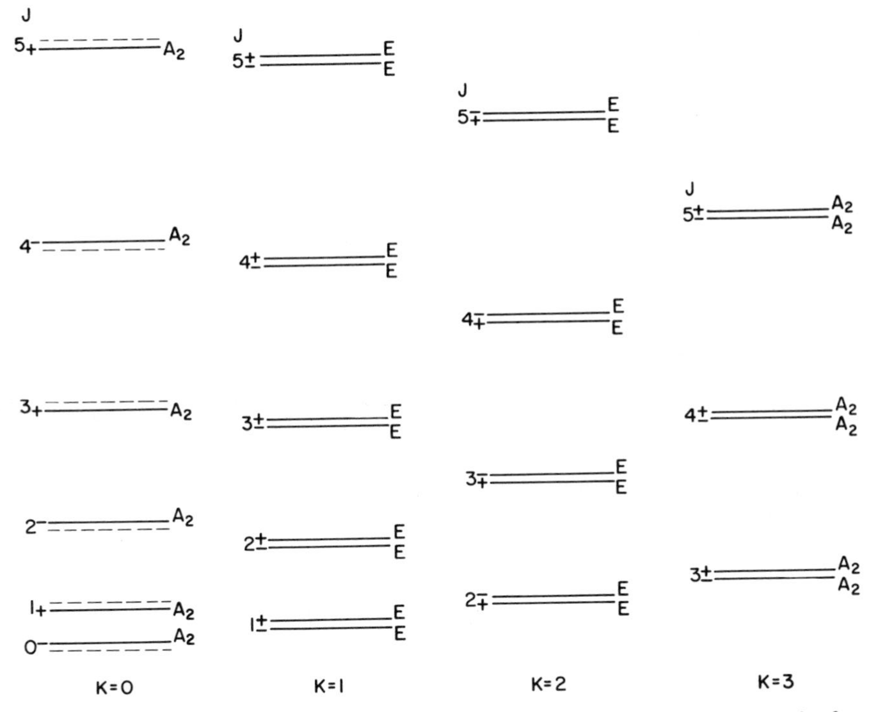

FIGURE 12 Rotational energy levels of NH_3 showing inversion doubling. The magnitude of the doubling is greatly exaggerated. The broken line levels with $K = 0$ are absent in NH_3 but are present in ND_3. The symbols A_2, E refer to the symmetry of the overall eigenfunction (see Ref. 5), the + and − as in Fig. 4 is to the symmetry of this function with respect to inversion.

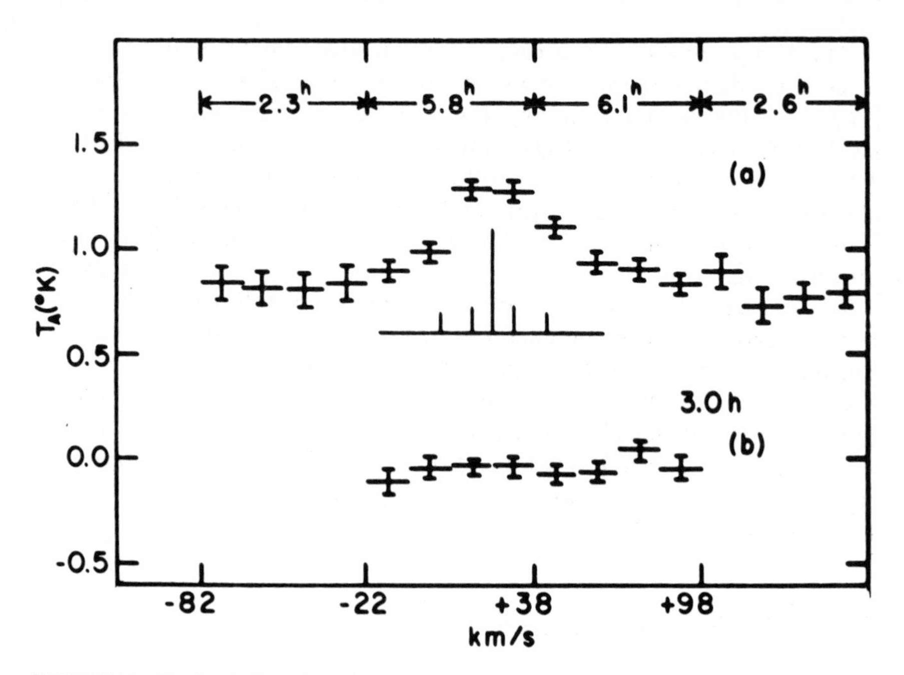

FIGURE 13 The $J = 1$, $K = 1$ inversion line of NH_3 as observed in interstellar emission from the galactic center after Cheung *et al.*[22] The crosses represent the scatter of the observations both in intensity and frequency. Curve (b) represents the background. The expected hyperfine splitting is shown by the vertical lines.

III. NEAR-INFRARED SPECTRA, ROTATION-VIBRATION SPECTRA

In the near infrared, the absorption spectra of diatomic and polyatomic molecules consist at low resolution of fairly strong "bands," often followed at twice (or three times) the frequency by much weaker bands. These bands were recognized very early as corresponding to the fundamental and overtone vibrations of the molecule. For diatomic molecules there is only one vibration, which, for heteronuclear molecules, gives rise to a single strong absorption band (ν_0), followed at sufficiently long absorbing paths by a progression of bands at nearly $2\nu_0$, $3\nu_0$, . . . with rapidly decreasing intensity. The frequency is of the order 10^3 cm^{-1} or 3×10^{13} Hz, which is about 1000 times greater than the (lowest) rotational frequencies.

If the vibrations of a diatomic molecule are considered as harmonic, the potential function is a parabola

$$V = \frac{1}{2}kx^2, \tag{11}$$

where x is the displacement from the equilibrium position and k is a constant giving the restoring force for unit displacement. The energy levels of such a harmonic oscillator according to wave mechanics are given by

$$E_v = h\nu_{osc}(v + \tfrac{1}{2}), \quad v = 0, 1, 2, \ldots, \tag{12}$$

or, in term values,

$$G(v) = \omega(v + \tfrac{1}{2}), \quad \omega = \nu_{osc}/c. \tag{13}$$

The vibrational frequency ν_{osc} is (classically) determined by the force constant k from the relation

$$\nu_{osc} = (\tfrac{1}{2}\pi)(k/\mu)^{1/2}, \tag{14}$$

where μ is the (reduced) mass of the oscillator.

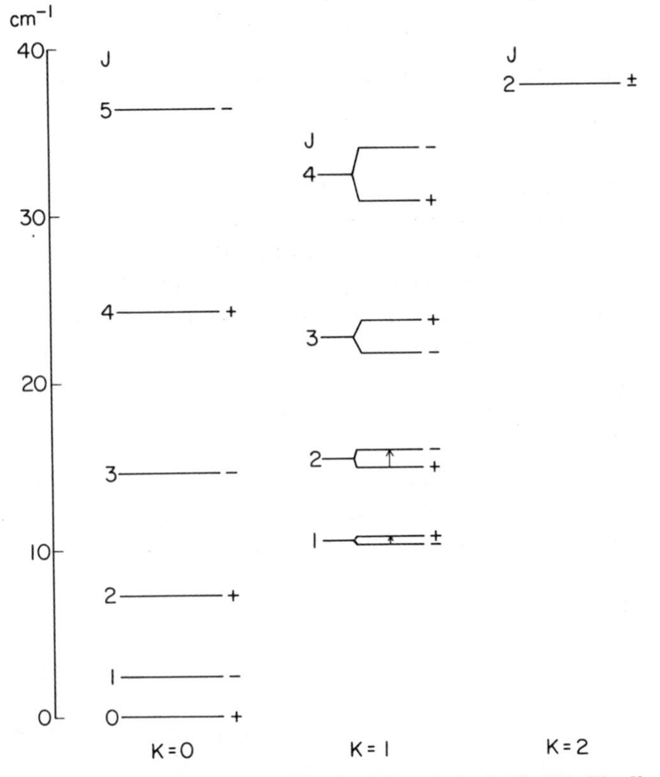

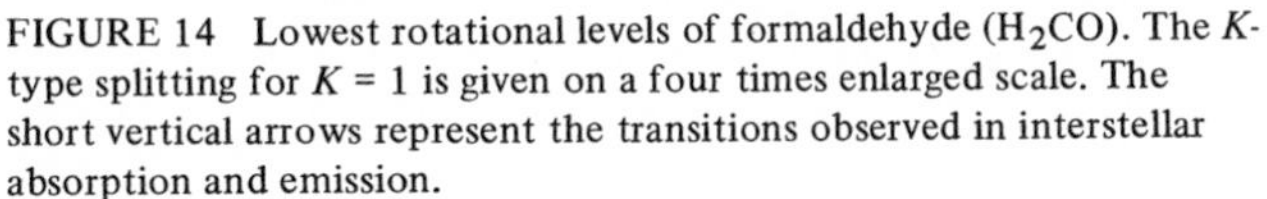
FIGURE 14 Lowest rotational levels of formaldehyde (H_2CO). The K-type splitting for $K = 1$ is given on a four times enlarged scale. The short vertical arrows represent the transitions observed in interstellar absorption and emission.

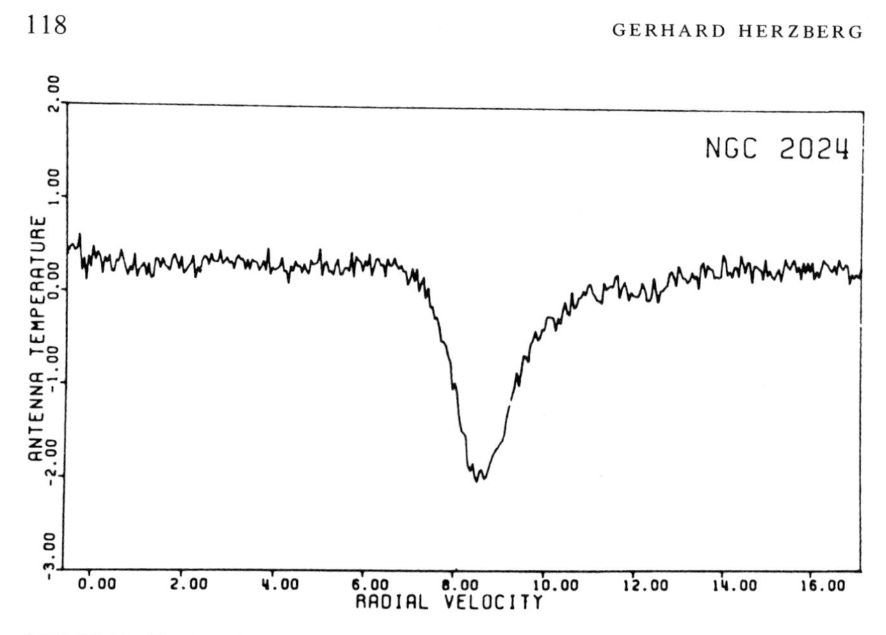

FIGURE 15 The $J = 1$ K-type doubling line of H_2CO at 4593.1 MHz as observed in interstellar absorption in the nebula NGC 2024 after Zuckerman *et al.*[23] The radial velocity scale on the abscissa (in km/sec) is a frequency scale.

According to Eq. (12) or (13), the vibrational levels of the harmonic oscillator are equidistant; the lowest level occurs at $\frac{1}{2}h\nu_{osc}$ corresponding to the zero-point vibrational energy. In the old quantum theory there was no zero-point vibration (one had $E_v = h\nu_{osc}\, v$). The existence of the zero-point vibration is difficult to establish from the infrared spectrum but has been established by a study of the vibrational structure of electronic transitions of isotopic molecules (see Section IV), first by Mulliken[24] just prior to the advent of quantum mechanics.

For the harmonic oscillator, the selection rule is $\Delta v = \pm 1$, and therefore only one absorption band is expected even if several vibrational levels are populated. For the actual molecule the potential function is not a parabola but a curve of the form shown in Fig. 16, i.e., it can be represented by the model of the *anharmonic oscillator*. Its energy levels are given by

$$G(v) = E_v/hc = \omega_e(v+\tfrac{1}{2}) - \omega_e x_e(v+\tfrac{1}{2})^2 \ldots, \quad \omega_e x_e \ll \omega_e, \tag{15}$$

that is, the spacing of the energy levels decreases slowly with increasing v (see Fig. 16), and since the selection rule is now

$$\Delta v = \pm 1, +2, \ldots$$

we have in absorption a progression of bands with slowly decreasing spacing and rapidly decreasing intensity, in agreement with observation. In the classical description they correspond to fundamental and first, second, . . . overtones.

The intensity of the fundamental depends primarily on the rate of change of the dipole moment with internuclear distance. Homonuclear molecules for which the dipole moment vanishes for all internuclear distances have no ordinary (dipole) near-infrared spectrum. They do, however, have (like the heteronuclear molecules) a vibrational Raman

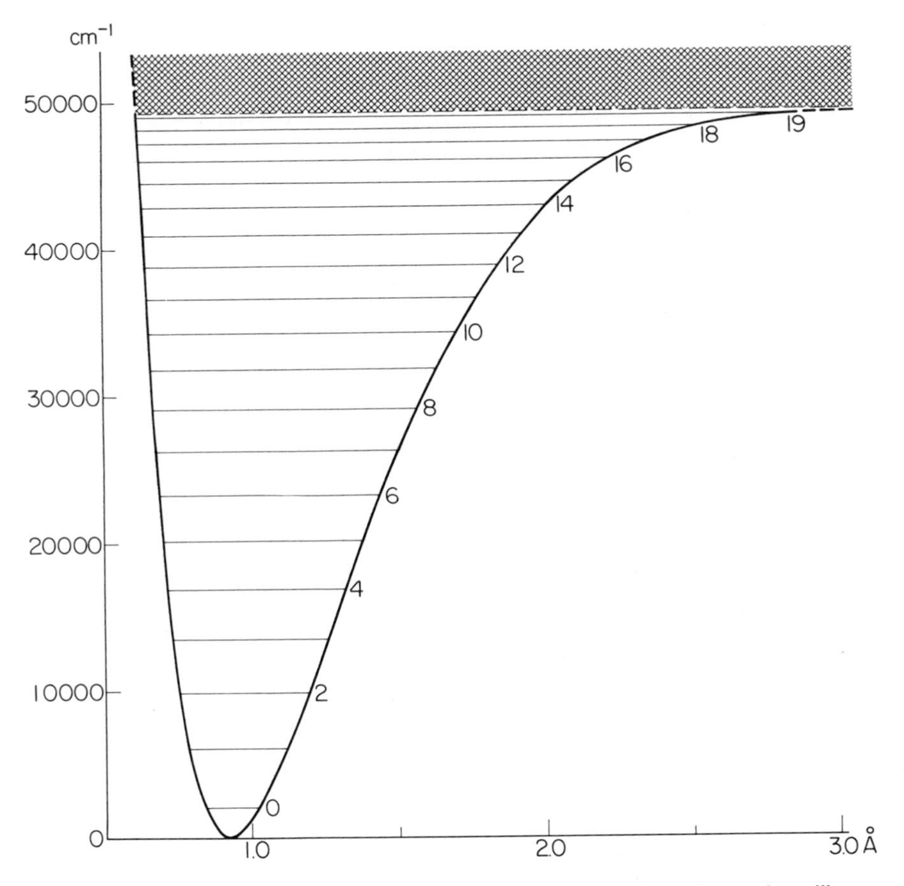

FIGURE 16 Potential function of the HF molecule as an example of an anharmonic oscillator after Di Lonardo and Douglas.[25] The observed levels are shown by the horizontal lines; the vibrational quantum number v goes from 0 to 19. Above the asymptote of the potential curve there is a continuous-term spectrum (indicated by cross-hatching), which corresponds to various amounts of kinetic energy of the two atoms (here H and F) flying apart or coming together.

spectrum, which, for molecules like H_2, N_2, has supplied the most accurate vibrational intervals (frequencies) in the ground state. Table 2 lists the lowest vibrational intervals $\Delta G(\frac{1}{2}) = G(1)-G(0)$ and the vibrational frequencies ω_e of some of the more important diatomic molecules as obtained from the infrared and Raman spectra. The vibrational frequencies ω_e give directly the restoring force constants k_e in the molecule for infinitesimal amplitudes, since [see Eq. (14)],

$$\begin{aligned} k_e &= 4\pi^2 \mu \nu_{\text{osc}}{}^2 \\ &= 4\pi^2 \mu c^2 \omega_e{}^2. \end{aligned} \tag{16}$$

For the hydrogen molecule and its isotopes the vibrational intervals ΔG have been derived theoretically directly from the wave equation by *ab initio* calculations. Table 3 compares the observed and calculated ΔG values. The very slight systematic discrepancies are believed to be due to the neglect of nonadiabatic corrections in the theoretical calculations (see Ref. 26).

While homonuclear molecules like H_2 have no electric dipole moment they do have a quadrupole moment whose variation during the vibration of the molecule leads to extremely weak absorption bands (10^{-9} times the intensity of dipole absorption bands of heteropolar molecules). Such quadrupole vibration bands have so far only been observed for H_2 with an absorbing path of the order of 1000 m. These observations are responsible for the first three ΔG values listed in Table 3.

For *polyatomic molecules* there are $3N - 6$ vibrational degrees of freedom (or $3N - 5$ when the molecule is linear) and therefore an equal number of fundamental frequencies. Figure 17 gives the form of the vibrational motion for two triatomic molecules, a linear one and a bent

TABLE 2 Vibrational Frequencies (in cm^{-1}) of Some Diatomic Molecules

Molecule	$\Delta G(\frac{1}{2})$	ω_e
H_2	4161.165	4404
N_2	2329.70	2358.03
O_2	1556.386	1580.36
F_2	893.95	917.8
OH	3569.63	3739.94
HF	3961.60	4138.32
HCl	2964.44	2990.95
CO	2143.27	2169.82
NO	1876.18	1904.12

TABLE 3 Vibrational Quanta in the Ground State of H_2

v	$\Delta G(v+\frac{1}{2})$ Obs.	Theor.
0	4161.14	4162.06
1	3925.98	3926.64
2	3695.24	3696.14
3	3468.01	3468.68
4	3241.56	3242.24
5	3013.73	3014.49
6	2782.18	2782.82
7	2543.14	2543.89
8	2292.96	2293.65
9	2026.26	2026.81
10	1736.66	1737.13
11	1414.98	1415.54
12	1049.18	1048.98
13	621.96	620.16

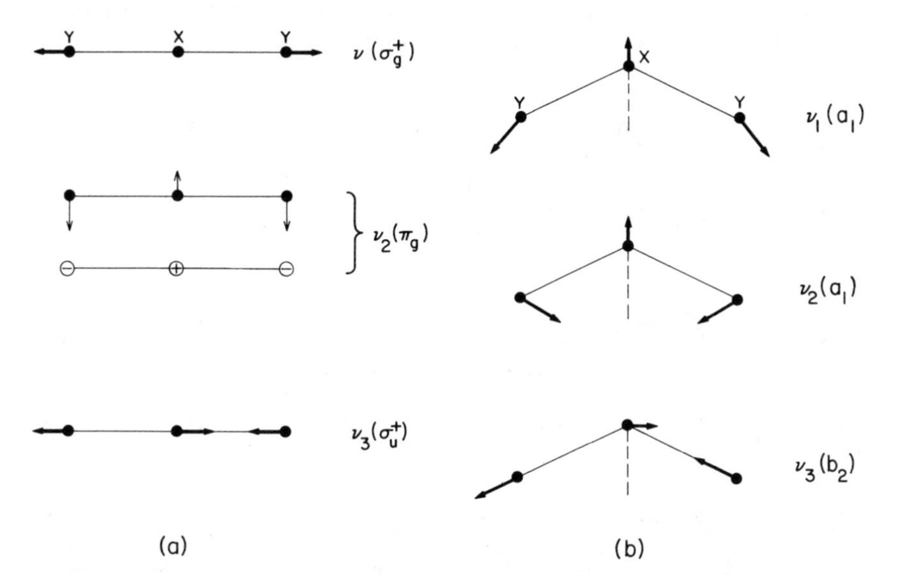

FIGURE 17 Normal vibrations of (a) linear XY_2 and (b) bent XY_2 molecules. The arrows show approximately the amplitudes of the vibration of the nuclei to which they are attached. For the vibration ν_2 of linear XY_2 there are two modes, in the second of which the motions are at right angles to the plane of the paper indicated by plus or minus. For an explanation of the symbols $\sigma_g{}^+$, π_g, . . . , see Ref. 5.

one (CO_2 and H_2O, respectively, are examples. The vibration ν_2 is a bending vibration, which in the linear case is doubly degenerate since the vibration may occur in any of the planes through the internuclear axis (thus there are four vibrational degrees of freedom for linear triatomic molecules but only three for bent triatomic molecules). Figure 18 illustrates the normal vibrations of a planar X_2Y_4 molecule like ethylene with $3N - 6 = 12$ vibrational degrees of freedom. The vibrational frequencies of polyatomic molecules are determined by the force constants, which conversely can be determined from the observed frequencies, but this is now a much more difficult task than for diatomic molecules.

For nonlinear triatomic molecules in general, during each of the three vibrations the dipole moment changes. The three vibrations are therefore infrared active. For linear symmetric triatomic molecules, the vibration ν_1 preserves the symmetry of the molecule during the whole vibration (it is totally symmetric), and therefore there is no change of the dipole moment: the vibration is infrared inactive. During the vibrations ν_2 and ν_3 the symmetry of the molecule is not preserved and an oscillating dipole moment develops: these vibrations are infrared active and have been observed for many such molecules. On the other hand, in the Raman effect ν_1 is the only vibration that occurs. Thus, conversely, from qualitative observations of the infrared and Raman spectra it is possible to

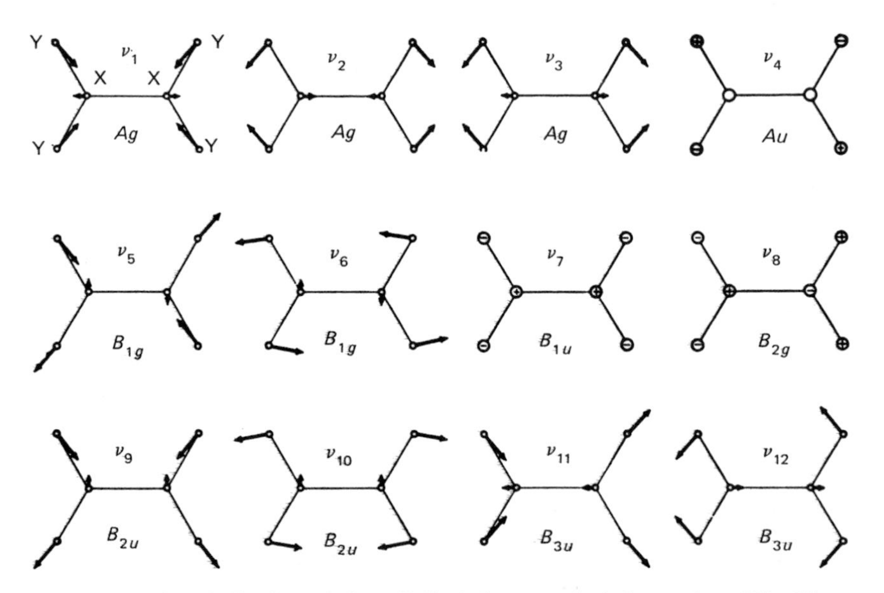

FIGURE 18 Normal vibrations of planar X_2Y_4 (point group D_{2h}). See caption of Fig. 17.

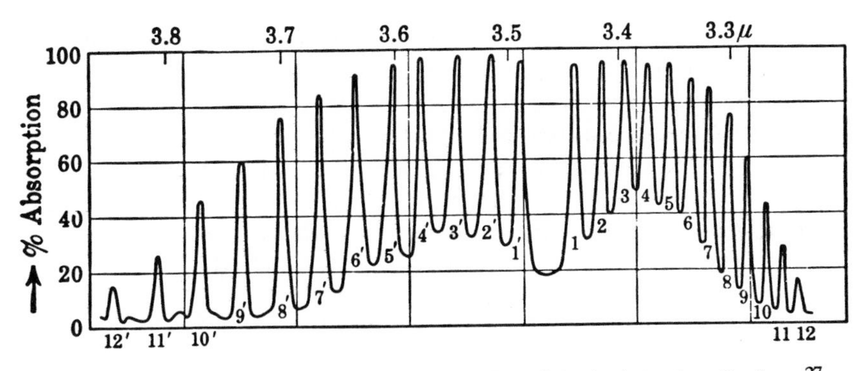

FIGURE 19 Fine structure of the fundamental of HCl at 3.5 μ in absorption after Imes.[27] The numbers attached to the peaks are the m values [see Eq. (17)]. The P branch is to the left, the R branch to the right. With higher resolution, each line is found to consist of two components corresponding to the two isotopes $H^{35}Cl$ and $H^{37}Cl$.

draw conclusions about the structure of the molecule (e.g., whether it is linear or nonlinear). For example, in the spectrum of C_2H_4 only the vibrations ν_7, ν_9, ν_{10}, ν_{11}, and ν_{12} are infrared active (see Fig. 18), while ν_1, ν_2, ν_3, ν_5, ν_6, and ν_8 are Raman active, an observation that supplied the first direct evidence that C_2H_4 has the planar, symmetrical structure.

In recent years, a good deal of work has been done on torsional vibrations (like ν_4 of C_2H_4), which, particularly for single bonds (C_2H_6), have rather low frequencies. Their analogues in ring molecules (ring puckering vibrations) give interesting information about the forces that keep a molecule planar or prevent it from being planar.

When the near-infrared bands are studied under high resolution, a *fine structure* is observed which is found to correspond to rotational transitions that take place simultaneously with the vibrational transition. One of the first spectra so resolved was that of HCl by Imes[27] just over 50 years ago. It is shown in Fig. 19. There are two simple series of lines called branches, an R branch and a P branch, which start out from the zero gap to shorter and longer wavelengths, respectively. Alternatively, one may also say that there is one single series of lines with a separation that increases from short to long wavelengths with one line missing in the center (zero gap). In a first approximation, this series can be represented by the simple formula

$$\nu = \nu_0 + dm + em^2, \tag{17}$$

where $m = +1, +2, \ldots$ for the R branch and $m = -1, -2, \ldots$ for the P branch and ν_0 is the wavenumber of the missing line.

We must now try to understand the constants d and e in terms of molecular constants. For this purpose we introduce the model of the *rotating vibrator* (or vibrating rotator), that is, a system that can carry out simultaneously rotations and vibrations. In a first approximation, the energy levels of such a system are obtained by taking the sum of the vibrational energy $G(v)$ of the anharmonic oscillator and the rotational energy $F(J)$ of the rotator in which, however, the rotational constant has been averaged over the vibration, i.e., B_0 is replaced by

$$B_v = (h/8\pi^2 cI_v)_{\text{average}}. \tag{18}$$

Thus we have the levels

$$T_{vr} = G(v) + F_v(J) = \omega_e(v+\tfrac{1}{2}) - \omega_e x_e(v+\tfrac{1}{2})^2 + \ldots + B_v J(J+1) + \ldots. \tag{19}$$

These levels are represented graphically in Fig. 20. The interaction of rotation and vibration is taken into account by the averaging in Eq. (18). One finds for a diatomic system (with only one vibration)

$$B_v = B_e - \alpha(v+\tfrac{1}{2}) + \ldots, \tag{20}$$

$$B_e = h/8\pi^2 c\mu r_e^2, \tag{21}$$

where r_e is the equilibrium internuclear distance. Thus with increasing v the rotational constant decreases gradually and even in the lowest vibrational level it is slightly smaller (about 0.5%) than the value B_e corresponding to the equilibrium position. For example, the B_0 and B_e values of CO in its ground state are

$$B_0 = 1.92253, \quad B_e = 1.93127 \text{ cm}^{-1},$$

yielding

$$r_0 = 1.13089, \quad r_e = 1.12832 \text{ Å}.$$

The difference between B_0 and B_e must be remembered when the rotational constants from the microwave or rotational Raman spectrum are used.

If we now consider a vibrational transition, e.g., the fundamental band $v = 1 \leftarrow 0$, we obtain the diagram in Fig. 21, where the spacing of the rotational levels in the upper state is slightly smaller than in the lower state. If there were no such difference, the two line series corresponding

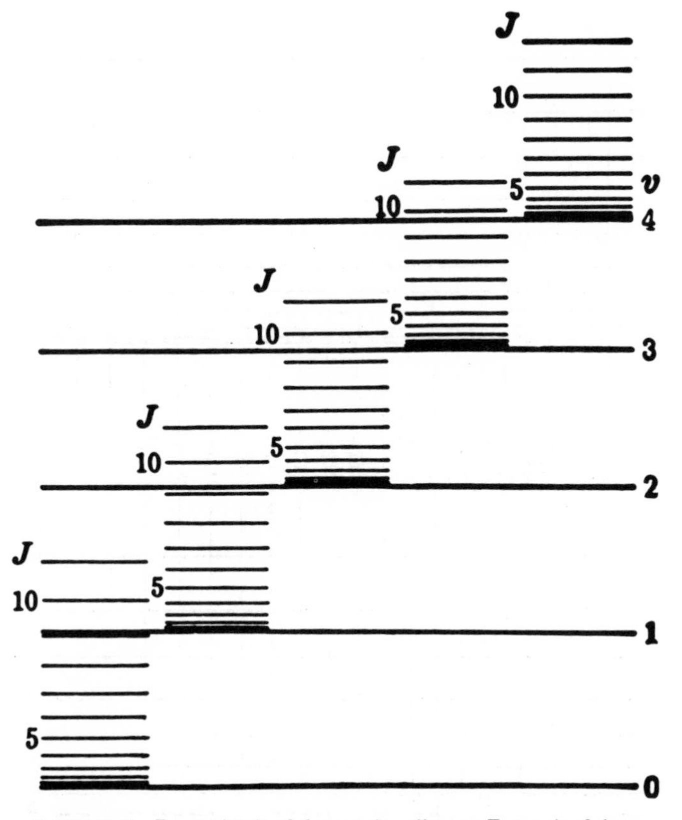

FIGURE 20 Energy levels of the rotating vibrator. For each of the first five vibrational levels the lowest twelve rotational levels are shown, indicated by the short horizontal lines.

to the selection rules $\Delta J = +1$ and -1 (*R* and *P* branch) would be equidistant with a spacing $2B$, i.e., the same as the spacing in the far infrared, but because of this difference the line spacing decreases in the *R* branch, increases in the *P* branch. From

$$\nu = T' - T'' = \nu_0 + F_v'(J') - F_v''(J'') \qquad (22)$$

with

$$\Delta J = J' - J'' = \pm 1$$

we find for the *R* branch

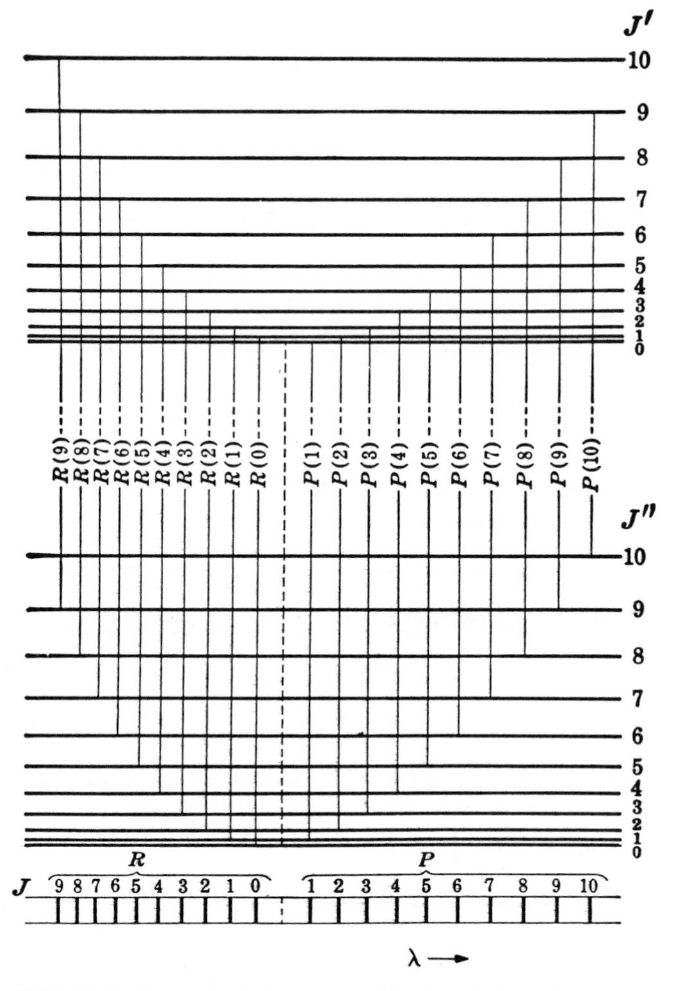

FIGURE 21 Energy-level diagram explaining the rotational fine structure of a rotation–vibration band. The schematic spectrogram below gives the appearance of the spectrum. The broken-line transition is forbidden by the selection rule; it corresponds to the zero-line ν_0.

$$\nu_R = \nu_0 + 2B_v' + (3B_v' - B_v'')J'' + (B_v' - B_v'')J''^2,$$

$$J'' = 0, 1, \ldots, \qquad (23)$$

and for the P branch

$$\nu_P = \nu_0 - (B_v' + B_v'')J'' + (B_v' - B_v'')J''^2,$$

$$J'' = 1,2,\ldots, \qquad (24)$$

or in a single formula

$$\nu = \nu_0 + (B_v' + B_v'')\,m + (B_v' - B_v'')\,m^2,$$

$$m = J+1 \text{ for } R \text{ and } = -J \text{ for } P.$$

Equation (25) is of the same form as the empirical Eq. (17). In other words, we can obtain from the near-infrared bands the same rotational constants B_0 (and D_0) as from the far-infrared bands. But now we obtain in addition the rotational constants of the upper state $v = 1$, or 2, . . . , that is, B_1, or B_2 . . . (and D_1, D_2, . . .) and thus are able to evaluate the rotational constant α and therefore the values of B_e and r_e for the equilibrium position.

In Table 4 we give the B_v values for HCl in the six lowest vibrational levels, and in Table 5 we give B_e, α_e, and r_e resulting from the data of Table 4 for HCl and from similar data for a number of other diatomic molecules and radicals. The data for H_2 are from the quadrupole spectrum mentioned earlier. For it, the selection rule is $\Delta J = 0, \pm 2$ leading to Q, S, and O branches from which B_v can be determined in essentially the same way as described before. In Fig. 22 a few of the lines of H_2 in the photographic infrared are shown. These lines were obtained with an absorbing path of several kilometers. They are of astronomical interest since their observation in the spectra of the outer planets allows a determination of the amount of H_2 in their atmospheres.

The near infrared spectrum of OH is of considerable interest in the study of the upper atmosphere. It appears strongly in emission in the light of the night sky. Because of the nature of the ground state of

TABLE 4 Rotational Constants of HCl in Its Ground State

v	B_v	ΔB_v
0	10.44025	
		0.30402
1	10.13623	
		0.30156
2	9.83467	
		0.29982
3	9.53485	
		0.29865
4	9.2362	
		0.2977
5	8.9385	

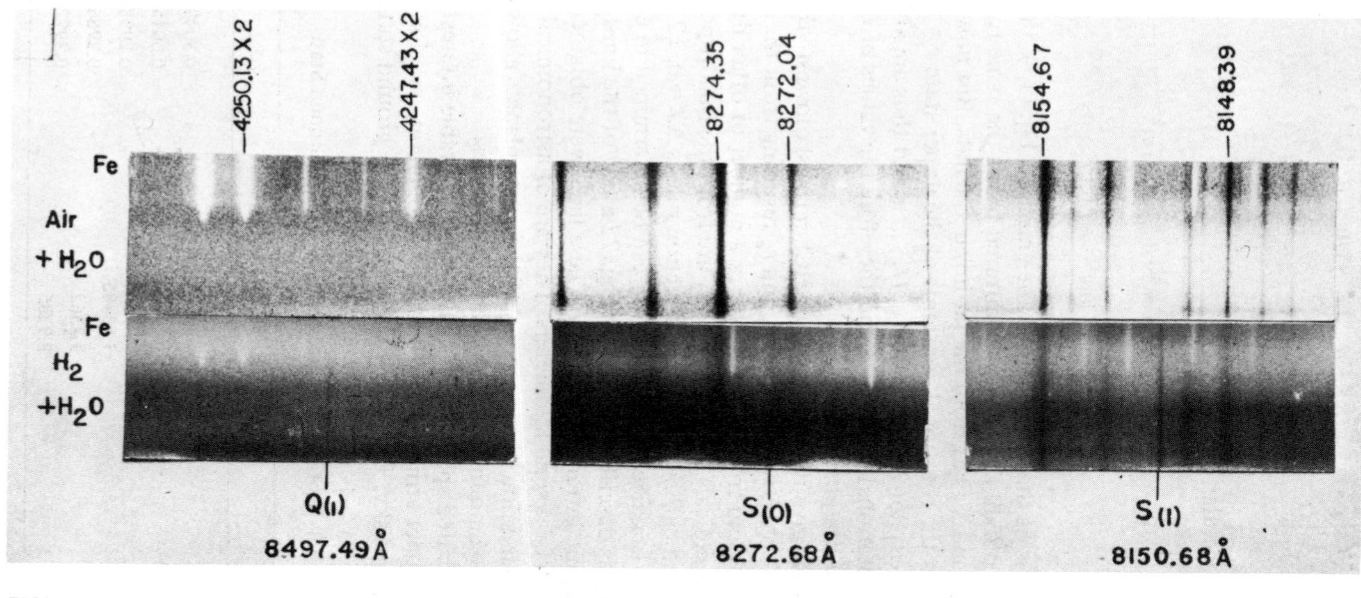

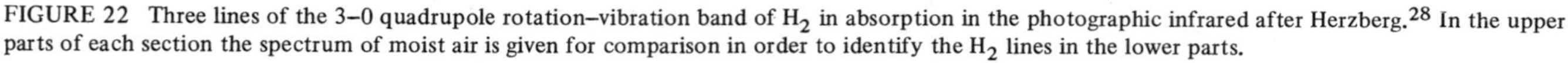
FIGURE 22 Three lines of the 3–0 quadrupole rotation–vibration band of H_2 in absorption in the photographic infrared after Herzberg.[28] In the upper parts of each section the spectrum of moist air is given for comparison in order to identify the H_2 lines in the lower parts.

TABLE 5 Rotational Constants and Internuclear Distances of Some Diatomic Molecules

Molecule	B_e (cm^{-1})	α_e (cm^{-1})	r_e (Å)
H_2	60.853	3.0622	0.7414$_4$
N_2	1.9980	0.01772	1.0978
O_2	1.44557	0.01579	1.20754
F_2	0.88925	0.0131	1.4127
OH	18.867	0.708	0.9708
HF	20.9557	0.798	1.02219
HCl	10.59342	0.30718	0.27455
CO	1.93127	0.017513	1.12832
NO	1.7046	0.0177	1.1508

OH ($^2\Pi$) the band structure is somewhat different from that described earlier for diatomic molecules without spin and electronic angular momentum in the ground state: the OH bands have a Q branch corresponding to $\Delta J = 0$, in addition to a P and R branch. Figure 23 shows part of the spectrum in the photographic infrared as obtained by Meinel.[29] The vibrationally excited OH radicals are formed in the upper atmosphere in the process of destruction of O_3 by H atoms according to

$$H + O_3 \rightarrow OH(v \leqq 9) + O_2. \tag{26}$$

For *linear polyatomic molecules* the fine structure of near-infrared bands is entirely similar to that of diatomic molecules except that for perpendicular fundamentals in addition to $\Delta J = \pm 1$ also $\Delta J = 0$ becomes possible, giving rise to a Q branch whose lines, when $B' \approx B''$, nearly coincide at ν_0. Figures 24, 25, and 26 show infrared bands of C_2H_2 and HCN as examples.

In the C_2H_2 spectrum (Fig. 24) the intensity alternation characteristic of symmetric molecules with identical nuclei is clearly present. The simple P, R structure immediately proves that the molecule is linear, and the intensity alternation shows that it must be symmetrical. A determination of the spacing of the lines gives directly the rotational constants, i.e., the moments of inertia in the upper and lower vibrational level. Since there are two parameters that determine the molecular structure (i.e., the C–H and the C–C distance), we cannot determine it from one moment of inertia. But if the moment of inertia (B value) is also determined for an isotopic molecule, C_2HD or C_2D_2, both parameters can be determined if it is assumed that these parameters are the same for isotopic molecules. The latter assumption is not rigorously fulfilled for the r_0 values but is certainly fulfilled for the r_e values.

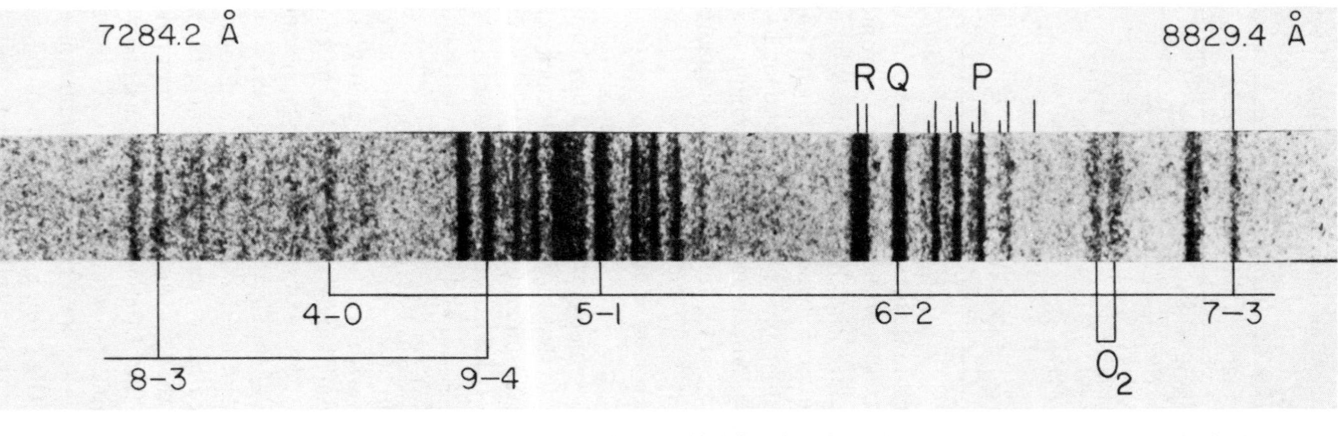

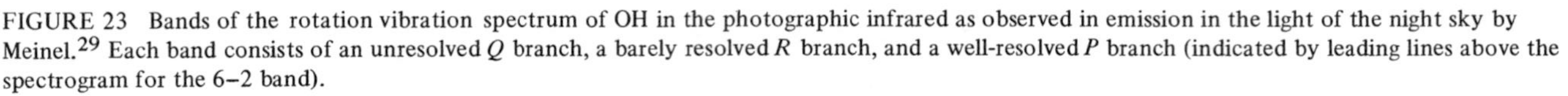

FIGURE 23 Bands of the rotation vibration spectrum of OH in the photographic infrared as observed in emission in the light of the night sky by Meinel.[29] Each band consists of an unresolved *Q* branch, a barely resolved *R* branch, and a well-resolved *P* branch (indicated by leading lines above the spectrogram for the 6–2 band).

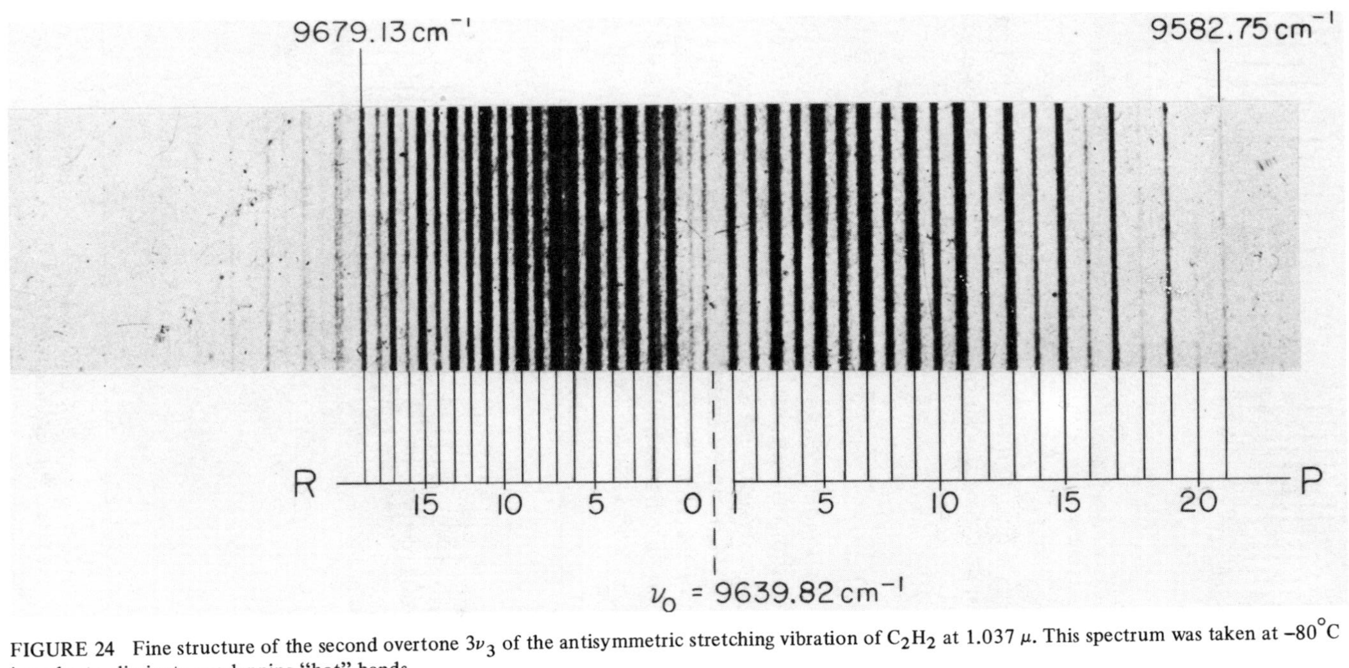

FIGURE 24 Fine structure of the second overtone $3\nu_3$ of the antisymmetric stretching vibration of C_2H_2 at 1.037 μ. This spectrum was taken at $-80^\circ C$ in order to eliminate overlapping "hot" bands.

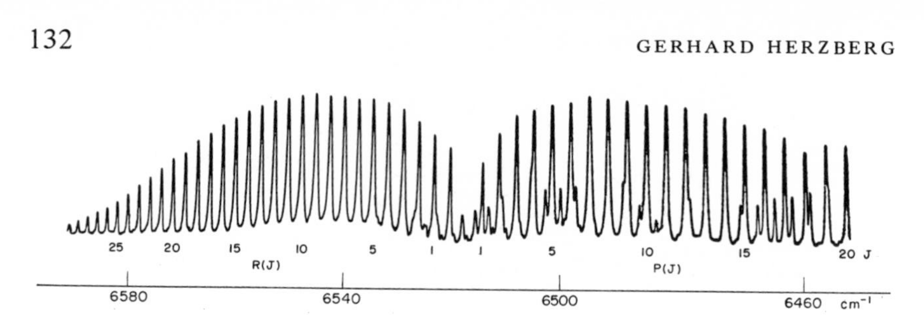

FIGURE 25 Fine structure of the overtone $2\nu_3$ of HCN near 1.53 μ. The weak lines overlapping the P branch and zero-gap belong to a "hot" band $2\nu_3 + \nu_2 - \nu_2$.

The r_e values are more difficult to determine for linear polyatomic than for diatomic molecules since $B_{[v]}$ now depends on all normal vibrations:

$$B_{[v]} = B_{v_1 v_2} \ldots = B_e - \Sigma\alpha_i\,[v_i + (d_i/2)] + \ldots d_i = 1,2, \tag{27}$$

where $d_i = 1$ for nondegenerate (stretching) vibrations, $d_i = 2$ for doubly degenerate (bending) vibrations. Only if $B_{[v]}$ has been determined for a sufficient number of different vibrational levels such that all α_i (three for HCN, five for C_2H_2) can be evaluated is it possible to obtain a reliable value for B_e and therefore I^e. This has been done for HCN and C_2H_2; the resulting data are listed in Table 6 together with the data referring to the lowest vibrational level, i.e., B_0, r_0. The differences are not large

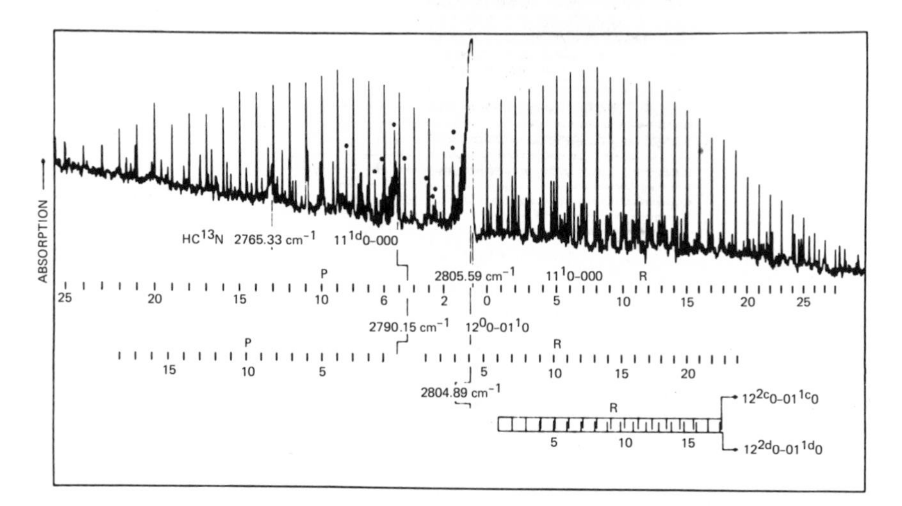

FIGURE 26 Fine structure of the perpendicular band $\nu_1 + \nu_2$ of HCN showing P, Q, and R branches after Maki and Blaine.[30] In addition to the main band a few weaker bands, mostly hot bands, are present.

TABLE 6 Rotational Constants and Internuclear Distances in HCN and C_2H_2

Molecule	B_0, B_e	r_0, r_e
HCN	1.47822 1.4849	r_0 (CH) = 1.064, r_0 (CH) = 1.156 Å r_e (CH) = 1.065_7, r_e (CH) = 1.153_0 Å
DCN	1.20775 1.2118	
C_2H_2	1.17660 1.18246	r_0 (CH) = 1.058, r_0 (CC) = 1.208 Å r_e (CH) = 1.060_4, r_e (CC) = 1.203_3 Å
C_2D_2	0.84794 0.85069	

but nevertheless significant. Data similar to those in Table 6 are available for a number of linear polyatomic molecules.

For *nonlinear polyatomic molecules*, if two of the principal moments of inertia are equal (symmetric top) we have the energy formula

$$T = G(v_1, v_2, \ldots) + F_{[v]}(J,K), \tag{28}$$

$$F_{[v]}(J,K) = B_{[v]} J(J+1) + (A_{[v]} - B_{[v]}) K^2 + \ldots, \tag{29}$$

where

$$\begin{aligned} B_{[v]} &= B_e - \Sigma\alpha_i^B [v_i + (d_i/2)], \\ A_{[v]} &= A_e - \Sigma\alpha_i^A [v_i + (d_i/2)]. \end{aligned} \tag{30}$$

The selection rules are different depending on whether the vibration has its alternating dipole moment parallel or perpendicular to the figure axis of the top, namely,

$$\| \qquad \Delta K = 0, \qquad \Delta J = 0, \pm 1; \tag{31}$$

$$\perp \qquad \Delta K = \pm 1, \qquad \Delta J = 0, \pm 1. \tag{32}$$

Correspondingly, there are two different band types, $\|$ bands and $\perp$ bands, each with a series of subbands. In the first type ($\|$) often all subbands nearly coincide, giving rise to an apparently very simple band structure exemplified by the CH_3CCH band in Fig. 27, while in the second type ($\perp$) the subbands are spaced $2(A_v - B_v)$ apart giving rise to the predominance of a series of linelike Q branches as shown by the example of the CH_3Br band in Fig. 28. In this example we also notice

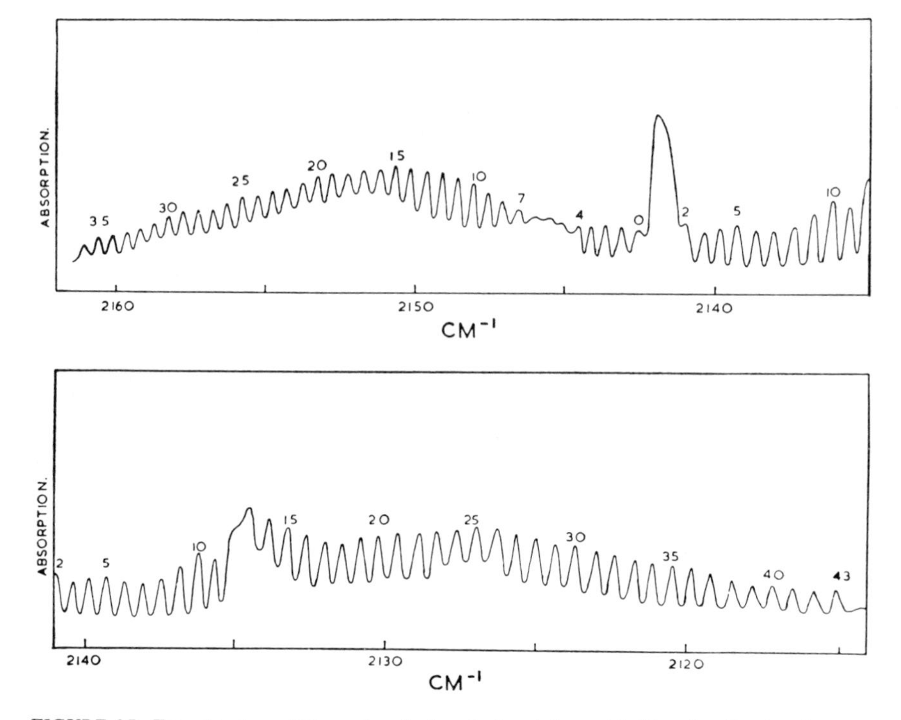

FIGURE 27 Fine structure of the ν_3 band of methyl acetylene at 4.67 μ after Boyd and Thompson.[31] The peak of 2134 cm^{-1} and the blurring of the structure of the main band near 2145 cm^{-1} are due to overlapping hot bands.

the effect of the identity of the three nuclei surrounding the symmetry axis: every third "line" is twice as strong as the intermediate "lines." For zero nuclear spin the intermediate lines would be missing.

From the spacing of the lines in the subbands of || or ⊥ bands the rotational constants $B_{[v]}$ can be obtained in the same way as for diatomic or linear polyatomic molecules, while from the spacing of the subbands in a ⊥ band the values of $A - B$ can in principle be obtained. However, for symmetrical molecules a complication arises on account of the Coriolis interaction between the ⊥ vibration and the rotation about the symmetry axis leading to a spacing of $2[A_v(1-\zeta_v)-B_v]$ instead of $2(A_v - B_v)$, where ζ is a Coriolis parameter that is somewhat difficult to evaluate.

The structures of many symmetric top molecules have been evaluated by such studies of rotation–vibration spectra. This method is particularly important for molecules like C_2H_6, which because of the absence of a permanent dipole moment cannot be studied in the microwave

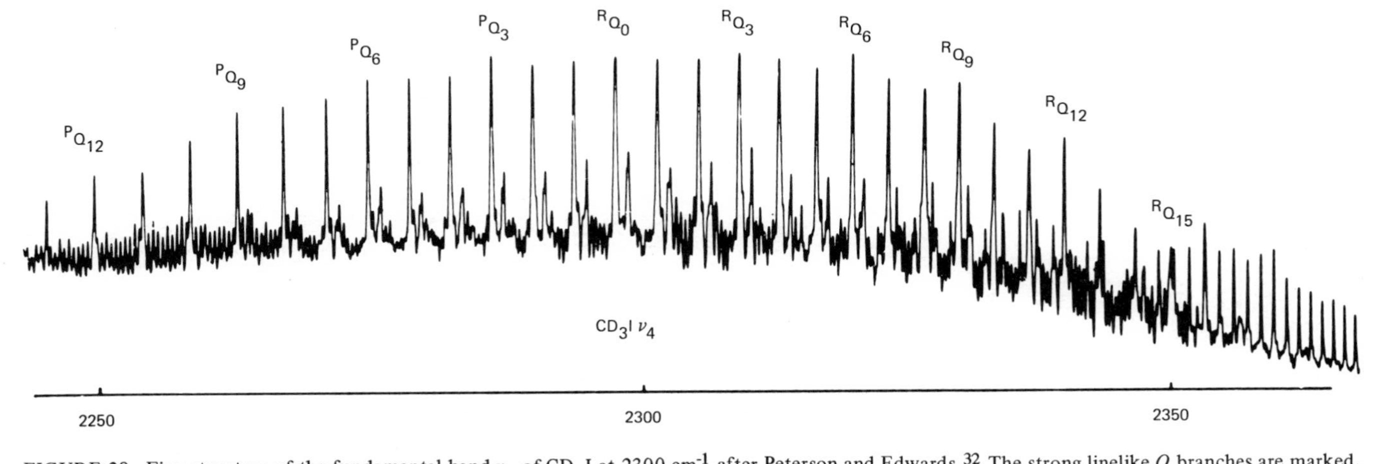

FIGURE 28 Fine structure of the fundamental band ν_4 of CD_3I at 2300 cm^{-1} after Peterson and Edwards.[32] The strong linelike Q branches are marked. The intensity alternation strong weak, weak strong, . . . is apparent.

TABLE 7 Rotational Parameters of Some Symmetric Top Molecules

Molecule	Rotational Constants (cm^{-1})	Geometric Parameters
C_2H_6	A_0 = 2.681 B_0 = 0.6622	r_0 (CC) = 1.536 Å r_0 (CH) = 1.091 Å <HCH = 108.0°
C_6H_6	B_0 = 0.1896	r_0 (CC) = 1.397 Å r_0 (CH) = 1.084 Å
NH_3	B_0 = 9.4443 C_0 = 6.196	r_0 (NH) = 1.0173 Å <HNH = 107.8°
CH_3Cl	A_0 = 5.097 B_0 = 0.44340	r_0 (CCl) = 1.781 Å r_0 (CH) = 1.113 Å <HCH = 110.5°
CH_2CCH_2	A_0 = 4.81 B_0 = 0.29632	r_0 (CC) = 1.308 Å r_0 (CH) = 1.087 Å <HCH = 118.2°

region. Again, the study of isotopic molecules is necessary in order to obtain all the necessary parameters. Table 7 gives a few examples.

When all three principal moments of inertia of a molecule are equal, i.e., for the model of the spherical top, of which CH_4 is an example, the energy levels are again given by the simple formula

$$F_{[v]}(J) = B_{[v]} J(J+1) + \dots \qquad (33)$$

[since the second term in the symmetric top formula (29) vanishes]. In a first approximation a simple band structure arises as for a linear molecule. This is illustrated by the CH_4 band in Fig. 29. However, because of the Coriolis interactions between rotation and vibration, splittings

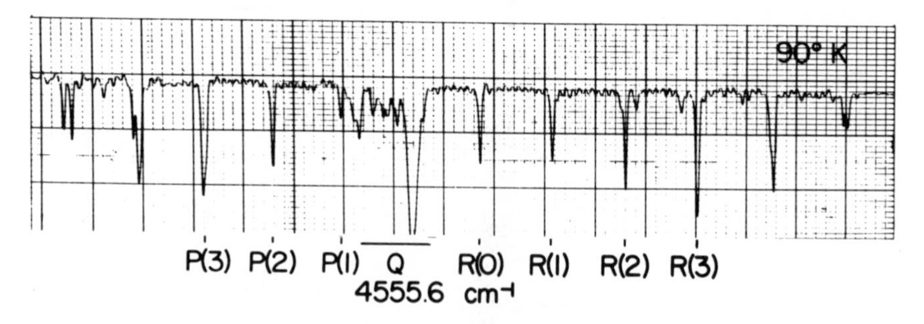

FIGURE 29 Fine structure of the band $\nu_2 + \nu_3$ of CH_4 near 2.20 μ. The spectrum was taken at liquid N_2 temperature.

occur for higher J values as seen in Fig. 29. Only if these structures are fully understood and evaluated can precise values for the molecular constants be obtained. The result for CH_4 is

$$B_0 = 5.2412\ \mathrm{cm}^{-1},\ r_0 = 1.0940\ \text{Å}.$$

It is based partly on the study of the fine structure of a vibrational Raman band (see Hecht[33] and Herranz and Stoicheff[34]).

The presence of CH_4 in the atmosphere of the earth was first established by the observation of its infrared bands in the solar spectrum. These bands will afford also the only way to detect CH_4 in interstellar space since CH_4 has no dipole moment and therefore no ordinary microwave spectrum.* A high-resolution infrared spectrum of the star α Orionis obtained by Connes (unpublished) does show indications of interstellar CH_4 (see Ref. 35).

If all three principal moments of inertia are different, as for example for H_2O, the band structure becomes quite complicated, as shown by the example in Fig. 30. I shall not attempt a more detailed discussion but only mention that quite a number of such spectra have been analyzed and from them the molecular structures have been determined. When two of the three principal moments of inertia are nearly the same, we have, of course, an approach to the much simpler band structure of a symmetric top. Table 8 gives results for a few asymmetric top molecules.

IV. ELECTRONIC BAND SPECTRA IN THE VISIBLE AND ULTRAVIOLET REGIONS

All transitions so far discussed take place in one and the same electronic state, usually the ground state. A molecule, just as an atom, has many electronic states corresponding to different distributions of the electrons over their various orbitals and to different orientations of the electronic angular momenta. The spectra that arise in transitions from one electronic state to another are similar to infrared spectra except that now the upper and lower vibrational and rotational terms belong to different electronic states and may differ considerably.

The energy of the molecule may be written as the sum of three contributions: electronic, vibrational, and rotational, that is,

$$E = E_e + E_v + E_r, \tag{34}$$

*An extremely weak far-infrared spectrum caused by rotation–vibration interaction has recently been reported by Rosenberg *et al.*[36]

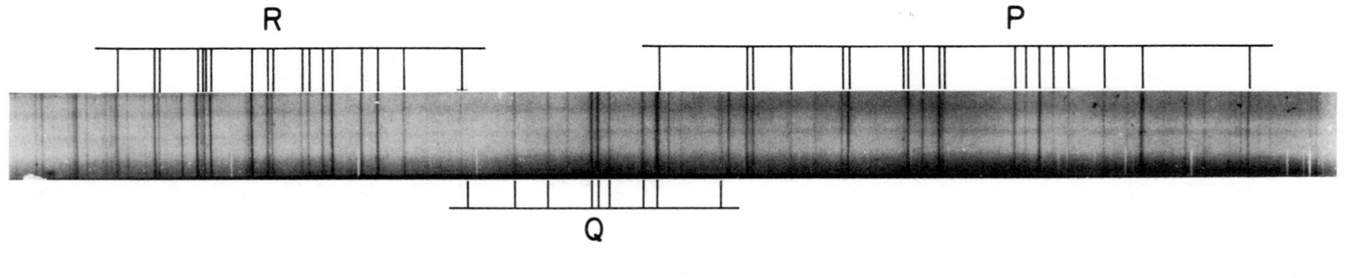
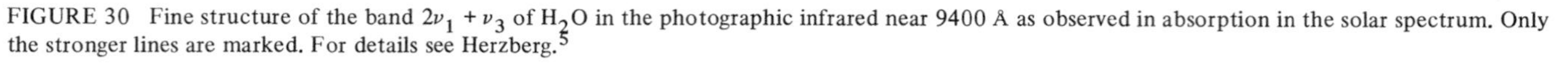

FIGURE 30 Fine structure of the band $2\nu_1 + \nu_3$ of H_2O in the photographic infrared near 9400 Å as observed in absorption in the solar spectrum. Only the stronger lines are marked. For details see Herzberg.[5]

TABLE 8 Rotational Parameters of Some Asymmetric Top Molecules

Molecule	Rotational Constants	Geometric Parameters
H_2O	$A_0 = 27.877$ $B_0 = 14.512$ $C_0 = 9.285$	r_0 (OH) = 0.956 Å <HOH = 105.2°
H_2CO	$A_0 = 9.4053$ $B_0 = 1.2954$ $C_0 = 1.1343$	r_0 (CH) = 1.102 Å r_0 (CO) = 1.210 Å <HCH = 121.1°
C_2H_4	$A_0 = 4.828$ $B_0 = 1.0012$ $C_0 = 0.8282$	r_0 (CH) = 1.086 Å r_0 (CC) = 1.339 Å <HCH = 117.6°
N_3H	$A_0 = 20.616$ $B_0 = 0.40142$ $C_0 = 0.39299$	r_0 (NH) = 0.975 Å r_0 (NN) = {1.237 Å, 1.133 Å} <HNN = 114.1°

or, in term values,

$$T = E/hc = T_e + G(v) + F_v(J). \tag{35}$$

For an electronic transition of a *diatomic molecule* we have

$$\nu = T' - T'' = \nu_e + \nu_v + \nu_r, \tag{36}$$

where

$$\nu_e = T_e' - T_e'', \quad \nu_v = G'(v') - G''(v''),$$

$$\nu_r = F_v'(J') - F_v''(J'').$$

If we consider one particular electronic transition, ν_e is fixed: all possible values of ν_v and ν_r give rise to a *band system*. Under low resolution we can in a first approximation neglect ν_r and obtain for the bands of an electronic band system of a diatomic molecule

$$\begin{aligned}\nu = \nu_e &+ \omega_e'(v' + \tfrac{1}{2}) - \omega_e' x_e'(v' + \tfrac{1}{2})^2 + \ldots \\ &- [\omega_e''(v'' + \tfrac{1}{2}) - \omega_e'' x_e''(v'' + \tfrac{1}{2})^2 + \ldots].\end{aligned} \tag{37}$$

Figure 31 gives as an example a low-resolution spectrogram of a band system of the CN free radical as observed in the ordinary carbon arc.

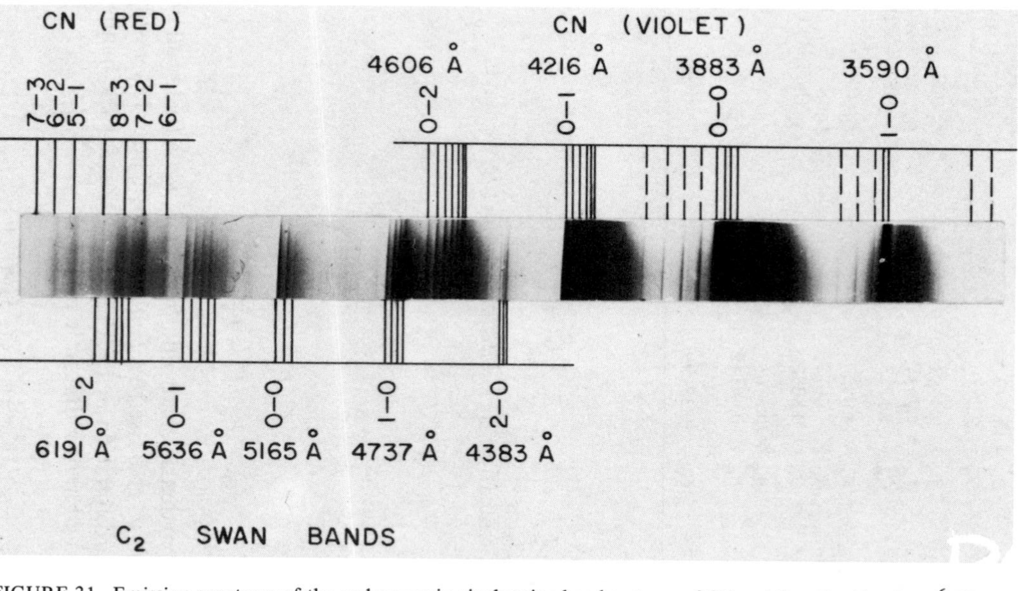

FIGURE 31 Emission spectrum of the carbon are in air showing band systems of CN and C_2 after Herzberg.[6] The bands of these band systems are grouped in sequences (Δv = const.). For both the violet CN and the C_2 Swan bands the vibrational assignments and the wavelengths of only the first member of each sequence are indicated. The broken leading lines refer to the so-called tail bands (see Ref. 6).

Figure 32 gives schematically the vibrational levels. The band system can be considered either as consisting of a number of v' progressions or of a number of v'' progressions. The spacings of appropriate bands give directly the spacings of the vibrational levels in the upper and lower state or, in other words, the vibrational frequencies $\omega_e{}'$ and $\omega_e{}''$ in the two states. These together with the anharmonicities (from the change of spacing with v) give a first rough approximation of the potential function of each state.

In some cases in absorption a band progression is observed up to the last vibrational level of the upper state and is followed by a continuous absorption spectrum, which represents transitions to the continuous range of energy levels above the asymptote of the potential function (see Fig. 16) and which corresponds to a dissociation into two atoms with various amounts of kinetic energy. This is illustrated by the potential diagram of H_2 in Fig. 33. The dissociation limit ν_{lim} is the convergence limit of the band progression corresponding to transitions from the ground state ($X\ {}^1\Sigma_g^+$) to the excited state ($B'\ {}^1\Sigma_u^+$). If the excitation energy of the resulting atoms is subtracted from the dissociation limit, a precise value for the dissociation energy of the molecule is obtained. The results for H_2, HD, and D_2 are shown in Table 9 and are compared with the values obtained by *ab initio* calculations from wave mechanics. The agreement is within one part in 50,000, confirming nicely the validity of wave mechanics for molecular systems.

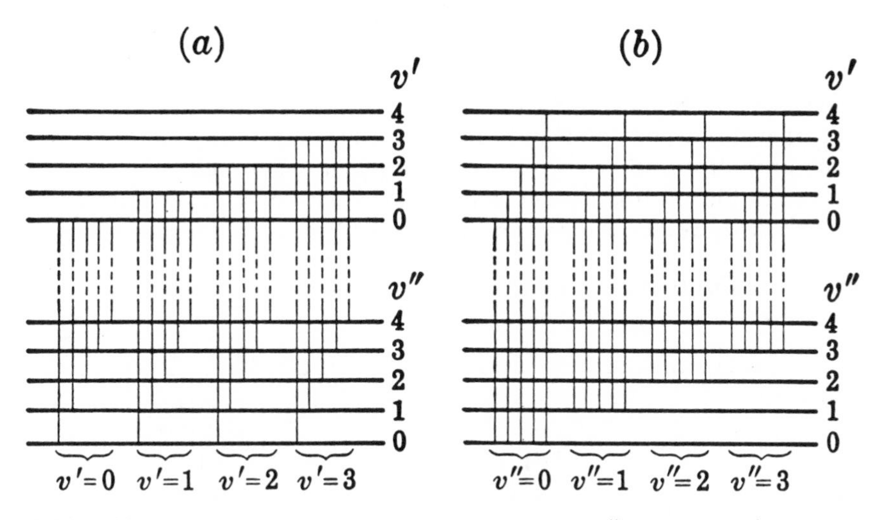

FIGURE 32 Energy-level diagrams for progressions of bands (a) v'' progressions (v' = const.) and (b) v' progressions (v'' = const.). A band system can be fully represented by either (a) or (b).

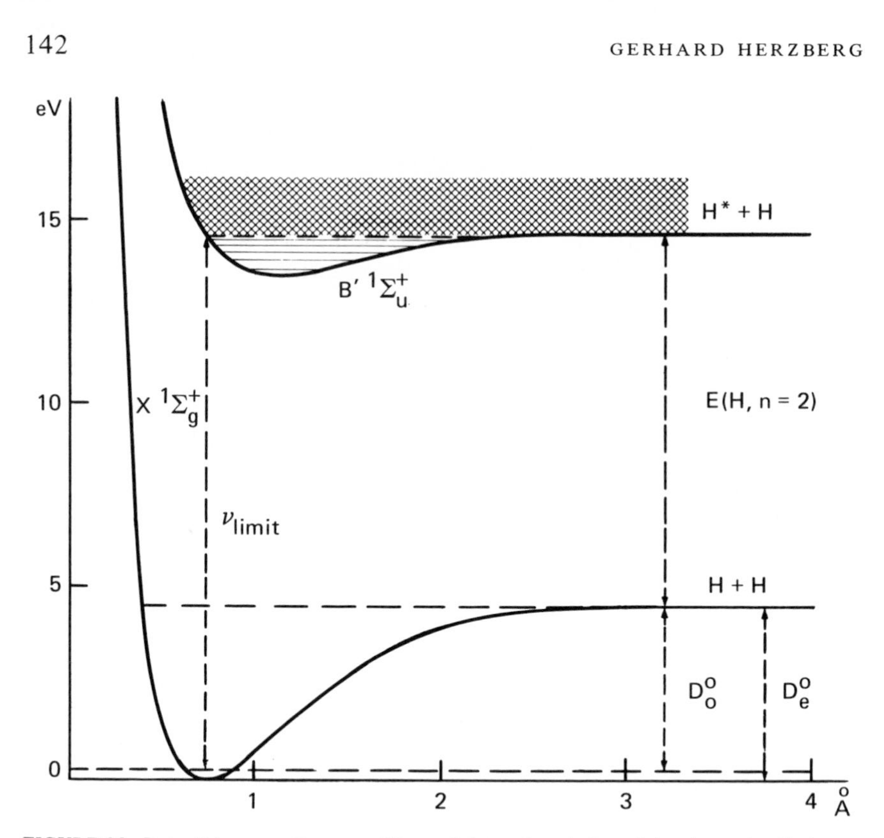

FIGURE 33 Potential energy diagram of H_2 explaining the relation of the absorption limit to the dissociation energy of the ground state. The observed absorption continuum joins onto the band progression corresponding to the transitions from the $\nu = 0$ ground state to the vibrational levels in the $B'\ ^1\Sigma_u^+$ state.

The characteristic bandlike structure of the CN bands in Fig. 31, a sharp edge at one end and a gradual fading out toward the other, is readily understood when the bands are observed under high resolution, as shown in Fig. 34 for the 0–0 band of CN. We have, as in the infrared, two branches given by the previous formulas (23) – (25); but here since B'

TABLE 9 Calculated and Observed Dissociation Energies of H_2, HD, and D_2[a]

	Theor. (cm^{-1})	Obs. (cm^{-1})
$D_0{}^0(H_2)$	36117.9	<36118.3 >36116.3
$D_0{}^0(HD)$	36405.5	36405.8
$D_0{}^0(D_2)$	36748.2	36748.9 ± 0.4

[a]From Refs. 26 and 37.

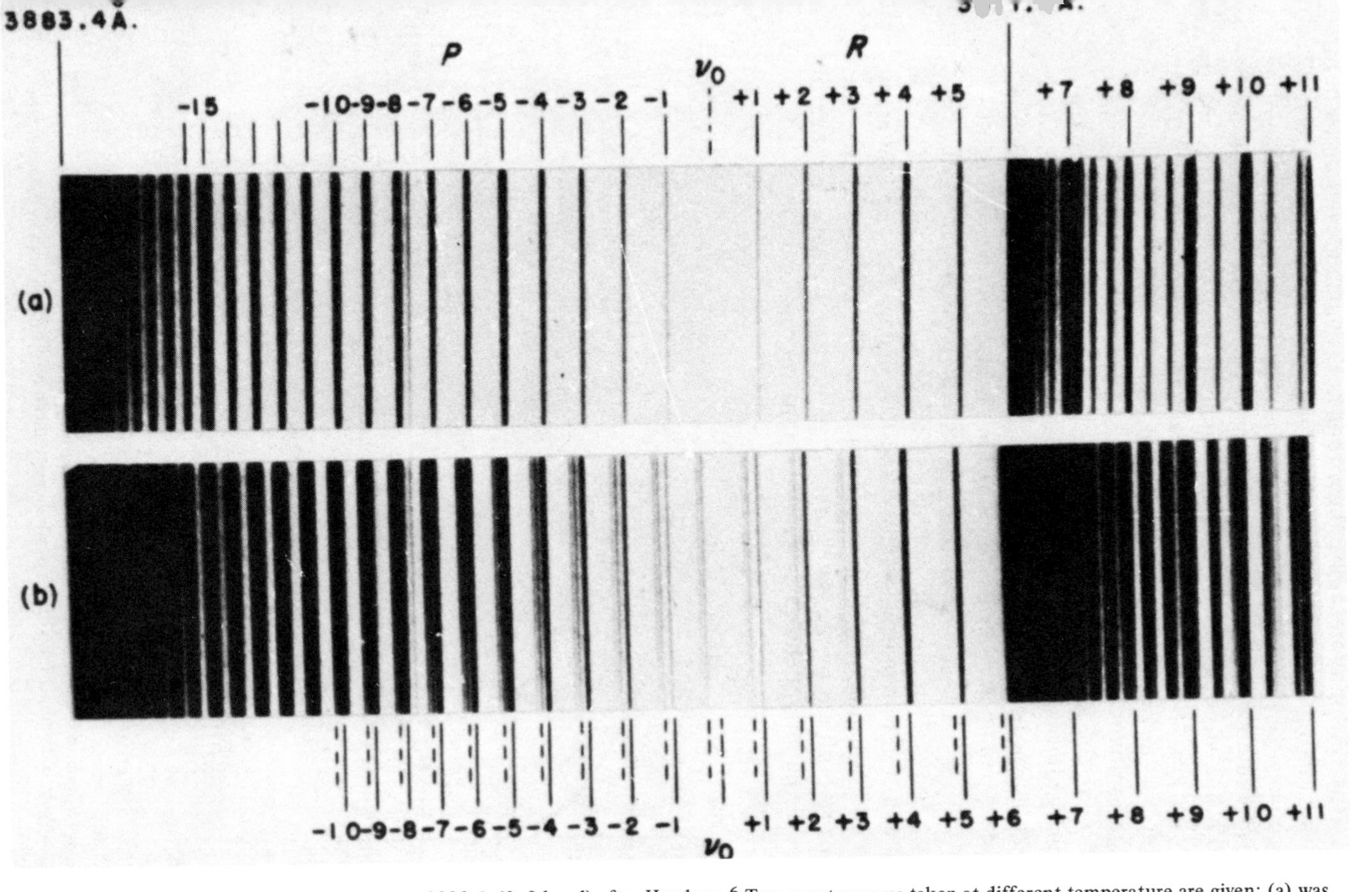

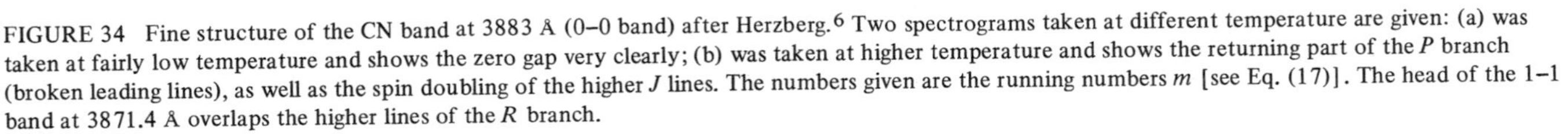
FIGURE 34 Fine structure of the CN band at 3883 Å (0–0 band) after Herzberg.[6] Two spectrograms taken at different temperature are given: (a) was taken at fairly low temperature and shows the zero gap very clearly; (b) was taken at higher temperature and shows the returning part of the P branch (broken leading lines), as well as the spin doubling of the higher J lines. The numbers given are the running numbers m [see Eq. (17)]. The head of the 1–1 band at 3871.4 Å overlaps the higher lines of the R branch.

and B'' may differ appreciably, the divergence is much greater and a head is formed in the P or R branch depending on whether $B' > B''$ or $B' < B''$, respectively. In Fig. 35 the value of m [see Eq. (25)] is plotted against the wavenumber (Fortrat diagram) showing clearly the formation of the band head. Although band heads are usually formed in electronic spectra there are also cases of bands without head. This arises when $B' \approx B''$. An example is provided by the CN^+ band in Fig. 36.

For homonuclear molecules the lines in the electronic bands, similar to those in the Raman spectrum discussed earlier, show a characteristic intensity alternation. Indeed, this phenomenon was first recognized in electronic spectra, particularly the violet bands of N_2^+ of which Fig. 37 shows an example. Alternate missing lines have been observed in all bands of O_2, C_2, S_2, and others for which the nuclear spin $I = 0$. Figure 38 shows as a recent example an absorption band of the C_2^- ion.

In addition to bands with only P and R branches there are also many examples of bands with P, Q, and R branches. The Q branch arises when in one or both electronic states there is an electronic angular momentum Λ about the internuclear axis (Π, Δ, . . . states correspond to $\Lambda = 1, 2, \ldots$,

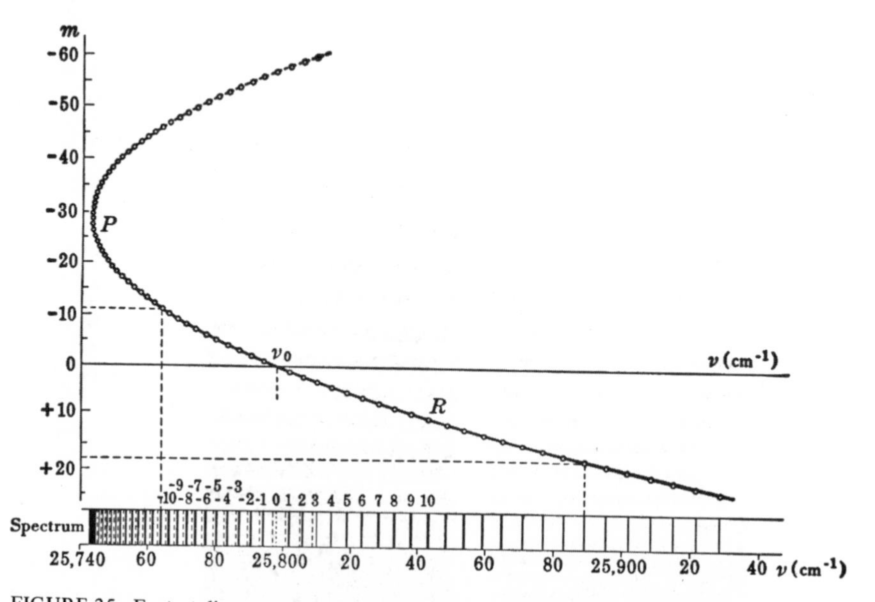

FIGURE 35 Fortrat diagram of the band 3883 Å of CN showing the formation of the band head. The schematic spectrum below is the projection of those points in the Fortrat parabola, which correspond to integral m. No line is observed for $m = 0$ (dotted line). The band lines corresponding to the upper limb of the parabola are indicated by broken lines in the schematic spectrum.

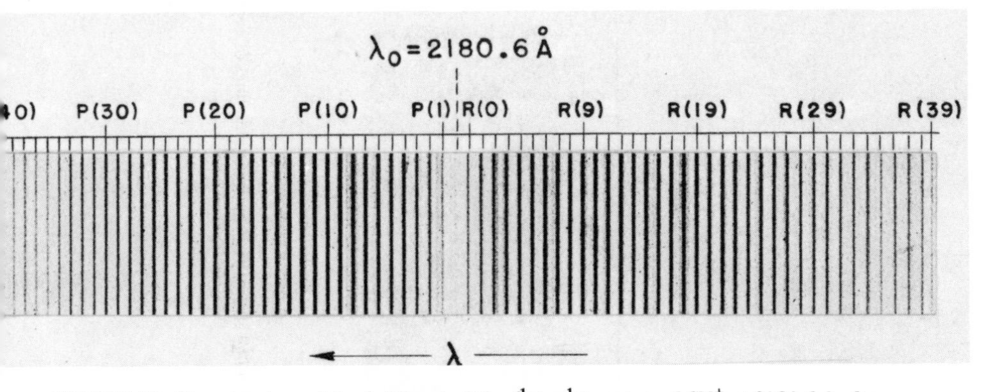

FIGURE 36 Fine structure of the 0–0 band of the $f^1\Sigma$–$a^1\Sigma$ system of CN^+ at 2180.6 Å after Douglas and Routley.[38] Since B′ ≈ B″ no band head is formed. The zero-gap is clearly shown.

while Σ states correspond to $\Lambda = 0$). Figure 39 shows an emission band of AlH, which shows a *Q* branch in addition to *P* and *R*.

In all these cases, an analysis of the band structure gives B' and B'' values and, if a sufficient number of bands has been observed, $B_e{}'$, $B_e{}''$, α', α'', and other constants. There are many diatomic molecules, including almost all short-lived molecules and homonuclear molecules, for which all structural information is based on the study of electronic transitions.

In all examples so far considered here, the effect of the *electron spin S* was negligible, because either *S* was zero or its coupling with the rest of the molecule was small. For $S \neq 0$ and strong spin-orbit coupling we have multiplet electronic states. We have then as many subbands as indicated by the multiplicity. For intermediate or small spin-orbit coupling often rather complicated band structures arise. For example, a band of a $^3\Sigma$–$^3\Pi$ transition has 27 branches. Figure 40 shows as an example a C_2 band. We shall not consider the multiplet structure further except to say that the interaction of rotation and electronic motion is well understood and this knowledge makes it possible to analyze even the more complicated band types, resulting in precise values for the rotational and vibrational constants of the particular electronic states. In addition, of course, the coupling parameters between spin and orbital angular momentum and spin and rotation are obtained.

The absorption and emission spectrum of the earth's atmosphere contains many examples of diatomic electronic spectra. Many of them are forbidden electronic transitions, like the infrared and red atmospheric oxygen bands ($^1\Delta_g$–$^3\Sigma_g^-$ and $^1\Sigma_g^+$–$^3\Sigma_g^-$), which are such a characteristic absorption feature of the spectrum of the sun as observed from the sur-

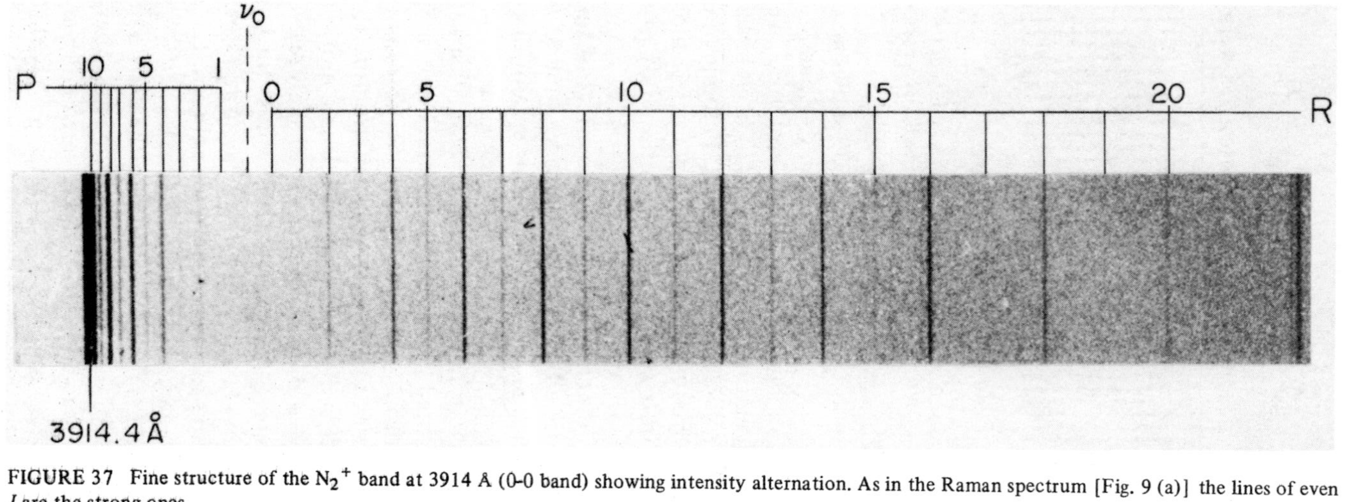

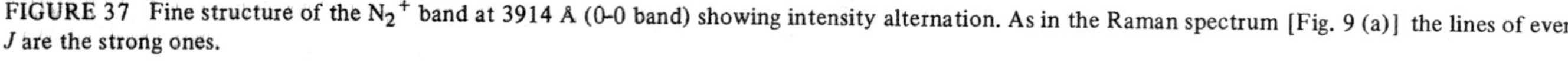

FIGURE 37 Fine structure of the N_2^+ band at 3914 Å (0-0 band) showing intensity alternation. As in the Raman spectrum [Fig. 9 (a)] the lines of even *J* are the strong ones.

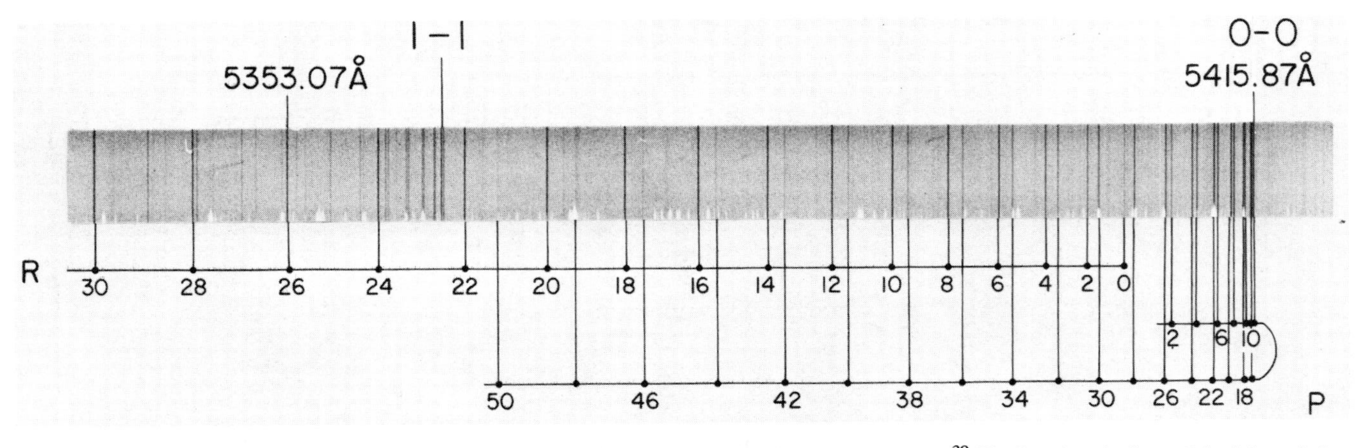

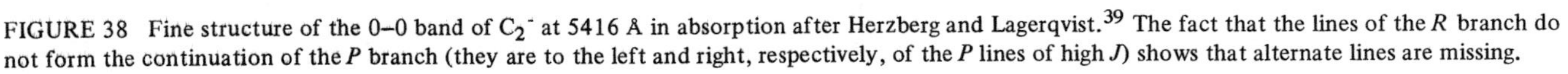

FIGURE 38 Fine structure of the 0–0 band of C_2^- at 5416 Å in absorption after Herzberg and Lagerqvist.[39] The fact that the lines of the *R* branch do not form the continuation of the *P* branch (they are to the left and right, respectively, of the *P* lines of high *J*) shows that alternate lines are missing.

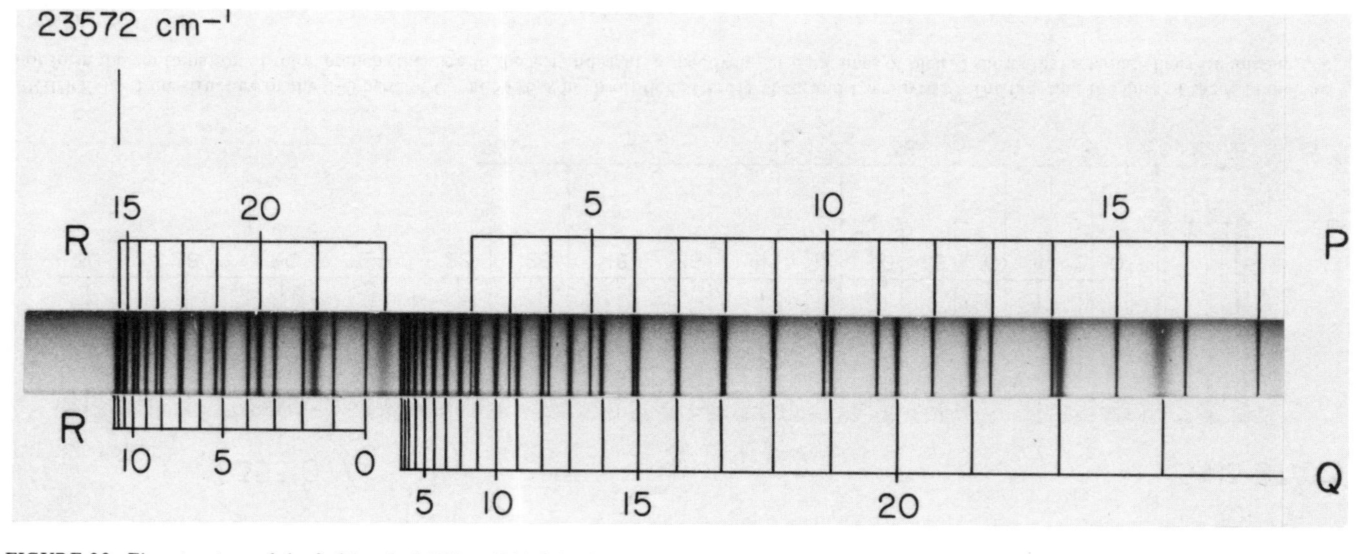

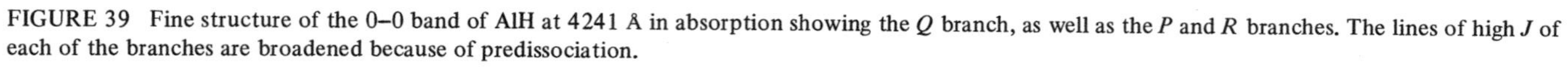

FIGURE 39 Fine structure of the 0–0 band of AlH at 4241 Å in absorption showing the Q branch, as well as the P and R branches. The lines of high J of each of the branches are broadened because of predissociation.

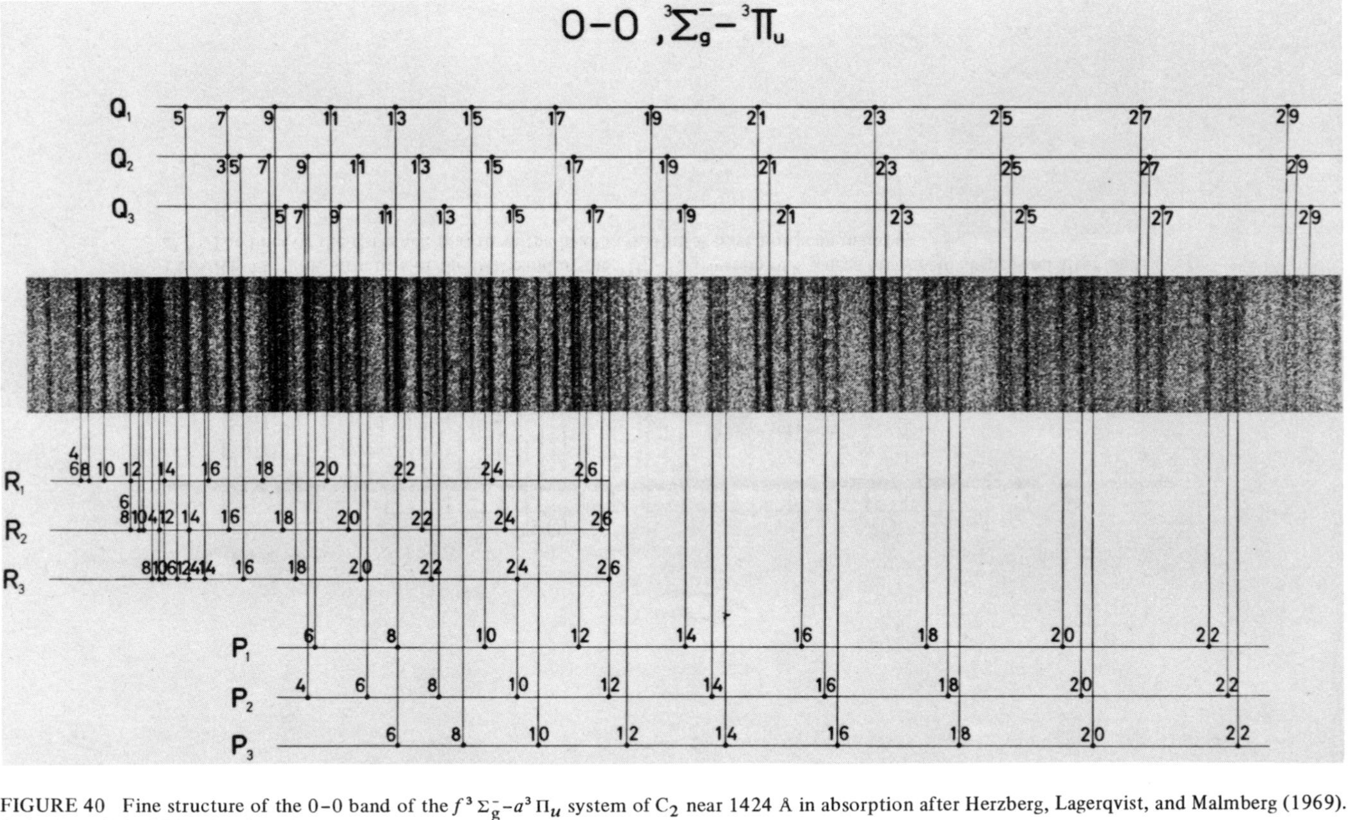

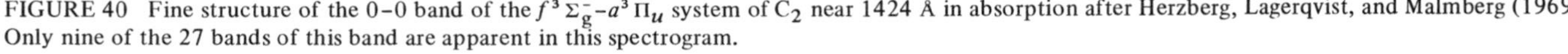

FIGURE 40 Fine structure of the 0–0 band of the $f\,^3\Sigma_g^- - a\,^3\Pi_u$ system of C_2 near 1424 Å in absorption after Herzberg, Lagerqvist, and Malmberg (1969). Only nine of the 27 bands of this band are apparent in this spectrogram.

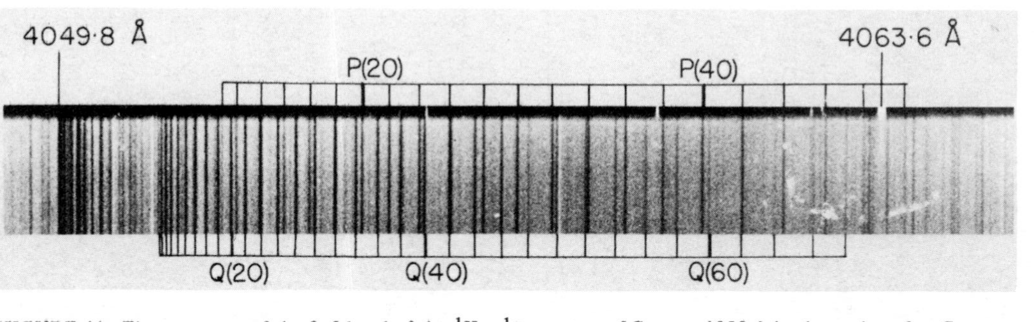

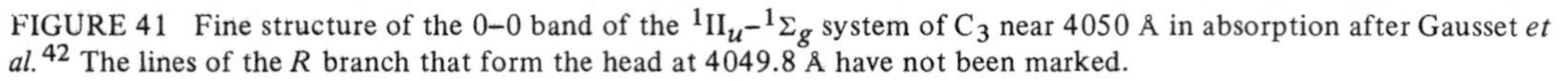

FIGURE 41 Fine structure of the 0–0 band of the ${}^1\Pi_u - {}^1\Sigma_g$ system of C_3 near 4050 Å in absorption after Gausset *et al.*[42] The lines of the *R* branch that form the head at 4049.8 Å have not been marked.

face of the earth, or the Vegard-Kaplan bands of N_2 ($^3\Sigma_u^+ - {}^1\Sigma_g^+$) observed in emission in the aurora. In the spectra of comets only allowed electronic transitions of diatomic molecules are observed: CN, C_2, OH, NH, CH, etc. Here, unlike the earth's upper atmosphere, the excitation mechanism is pure fluorescence as was proven by Swings[40] and McKellar[41] from the effect of solar Fraunhofer lines on the intensity distribution in the rotational fine structure. Studies of this intensity distribution also have supplied important information about the temperature in the upper atmosphere of the earth and in the heads of comets.

Electronic spectra of diatomic molecules are also observed in the spectra of the sun and particularly in sunspots, as well as in all late-type stars. The first molecules discovered in the interstellar medium—CH, CH^+, CN—were discovered by means of their electronic spectra in the visible and near-ultraviolet regions. Here, because of the low temperature (and the long time between collisions and excitations) instead of observing bands, only single lines are found—the $R(0)$ lines, which originate from the lowest rotational level. Only for CN has a line [$R(1)$] originating from the second lowest level also been observed. The intensity ratio of $R(1)$ and $R(0)$ indicates a temperature of 3 K in agreement with the much later discovery of the universal 3 K radiation.

The electronic spectra of *linear polyatomic molecules* are in most respects similar to those of diatomic molecules except that the vibrational structure is more complicated because of the presence of several vibrations. An additional complication arises when the bending vibrations are excited: there is often a strong coupling of the vibrational angular momentum with the electronic angular momentum (Renner-Teller effect). A striking example is provided by the $^1\Pi - {}^1\Sigma$ transition of the C_3 molecule first observed in cometary spectra. Figure 41 shows a high-resolution absorption spectrogram of the 0–0 band, which shows P, Q, and R branches. The vibrational structure remained a puzzle until it was recognized[42] that in the lower state the bending frequency is exceptionally small ($\nu_2 = 63\ \text{cm}^{-1}$), while in the excited state a strong Renner-Teller effect causes large splittings when the bending vibration is excited.

Several molecules that are linear in their ground states have been found to be nonlinear in some of their excited states. This change of shape is indeed a fairly common phenomenon. Usually in such a case the molecule in its nonlinear form, while strictly speaking an asymmetric top, is sufficiently close to a symmetric top that the rotational fine structure is still quite simple. Figure 42 shows an absorption band of HCN at 1816 Å. It has a simple P, Q, and R branch as if it belonged to a transition between two states in both of which the molecule is linear.

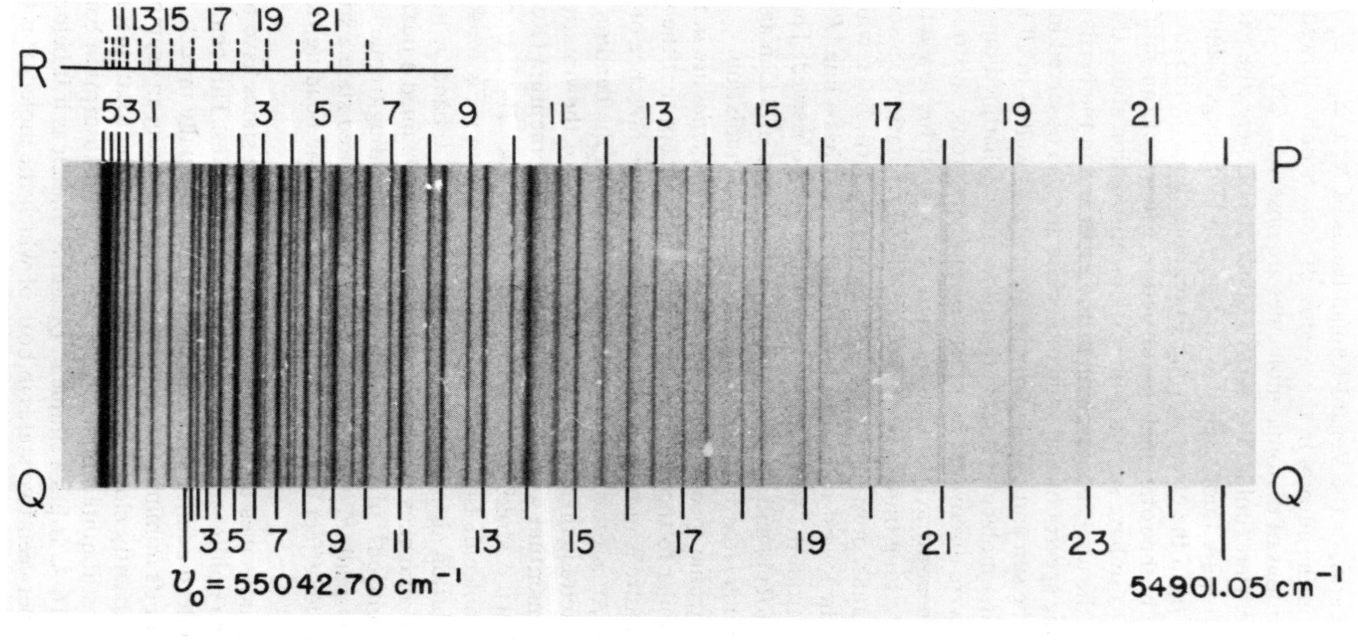

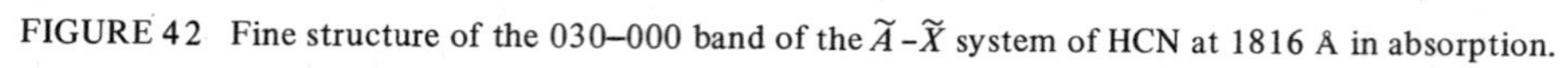

FIGURE 42 Fine structure of the 030–000 band of the $\tilde{A}-\tilde{X}$ system of HCN at 1816 Å in absorption.

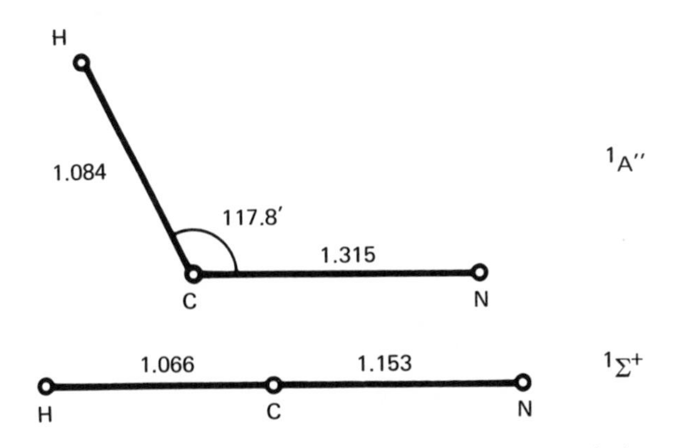

FIGURE 43 Structures of HCN in the electronic ground state $^1\Sigma^+$ and the first excited state $^1A''$. The geometrical parameters given refer to the lowest vibrational level, not to the equilibrium configuration.

Only by a more detailed analysis was it possible to show conclusively that the molecule is nonlinear in the excited state. Figure 43 illustrates the resulting structures.

The absorption band of HCO shown in Fig. 44 exemplifies the opposite case in which the molecule is nonlinear in the ground state and linear in the excited state. In such a transition, since levels with several K values are occupied in the ground state, and since several values of the vibrational angular momentum ℓ' occur in the upper state, one might have expected the presence of a number of subbands, but only a single one appears for a given vibrational transition. The reason for the absence of all subbands other than the one with $K' = \ell' = 0$ was found to be that there is a strong Auger process (here called predissociation) which broadens the subbands with $K' \neq 0$ so much that they are difficult to recognize. The structure derived from the spectrum is shown in Fig. 45.

There are many cases in which a molecule is a symmetric top (or a nearly symmetric top) in both upper and lower state. The band structure is then similar to that of infrared bands of (nearly) symmetric top molecules: we have $\parallel$ and $\perp$ bands as described earlier. Figure 46 shows as an example a $\perp$ absorption band of CH_3I, which shows the intensity alternation characteristic of molecules with three identical nuclei with nonzero spin.

As an example of a nearly symmetric top molecule we show in Fig. 47 one of the subbands of the 0–0 band of HNO, which shows clearly the effect of K-type doubling. Figure 48 shows the structures derived for HNO in the upper and lower states with the help of the corresponding

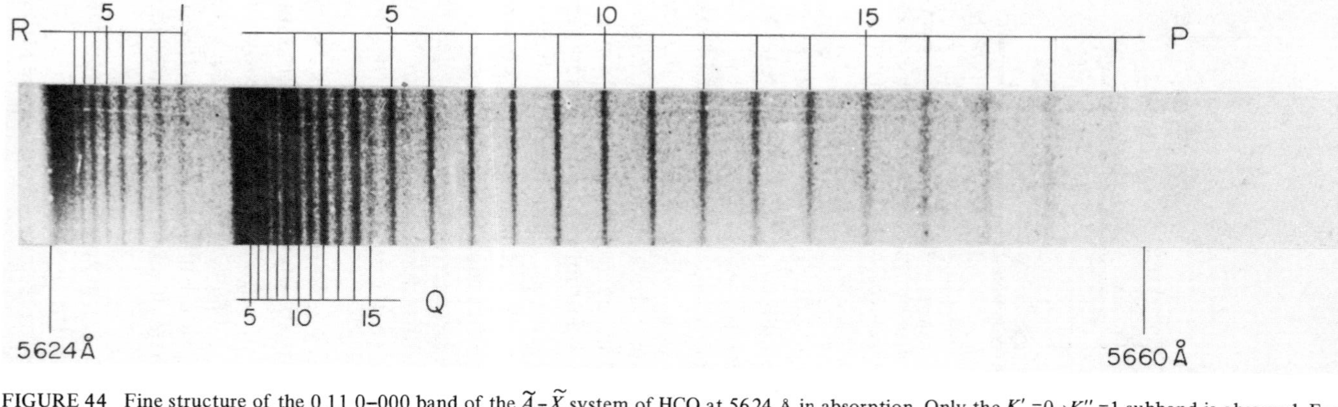
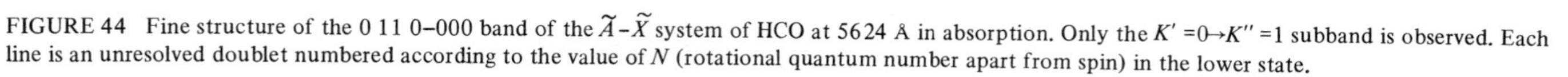

FIGURE 44 Fine structure of the 0 11 0–000 band of the $\tilde{A}-\tilde{X}$ system of HCO at 5624 Å in absorption. Only the $K'=0 \rightarrow K''=1$ subband is observed. Each line is an unresolved doublet numbered according to the value of N (rotational quantum number apart from spin) in the lower state.

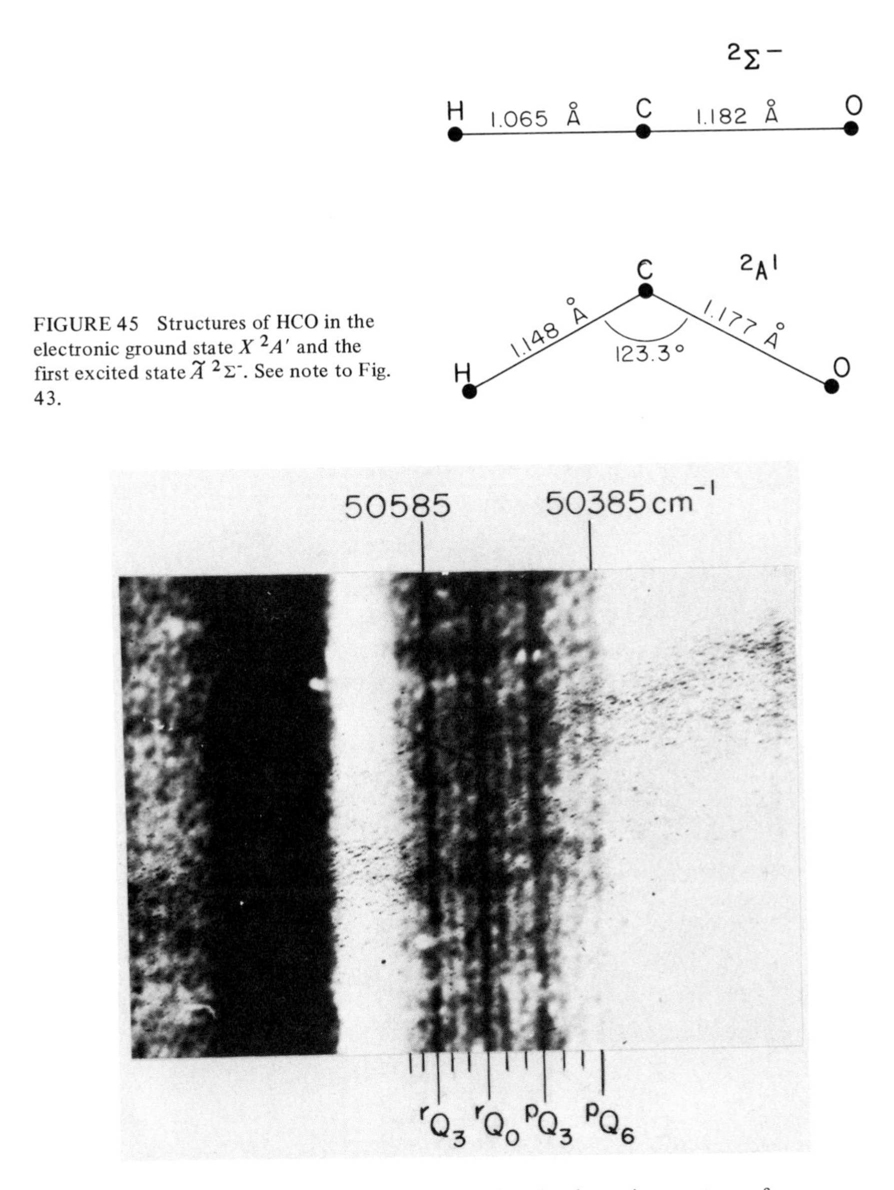

FIGURE 45 Structures of HCO in the electronic ground state $X\ ^2A'$ and the first excited state $\tilde{A}\ ^2\Sigma^-$. See note to Fig. 43.

FIGURE 46 Fine structure of a ⊥ band in the absorption spectrum of CH_3I near 1980 Å. The "lines" are the unresolved Q branches of the subbands.

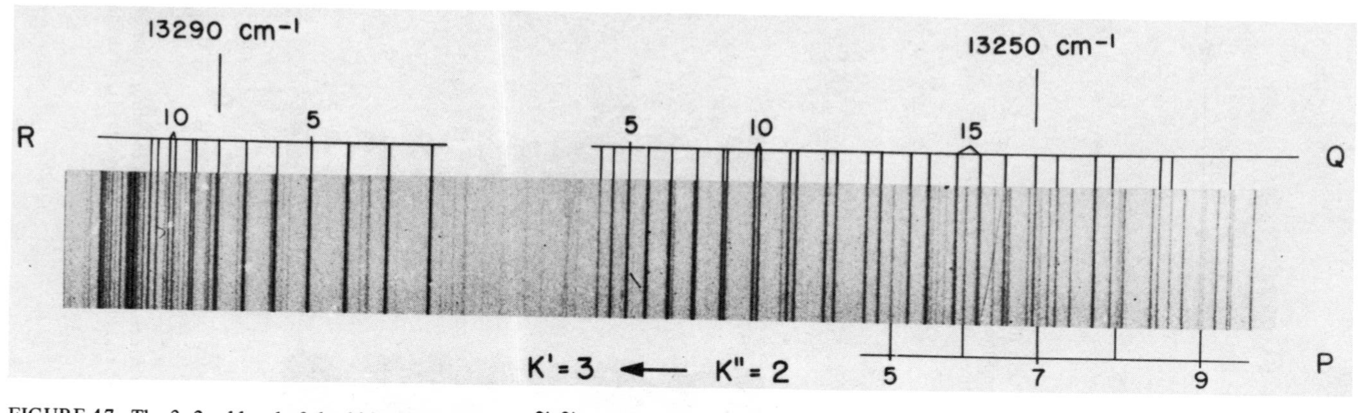

FIGURE 47 The 3–2 subband of the 000–000 band of the $\tilde{A}$–$\tilde{X}$ system of HNO near 7534 Å in absorption after Dalby.[43] Note the *K*-type doubling at higher *J* values.

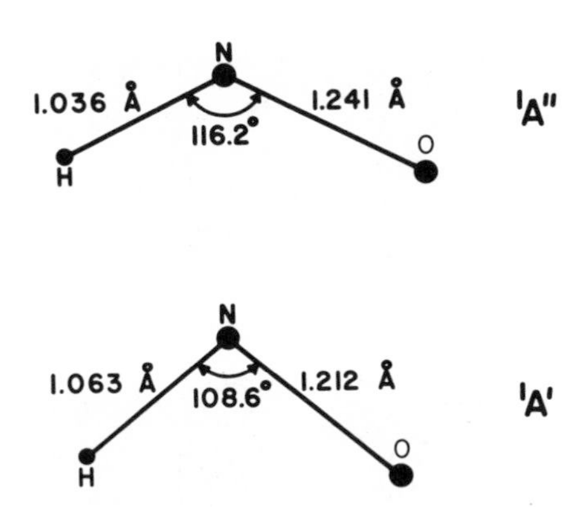

FIGURE 48 Structures of HNO in the electronic ground state $\tilde{X}\ ^1A'$ and the first excited state $\tilde{A}\ ^1A''$. See note to Fig. 43.

spectrum of DNO (since there are three geometrical parameters, a determination from HNO alone is not possible).

Another important example is the absorption spectrum of the CH_2 radical. In the vacuum ultraviolet it shows, at 1415 Å, a strong absorption band that, while rather diffuse for CH_2, is quite sharp for CHD and CD_2. The appearance of this band, shown in Fig. 49, is certainly that of a band of a linear molecule. However, here also, various pieces of evidence have recently led to the conclusion[45] that this is only one subband ($K = 0$) of a $\parallel$ band of a nearly symmetric top, that is, of bent CH_2. For reasons too involved to be discussed here, it was established that the CH_2 band of Fig. 49 is a triplet transition. In the red region of the spectrum a second absorption of CH_2 has been found which is a singlet transition between states in which the molecule is an asymmetric top. The analysis of both these spectra has led to the structural data represented in Fig. 50. The ground state of the molecule is $^3B_1$. The lowest singlet state $^1A_1$ lies above this state by about 0.8 eV. However, an experimental verification of this number has not yet been achieved.

For a number of diatomic and polyatomic molecules, series of electronic states very similar to the Rydberg series of atoms have been observed. They can be represented by the simple formula

$$\nu = \nu_{\text{limit}} - [R/(n-\delta)^2],$$

where n is the principal quantum number and δ the quantum defect. The limit of such a series, ν_{limit}, corresponds to an ionization potential of the molecule. If members of a Rydberg series up to high values of n have been observed, very precise values of the ionization potential can be obtained. In Fig. 51, as an example two Rydberg series of H_2 are shown near the series limit. This spectrum was taken at liquid nitrogen

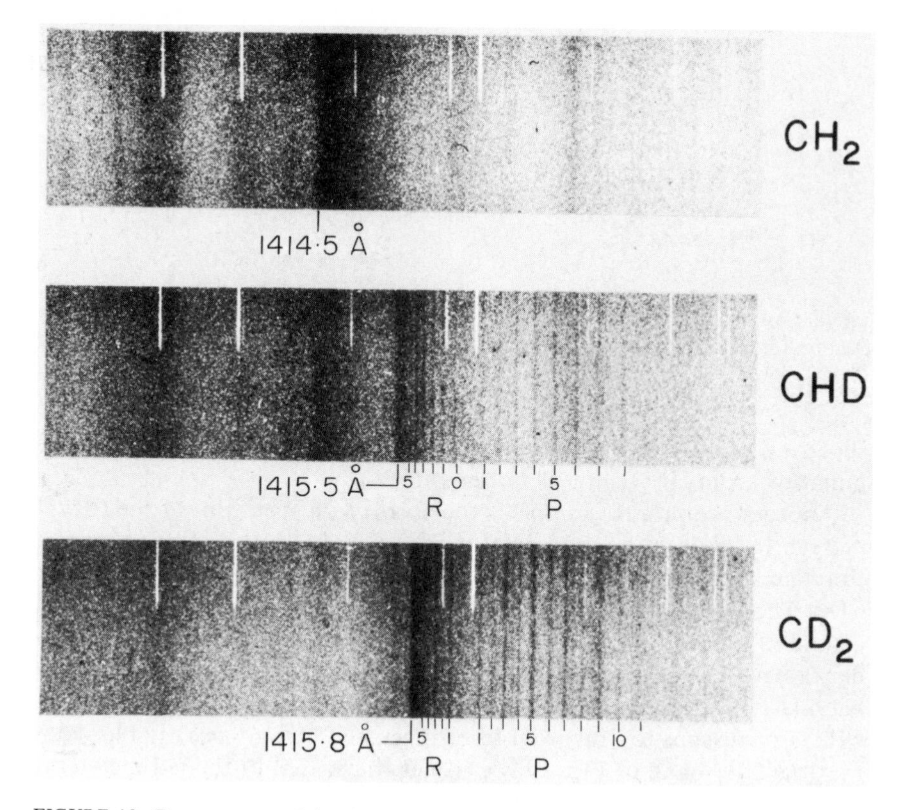

FIGURE 49 Fine structure of the ultraviolet absorption bands of CH_2, CHD, and CD_2 near 1415 Å after Herzberg.[44] These spectra have been obtained by the flash photolysis of diazomethane.

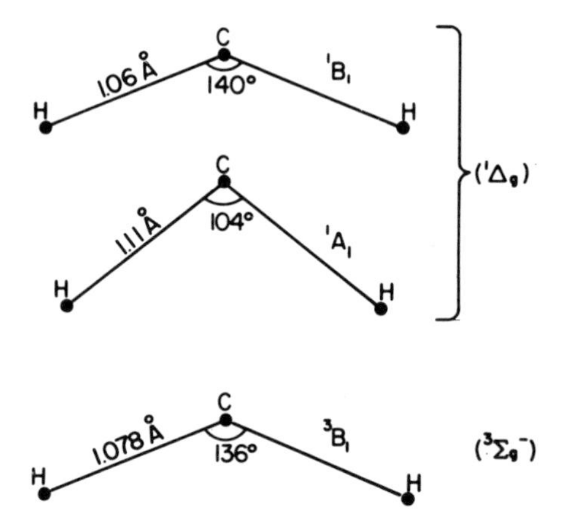

FIGURE 50 Structure of CH_2 in the electronic ground state $X\ ^3B_1$ and the first two excited states $\tilde{a}\ ^1A_1$ and $\tilde{b}\ ^1B_1$. See note to Fig. 43. The designations for the corresponding linear conformations are added at the right.

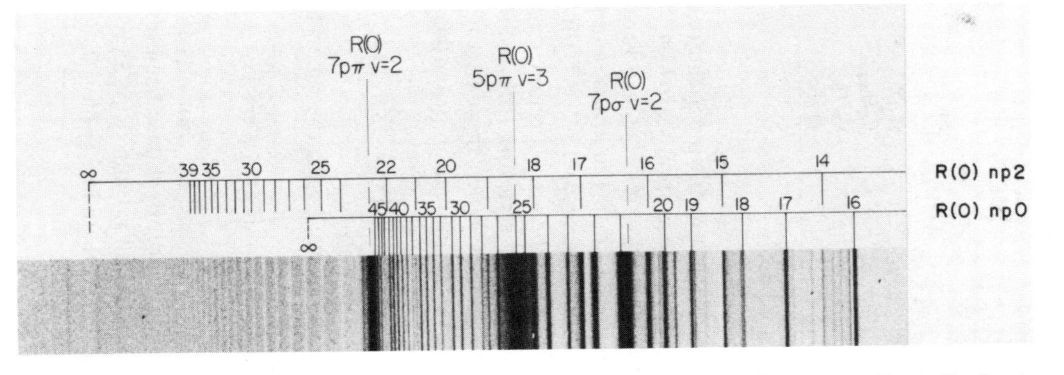

FIGURE 51 Rydberg series of *para*-H_2 for $\nu' = 1$ photographed at 80 K. The two series corresponding to $N = 0$ and $N = 2$ of the H_2^+ ion are marked. The latter appears, above the limit of the former, as an apparent emission series on account of preionization. The three strong lines marked belong to Rydberg series with different ν'.

TABLE 10 Ionization Potentials from Rydberg Series

Molecule	Ionization Potential (eV)	Molecule	Ionization Potential (eV)
H_2	15.42541	CH_2	10.396
HD	15.4441_7	H_2O	12.618
D	15.4661_3	N_2O	12.893
N_2	15.5803	CO_2	13.769
O_2	12.059	CH_3	9.843
CH	10.64	C_2H_2	11.41
OH	13.18	H_2CO	10.88
HF	16.06	C_2H_4	10.507
CO	14.013	C_6H_6	9.247
NO	9.2639		

temperature using *para*-H_2 and as a result practically all molecules are in the lowest rotational level of the ground state ($J = 0$), thus greatly simplifying the spectrum. The two series observed for each v' have two close-lying limits, which correspond to the lowest and the second lowest rotational level of the H_2^+ ion. The second series is diffuse above the first limit on account of an Auger process (preionization). Below this limit the second series and the first series perturb each other, leading to a number of apparent irregularities. When these perturbations have been taken into account, a value of the ionization potential,

$$\mathrm{I.P.}(H_2) = 124417.2\ \mathrm{cm}^{-1} = 15.4254_1\ \mathrm{eV},$$

is obtained that agrees extremely well with the theoretical value

$$\mathrm{I.P.}(H_2)_{\mathrm{theor.}} = 124417.3\ \mathrm{cm}^{-1}.$$

Similar agreement is found for HD and D_2. For no other molecule have such precise experimental and theoretical data been obtained. The spectroscopically determined ionization potentials of a number of diatomic and polyatomic molecules are collected in Table 10.

V. CONCLUSION

In the preceding discussion it has not been possible to describe any of the methods used for obtaining spectra. Nor was it possible to compare the results of other methods of molecular structure determination like electron diffraction with the spectroscopic results. Many parts of molecular spectroscopy like nuclear magnetic resonance and electron spin resonance have been omitted. Active work is going on in all these fields.

Much new information about the electronic structure of molecules is being obtained by the new method of photoelectron spectroscopy. This information refers to molecular ions and tells us what energies are required to remove various electrons from the neutral molecule. At the National Research Council of Canada we are engaged in studying directly the spectra, particularly absorption spectra, of molecular ions in order to obtain more detailed information about the various electronic states of molecular ions. This information may prove to be of astrophysical interest just as the information about spectra of neutral molecules and radicals did.

I hope I have succeeded in showing that molecular spectroscopy, which started, like the Union, 50 years ago, is still an active field of work in which many further interesting new results are to be expected in the future.

REFERENCES

1. K. Schwarzschild, *Sitzber. Königl. Preuss. Akad. Wiss. Berlin*, 548 (1916).
2. T. Heurlinger, *Phys. Z. 20*, 18 (1919); also *Diss. Lund* (1918).
3. W. Lenz, *Verk. Deut. Phys. Ges. 20*, 188 (1919).
4. A. Kratzer, *Z. Phys. 3*, 289 (1920).
5. G. Herzberg, *Molecular Spectra and Molecular Structure, Vol. II, Infrared and Raman Spectra of Polyatomic Molecules* (Van Nostrand-Reinhold, New York, 1945).
6. G. Herzberg, *Molecular Spectra and Molecular Structure, Vol. I, Spectra of Diatomic Molecules* (Van Nostrand-Reinhold, New York, 1950).
7. G. Herzberg, *Molecular Spectra and Molecular Structure, Vol. III, Electronic Spectra and Electronic Structure of Polyatomic Molecules* (Van Nostrand-Reinhold, New York, 1966).
8. G. Herzberg, *The Spectra and Structures of Simple Free Radicals: An Introduction to Molecular Spectroscopy* (Cornell U.P., Ithaca, N.Y., 1971).
9. M. Czerny, *Z. Phys. 34*, 227 (1925).
10. E. V. Loewenstein, *J. Opt. Soc. Am. 50*, 1163 (1960).
11. O. R. Gilliam, C. M. Johnson, and W. Gordy, *Phys. Rev. 78*, 140 (1950).
12. R. W. Wilson, K. B. Jefferts, and A. A. Penzias, *Astrophys. J. 161*, L43 (1970).
13. B. Starck, *Landolt-Börnstein New Series II*, 4 (1967).
14. W. S. Benedict, H. H. Claessen, and J. H. Shaw, *J. Res. Nat. Bur. Stand. U.S. 49*, 91 (1952).
15. W. Kiefer and H. J. Bernstein, unpublished (1972).
16. F. Rasetti, *Z. Phys. 61*, 598 (1930).
17. B. Stoicheff, *Adv. Spectrosc. 1*, 91 (1959).
18. B. Stoicheff, *Can. J. Phys. 32*, 339 (1954).
19. W. Heisenberg, *Z. Phys. 41*, 239 (1927).
20. F. Hund, *Z. Phys. 42*, 93 (1927).

21. W. Heitler and G. Herzberg, *Naturwissenschaften 17*, 673 (1929).
22. A. C. Cheung, D. M. Rank, C. H. Townes, D. D. Thornton, and W. J. Welch, *Phys. Rev. Lett. 21*, 1701 (1968).
23. B. Zuckerman, D. Buhl, P. Palmer, and L. E. Snyder, *Astrophys. J. 160*, 485 (1970).
24. R. S. Mulliken, *Phys. Rev. 25*, 119, 259 (1925).
25. G. Di Lonardo and A. E. Douglas, to be published.
26. G. Herzberg, in *Fundamental and Applied Laser Physics: Proceedings of the Esfahan Symposium,* M. Feld, A. Javan, and N. Kurnit, eds. (John Wiley & Sons, Inc., New York, 1973); also *Commentarii, Pontif. Acad. Sci.,* to be published.
27. E. S. Imes, *Astrophys. J. 50,* 251 (1919).
28. G. Herzberg, *Can. J. Res. A 28*, 144 (1950).
29. A. D. Meinel, *Astrophys. J. 111*, 555; *112,* 120 (1950).
30. A. G. Maki and L. R. Blaine, *J. Mol. Spectrosc. 12*, 45 (1964).
31. D. R. J. Boyd and H. W. Thompson, *Trans. Faraday Soc. 49*, 141 (1953).
32. R. W. Peterson and T. H. Edwards, *J. Mol. Spectrosc. 38*, 1 (1971).
33. K. T. Hecht, *J. Mol Spectrosc. 5*, 390 (1960).
34. J. Herranz and B. P. Stoicheff, *J. Mol. Spectrosc. 10*, 448 (1963).
35. G. Herzberg, *Highlights of Astronomy* (D. Reidel Publ. Co., Dordrecht, Netherlands, 1971), p. 145.
36. A. Rosenberg, I. Ozier, and A. K. Kudian, to be published.
37. G. Herzberg, *J. Mol. Spectrosc. 33*, 147 (1970).
38. A. E. Douglas and P. M. Routley, *Astrophys. J. 119*, 303 (1954).
39. G. Herzberg and A. Lagerqvist, *Can. J. Phys. 46*, 2363 (1968).
40. P. Swings, *Lick Observatory Bull. 19*, 131 (1941).
41. A. McKellar, *Rev. Mod. Phys. 14*, 179 (1942); *Astrophys. J. 99*, 162; *100*, 69 (1944).
42. L. Gausset, G. Herzberg, A. Lagerqvist, and B. Rosen, *Astrophys. J. 142*, 45 (1965).
43. F. W. Dalby, *Can. J. Phys. 36*, 1336 (1958).
44. G. Herzberg, *Proc. Roy. Soc. 262A*, 291 (1961).
45. G. Herzberg and J. W. C. Johns, *J. Chem. Phys. 54*, 2276 (1971).

JOHN BARDEEN *is a professor of physics and electrical engineering at the University of Illinois. He has served as president of the American Physical Society and president of the IUPAP Commission on Very Low Temperatures. He was awarded the Nobel Prize in Physics in 1956 for his work on the discovery of the transistor.*

6 *Solid-State Physics: Accomplishments and Future Prospects* / JOHN BARDEEN

I. INTRODUCTION

Solid-state physics is concerned with understanding the properties of condensed matter in terms of its electronic and atomic structure. To include liquids as well as solids, the field is now often called the physics of condensed matter. Aside from that, some may feel that the term solid state has been taken over by the consumer electronics industry. However, being old fashioned, I will continue to use the term solid-state physics but interpret it broadly to include quantum liquids. The field has grown very rapidly since the end of World War II and is now the largest branch of physics. During much of this period, the number of workers doubled about every five years. Now there are more than 4000 PhD solid-state physicists in the United States and many more in other countries throughout the world.

The importance of solid-state physics is recognized by four Commissions of IUPAP engaged in various aspects. In addition to the Solid State Commission, there are Commissions on Semiconductors, Magnetism, and Low Temperatures.

Applications cover a broad range: semiconductors in electronics, magnetic materials for transformers, tapes and computer memories, luminescent materials for light sources and displays, photoconductors in xerography, superconductors in electromagnets, and many others. The spectacular growth of electronic communications and of the computer industry has been directly dependent on products from solid-state physics. In electric power, semiconductor rectifiers are in wide use, and there is promise of large-scale use of superconductors in electric generators and motors, underground transmission lines, and other applications. Products that have come from advances in solid-state physics run to tens of billions of dollars per year.

Equally spectacular have been the advances in scientific understanding, the discovery of new physical phenomena in solids, and the discovery of new materials with unusual properties. In this paper I will give only a few highlights and will stress two areas in which I have been concerned personally—semiconductors and superconductors. There are other fields of equal or greater importance that unfortunately will have to be omitted. I regret that we will not be able to say much about such important topics as electron and nuclear spin resonance, the Mössbauer effect, lasers and nonlinear optics, studies of the Fermi surface and other electronic properties of metals, quantum fluids and quantum crystals, phase transitions, and many others in the vast field of condensed-matter physics. Each of these fields has had great advances, and each could warrant a separate talk. Solid-state physics is a diverse field with a great variety of problems rather than a few major ones.

I will first describe some of the historical background to our understanding of solids and then say something about the methods, both experimental and theoretical, that have characterized research of the past 25 years. A few examples will then be given of research and applications in the fields of semiconductors, magnetism, and superconductivity. Finally, I will discuss future prospects.

II. EARLY DEVELOPMENTS

In any field there are golden ages during which advances are made at a rapid pace. In solid-state physics, three stand out. One, the early years of the present century, followed the discoveries of x rays, the electron, Planck's quantum of energy, and the nuclear atom—the discoveries that ushered in the atomic era. The Drude-Lorentz electron theory of metals and Einstein's applications of the quantum principle to lattice vibrations in solids and to the photoelectric effect date from this period. Von Laue's suggestion in 1912 that a crystal lattice should act as a diffraction grating for x rays and research of the W. H. and W. L. Bragg opened up the vast field of x-ray structure determination.

The foundations of the field were firmly established during a second very active period, from about 1928 until the mid-thirties, which followed the discovery of quantum mechanics. Many of the world's leading theorists were involved in this effort. The Bloch theory, based on the one-electron model, introduced the concept of energy bands and showed why solids, depending on the electronic structure, may be metals, insulators, or semiconductors. The fundamentals of the theory of transport of electricity and of heat in solids were established. In these same years, the importance for many crystal properties of the role of imperfections in the crystal lattice, such as vacant lattice sites, dislocations, and impurity atoms was beginning to be recognized. Some of the names prominent in the developments of solid-state theory during this period are Bloch, Brillouin, Frenkel, Landau, Mott, Peierls, Schottky, Seitz, Slater, A. H. Wilson, Wigner, and Van Vleck. The third golden age has been the rapid expansion in the post-World War II years, with not only great advances in understanding but also in technology and new products.

The accomplishments of the 1930's were summarized in a book by Frederick Seitz published in 1940. This book has been the bible to many generations of budding solid-state physicists and has had great influence. Although advances in understanding during the 1930's were considerable, there was generally only limited qualitative agreement between theory and experiment. This was partially due to the limitations of the theory but even more to the paucity of experiments on well-defined systems. Many of the most important properties of crystals are what is called structure-sensitive, that is, they depend very sensitively on the presence of very small concentrations of impurities or imperfections in the crystal lattice. Structure-sensitive properties include semiconductivity, mechanical strength of materials, photoconductivity, luminescence, and many others. With a few exceptions, most measurements were made on polycrystalline material of poorly defined composition. A notable exception during the 1930's was a beautiful series of experiments by Hilsch and Pohl in Germany on the properties of alkali halide crystals.

III. POST-WORLD WAR II PERIOD

It may be worthwhile to spend a few minutes describing the nature of experimental and theoretical work that has characterized the post-World War II period. There are now remarkably detailed quantitative checks between theory and experiment for many crystal properties. As a result of a great deal of careful work, the elusive structure-sensitive properties have been brought under control not only in the laboratory but on an industrial scale.

Very essential has been the development of methods for growing single crystals of various materials of exceptional purity or of well-defined composition, free of defects. The purity required usually goes well beyond the limits of ordinary chemical methods. Physicists often grow their own crystals in order to have control of the properties. Impurities present in concentrations much less than a part in a million can have a marked effect. In a heroic effort, R. N. Hall of General Electric has purified germanium for use in radiation detectors to such an extent that the concentration of impurity ions is less than one in 10^{12}, with great improvement in the energy resolution over lithium-drift detectors. Extensive efforts are made to determine the characteristics of impurities and other defects that affect the properties. Many years of effort are often required to understand and control the impurities that affect structure-sensitive properties in a given class of materials.

Facilities and methods used for experimental work have been vastly improved over the years. Many new techniques have been introduced or greatly developed, such as electron and nuclear spin resonance, the Mössbauer effect, the de Haas-van Alphen effect, and cyclotron resonance. Measurements can be made over the complete spectrum from static or low frequencies through microwaves, infrared, optical, and into the far ultraviolet. Most laboratories have liquid helium available to reach very low temperatures. With dilution refrigerators, temperatures within a few millidegrees of the absolute zero can be obtained. Superconducting magnets with fields up to and above 100 kg are readily available. With pressures as high as 500,000 atm available in some laboratories, one can create new phases with unusual properties. For example, at high pressures many insulators and semiconductors change into metals. Some crystal forms, such as diamond, require high pressures for their fabrication.

Many of the advances in experimental techniques are based on solid-state research. For example, by far the most sensitive detectors for voltages, currents, and magnetic fields make use of the Josephson effect in superconductors. Solid-state devices are used as sources and as detectors throughout the frequency spectrum. Advances in the field have made possible advances in many other areas of science. Much solid-state research is done in industry and is stimulated by requirements of technology.

IV. ELEMENTARY EXCITATIONS

From a theoretical point of view, one of the most powerful and fruitful methods of approach has come from the concept of elementary excitations. One tries to understand the ground state of the system and the elementary thermal excitations, such as phonons and quasi-particles.

There is a close analogy with particle physics. The ground state of the ideal crystal, the state present at the absolute zero of temperature, corresponds to the vacuum. The elementary excitations correspond to elementary particles. Elementary excitations are thermally excited as the temperature is raised. At very low temperatures, the gas of elementary excitations is dilute and interactions between them are infrequent. With increasing temperatures, the interactions become more and more important and must be taken into account.

Elementary excitations obey either Bose or Fermi statistics. Some of the more common ones are given in the list below, which is by no means complete.

Bose Excitations:

Phonons, longitudinal, transverse; $E(q) = \hbar\omega_q$
Spin waves in magnetic materials
Phonons and rotons in superfluid helium, He II
Excitons in semiconductors

Fermi-Dirac Excitations:

Conduction electrons and holes in semiconductors
Quasi-particles in normal metals, superconductors, liquid ^{3}He and ^{3}He–^{4}He mixtures

Because of translation invariances, the momentum or wave vector is usually a good quantum number. The elementary excitations are approximate eigenstates of the entire interacting many-body system and have long lifetimes at low temperatures. The Fermi excitations are generally quasi-particle excitations of Fermi systems.

It is fortunate that there are two isotopes of helium, ^{3}He and ^{4}He, because they give examples of both Fermi and Bose systems. Both remain liquid down to the absolute zero. Another example of a Fermi liquid is a solution of ^{3}He and ^{4}He. Helium-3 is soluble in ^{4}He up to a concentration of about 6% at $T = 0$ K, and the ^{3}He quasi-particles in ^{4}He form an almost ideal Fermi-Dirac gas. As you know, ^{4}He becomes superfluid below about 2.2 K.

In addition to the elementary excitations, there are collective excitations, which are coherent motions of large numbers of elementary excitations. Some examples are acoustic waves, acoustoelectric waves, and magnetohydrodynamical waves of various types.

The study of elementary excitations and interactions between them has been actively pursued both experimentally and theoretically. Green's function methods have been adopted from quantum field theory. Feynman diagrams, Dyson equations, etc., are now part of the language of solid-state physics. In the theory, temperature plays the role of an imaginary time.

One of the most powerful experimental methods for studying the excitation spectrum of Bose particles is from neutron scattering experiments. A neutron excites an elementary excitation in the solid or liquid. From the momentum and energy loss of the neutron one can obtain the corresponding values for the elementary excitation. Phonon spectra of many crystals have been determined in this way. Many spin-wave spectra have also been observed. I will give just one example, the famous Landau quasi-particle excitation spectrum of superfluid helium, He II. Landau, who introduced the term elementary excitations, suggested on phenomenological grounds a spectrum very similar to the one shown in Fig. 1. It was Feynman who first suggested that one should be able to observe the spectrum by neutron scattering. When measurements were made, the spectrum was found to be almost identical with the one postulated by Landau. For small wave vector or momentum, the excitations are density waves, longitudinal phonons. At high momentum, there is the roton minimum. From the elementary excitation spectrum determined by neutron scattering, one can calculate quantitatively many properties of the liquid, such as the temperature variation of the specific heat and of the superfluid density, ρ_s. In fact,

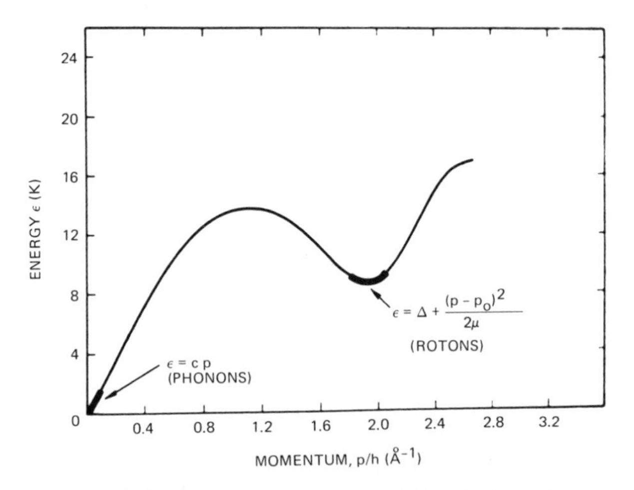

FIGURE 1 Spectrum of elementary excitations in liquid He II as postulated by Landau and measured by neutron scattering. [From A. B. Woods, in *Quantum Fluids*, D. F. Brewer, ed. (North-Holland Publishing Co., Amsterdam, 1966), p. 242.]

Landau originally suggested the spectrum to account for Andronikashvili's measurements of ρ_S as a function of temperature.

Another powerful method for investigating elementary excitations that is getting increasing use is the study of Brillouin and Raman scattering with laser beams. From such experiments, there is evidence that two rotons form a bound state.

The quasi-particle excitation spectrum of Fermi systems is what now replaces the Bloch one-electron model. There is a one-to-one correspondence between states of the interacting system and those of the non-interacting system. In the ground configuration, all states are occupied up to the Fermi level and higher states are unoccupied. Low-lying excitations correspond to excited particles above the Fermi surface and holes below. If the excitation energy is small compared with the Fermi energy, as is true of thermally excited electrons in metals, the quasi-particle can have a very long life. The mean free path of electrons in pure metals can be 1 cm or more at low temperatures. Even in ^3He, which one thinks of as a reasonably dense liquid, the quasi-particle mean free path can be tens of thousands of angstroms at very low temperatures.

Figures 2 and 3 illustrate the quasi-particle excitation spectrum of semiconductors according to the theory of A. H. Wilson. In the ground state of an ideal crystal at the absolute zero, there are just enough electrons to fill completely the highest occupied band of states. These are the electrons that one usually thinks of as forming the valence bonds. There is a relatively small energy gap (~1 eV) to the next higher unoccupied band—the conduction band. At high temperatures (Fig. 2), electrons can be thermally excited from the valence band to the conduction band, giving intrinsic conductivity. Both the electrons in the conduction band and the places from which electrons are missing in the valence band (the holes) are mobile and can contribute to the electrical conductivity. It is a consequence of quantum theory that holes near the top of the valence band behave in all respects like particles of positive charge and positive mass. In an intrinsic semiconductor the negative space charge of the conduction electron is compensated by the positive holes.

Electrons can also be excited from the valence to conduction bands by absorption of light quanta, giving photoconducticity. In the transistor, extra carriers, electrons, and holes are introduced into the semiconductor by current flow from an appropriate contact.

In extrinsic semiconductors (Fig. 3), space-charge balance is maintained by fixed positive or negative ions. In N-type material, positively charged impurity ions are called donors and their charge is balanced by negatively charged conduction electrons. In P-type, negatively charged

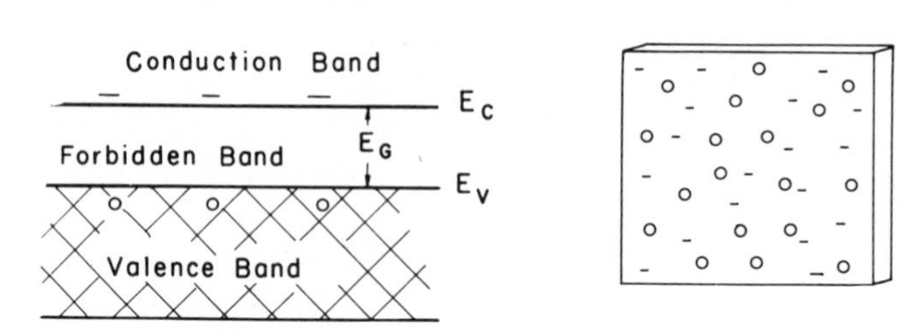

FIGURE 2 Quasi-particle energies of electrons in an intrinsic semiconductor according to the energy band picture of A. H. Wilson. The missing electrons or holes near the top of the valence band behave in all respects like particles of positive charge and positive mass. The negative charge of electrons in the conduction band is balanced by the positive charge of the holes.

acceptor ions are compensated by positively charged holes. Concentrations of parts per million or less of donor or acceptor ions can give rise to substantial conductivity by the compensating mobile charge carriers.

In 1957, Cooper, Schrieffer, and I gave a microscopic theory of superconductivity based on pairing of electrons of opposite spin and momentum in the ground state. The new ground state takes advantage of an effective attraction between electrons resulting from interaction with the field of phonons. We derived a quasi-particle excitation spectrum with a gap, as illustrated in Fig 4. The gap is very small compared with the Fermi energy; it is of the order of millivolts as compared with

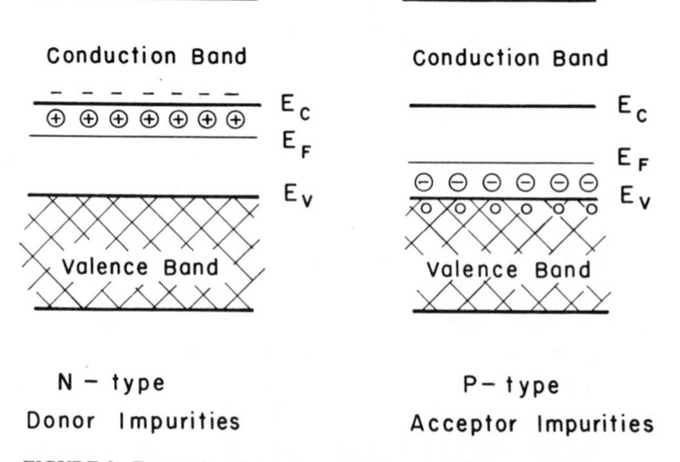

FIGURE 3 Energy band structure of N-type and P-type semiconductors. The charge density of the mobile carriers, electrons, or holes, is compensated by fixed impurity ions of opposite sign.

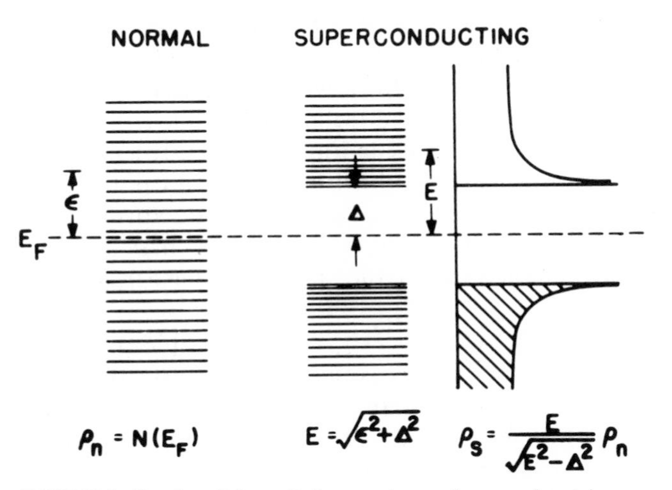

FIGURE 4 Quasi-particle excitation spectrum of a normal metal (left) and a superconductor with a gap (middle). The density of states in energy in the superconductor is shown on the right. The gap parameter, Δ, of the order of 10^{-3} eV is very small compared with the Fermi energy of several electron volts.

a Fermi energy of several volts. The difference in excitation spectrum accounts in part for the remarkable difference in properties between superconductors and normal metals. Calculations of the properties of superconductors are not much more difficult than for normal metals.

In some of the most exciting results reported at the Boulder Low Temperature Conference in August 1972, evidence was presented by Lee, Richardson, and others of Cornell University that liquid ^{3}He undergoes a phase transition when cooled below 2.7 millidegrees at high pressures. The suggestion was made that perhaps it is a pairing transition analogous to the superconductivity transition of electrons in metals. It would be very exciting indeed if a new superfluid system were found in ^{3}He. The measurements indicating the transition were made in a study of Pomeranchuk cooling. Several years ago, Pomeranchuk suggested that as ^{3}He is solidified under pressure (about 25 atm), cooling should result because of the higher spin entropy in the solid as compared with the liquid.

V. JOSEPHSON EFFECT

The past 10 or 15 years has seen a tremendous growth in interest in superconductivity. It was by far the most popular subject at the 1972

Boulder conference. This is in part because of the microscopic theory that can be used to interpret experiments and predict new effects and in part because of the discovery of new superconductors that withstand very high magnetic fields. The most remarkable new prediction from theory is the Josephson effect, which has had very important implications for both science and technology. While still a graduate student at Cambridge University in England, B. D. Josephson suggested on theoretical grounds in 1962 that a supercurrent can tunnel through a thin insulating barrier separating two superconductors and that the current should be proportional to the sine of the phase difference between the superconductors on either side of the barrier, as illustrated in Fig. 5.

Phase is normally a wave property, and the Josephson effect is an indication that superconductors exhibit quantum effects on a macroscopic scale. For mathematical convenience, Cooper, Schrieffer, and I used a wavefunction for the superconducting ground state that is a linear combination of states with differing numbers of particles

$$\psi_s = \Sigma_N a_N e^{i\chi N} \psi_N.$$

Here N is the number of ground state pairs, ψ_N is a ground-state wavefunction with exactly N pairs, and χ is the phase. P. W. Anderson later showed that in a quantum-mechanical sense χ and N are conjugate variables obeying an uncertainty relation:

$$N = i\partial/\partial\chi; \quad \Delta\chi\Delta N \sim 1.$$

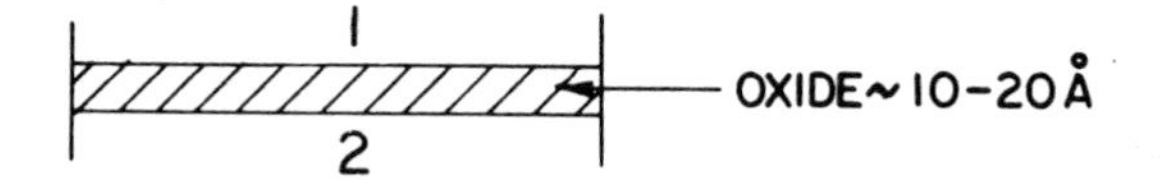

Josephson tunnel junction

$$J = J_1 \sin(\chi_1 - \chi_2) = \frac{2e}{\hbar} \frac{\partial W_{12}}{\partial(\chi_1 - \chi_2)}$$

$$W_{12} = -\frac{\hbar J_1}{2e} \cos(\chi_1 - \chi_2)$$

FIGURE 5 A Josephson tunnel junction formed by two superconductors separated by a very thin insulating barrier. A supercurrent proportional to the rise of the phase difference is related to the coupling energy W_{12} that results from pairs going back and forth across the barrier.

Josephson attended lectures given by Anderson at Cambridge and suggested the tunneling experiment (Fig. 5) to see whether the phase is actually a relevant variable for superconductors. He predicted that with no applied voltage a supercurrent should flow through the tunneling barrier and that the current should be proportional to the sine of the phase difference between the metals on either side of the barrier.

Josephson also predicted that if a voltage V is applied across the barrier, an alternating current should flow with a frequency given by

$$\nu = 2eV/h,$$

together with its harmonics. If V is the order of millivolts, the frequency is in the microwave range. The alternating current can be detected by applying a microwave signal across the barrier from an external source.

These predictions of Josephson were later verified experimentally. The Josephson effect has been used to make extremely sensitive detectors. It has also been used to make a precision measurement of the fundamental constants, $2e/h$. Langenberg and co-workers at Pennsylvania found that

$$2e/h = (483.593718 \pm 0.000060)\ \mathrm{MHz}/\mu\mathrm{V}_{\mathrm{NBS69}}.$$

The estimated error of 0.12 parts per million is due mainly to uncertainties in the standard voltage cell. This result has helped clarify values of the fundamental constants. Standards laboratories throughout the world are planning to use the ac Josephson effect to define the voltage standard.

VI. BACKGROUND OF TRANSISTOR DEVELOPMENT

I will discuss more about other applications of superconductivity later on. Before doing so I would like to say something about the background to developments in semiconductor electronics and other applications of semiconductors.

After World War II, the field of solid-state physics was ripe for exploitation. Firm foundations had been established in the 1930's but there was a large gap between theory and experiment particulary for the structure-sensitive properties. It was hoped that the gap could be overcome by a strong research effort. This was recognized among other places by the Bell Telephone Laboratories, where a research group in solid-state physics was initiated in late 1945, just at the end of the war. I joined the group at that time and got my first introduction to semiconductors through associations with Walter Brattain, William Shockley, Gerald Pearson, and others of the group.

None of us had worked on semiconductors during the war, and we had joint seminars and discussions in which we tried to get familiar with the literature. While general basic research activity in solid state and other fields essentially came to an end during the war, intensive research was carried out in a few areas of importance for the military effort. One of these was the development of silicon and germanium as detectors for radar. We decided to concentrate our efforts on these materials, because we thought that being elements they should be simpler to purify and understand theoretically than other semiconductors. The research program was one of basic research, directed toward understanding the properties of semiconductors rather than aimed toward any particular device. However, we were ready to exploit anything that we learned. One of the possibilities was to make an amplifier with a semiconductor. Shockley had suggested controlling the current in a semiconductor by use of a transverse electric field, the principle later used in metal-oxide-semiconductor (MOS) and other field-effect transistors. While trying to observe this effect on a germanium surface, Brattain and I found an entirely new way to modulate the conductivity of a semiconductor—by injection of minority carriers. It is this discovery that led to the invention of the transistor.

The invention was just the first step in opening up the vast new technology of semiconductor electronics. The technology has grown far beyond what any of us would have thought possible in our wildest dreams. Present-day technology is based on the ingenious efforts of many people. It seems that every year or two a new breakthrough comes along to give a new impetus to the field.

Listed below are just a few of the more important steps.

1.	Point-contact transistor	1947–48
2.	Growth of single crystals, zone refining	1949–
3.	Junction transistor	1950–
4.	Hearing aids, radios, computers	1952–
5.	Silicon transistors	1954–
6.	Planar technology	1960–
7.	Integrated circuits	1962–
8.	Large-scale integration	1968–

The twenty-fifth anniversary of the invention of the point-contact transistor is approaching; the invention was made in December 1947, and the first public announcement was in June 1948. When the transistor was invented, we had only polycrystalline material of moderate purity. The first single crystals of germanium were grown by Teal and Little. This made possible the fabrication of grown junction transistors, which had been suggested much earlier by Shockley on theoretical grounds. The first junction transistors were found to behave exactly as had been predicted, and most transistors made since have been of the junction variety. The first applications of transistors outside of the telephone system started appearing about 1952. Although it was recognized early that silicon should be better than germanium, it is more difficult to purify; and the first silicon transistors did not appear until 1954, when they were introduced by Texas Instruments. An exceedingly important step was the invention of the planar transistor by Hoerni, then of the Fairchild Co., about 1960. It was this discovery that opened the way to present-day integrated circuits.

Some of these steps are illustrated in the figures. The first demonstration model of the transistor, shown in Fig. 6, consisted of two pieces of gold foil in close proximity on the surface of a germanium block. This soon led to the point-contact transistor, Fig. 7, consisting of two cats whisker contacts very close together on a small block of germanium. The active region is only of order 10^{-3} cm across and is not much greater than that of present-day transistors. I would like to point out that except for the contacts it is planar in structure.

When single crystals of Ge were grown, it became possible to alter the composition so as to make a P–N junction. As illustrated in Fig. 8,

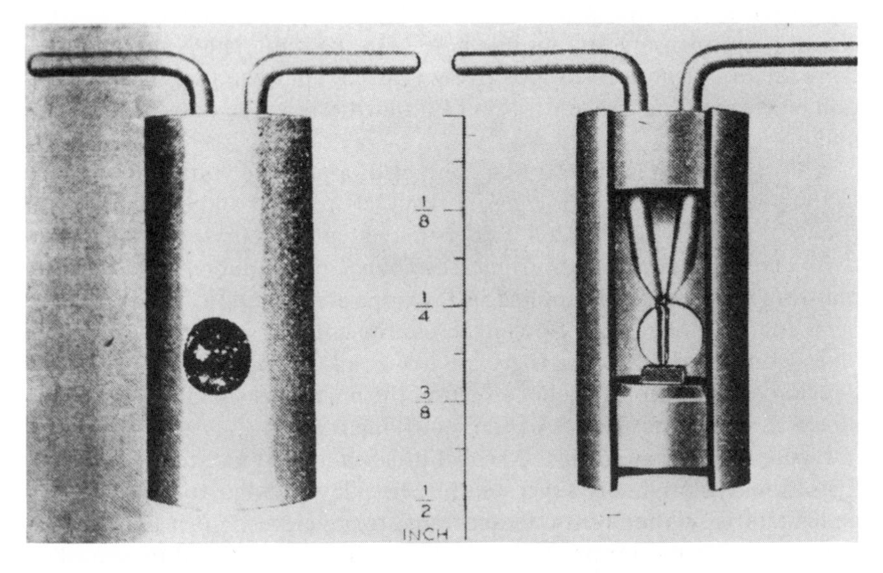

FIGURE 6 Early point-contact transistor. [From Bell Telephone Laboratories.]

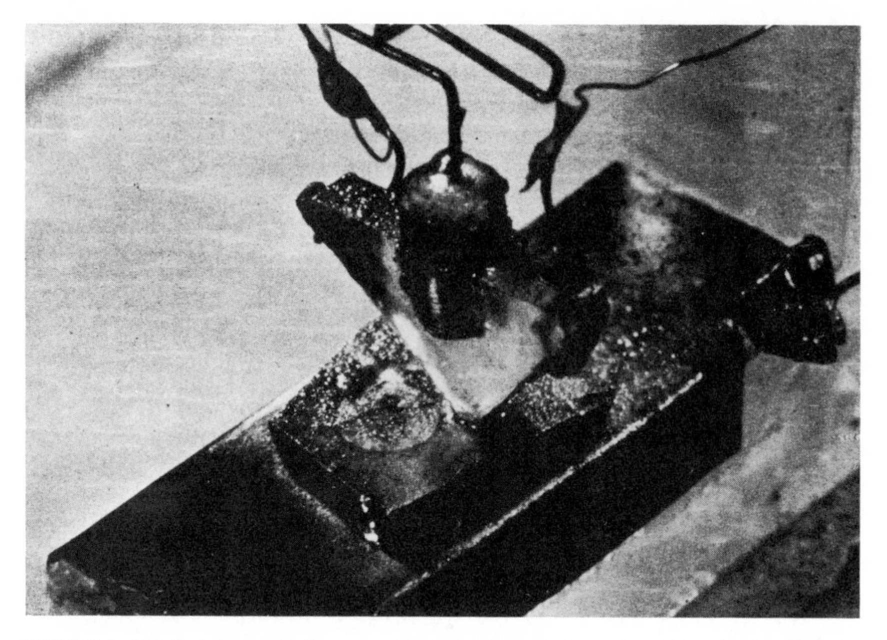

FIGURE 7 The first demonstration model showing the transistor effect. [From Bell Telephone Laboratories.]

there are two regions of the crystal, separated by a plane boundary, which differ only in the nature of the impurities present. In the P-type region on the left, there are acceptors, negatively ionized, whose space charge is compensated by mobile holes. On the right, the N-region, there are donor impurities, positively ionized, compensated by conduction electrons. The concentration of impurities may be only parts per million.

A P–N junction is a rectifying contact. If a positive voltage is applied to the P-side, holes may flow easily from left to right and electrons from right to left. The added carriers may greatly increase the conductivity. This is the transistor effect; the change in conductivity by current flow. If a voltage is applied in the opposite direction, the current must consist of electrons flowing across the contact from the left or holes from the right. Since there are few conduction electrons in the P-region and few holes in the N-region, the current is small; this is the direction of high resistance of the rectifying contact.

Figure 9 shows an idealized model of a junction transistor. This configuration was originally suggested by Shockley in order to better understand how the point-contact transistor operates, but it turned out to have much superior properties to the earlier varieties. All present-day

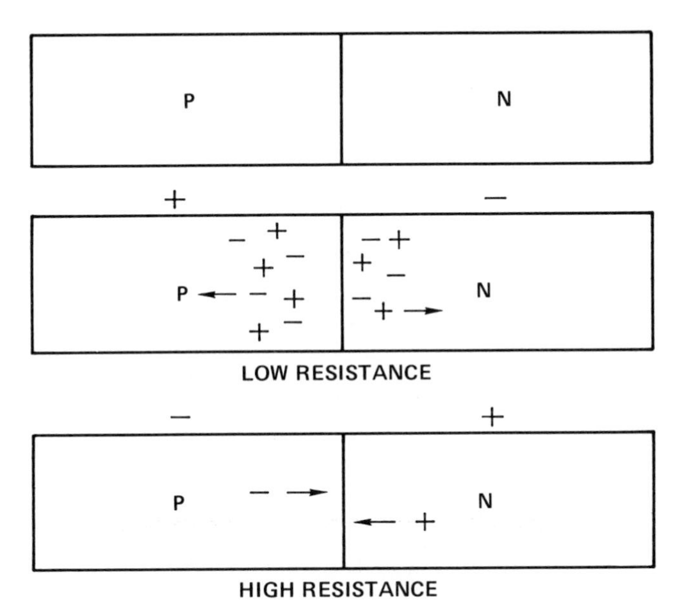

FIGURE 8 Schematic diagram of P–N junction in which the P- and N-regions of a single crystal differ only in the nature of the impurities present, acceptors on the left and donors on the right. The transistor effect is the great increase in conductivity that occurs when a positive voltage is applied to the P-side allowing holes to flow from left to right and conduction electrons from right to left. When voltage is applied in the opposite direction, little current flows, so that the junction is a rectifying contact.

bipolar transistors are of the junction variety. Originally, the impurities were introduced during the growth of the crystal, but later methods have been developed to make P–N junctions by high-temperature diffusion.

Figure 10 shows a simplified version of a modern planar transistor, which is fabricated by a series of diffusion and masking steps. They can

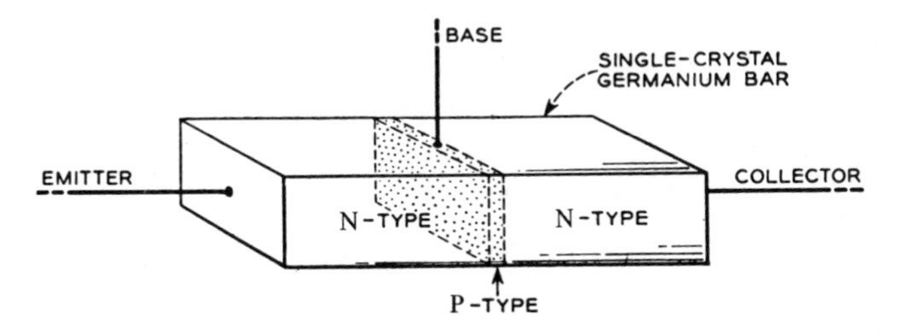

FIGURE 9 A schematic diagram of an early junction transistor in which two N-regions of a single crystal are separated by a narrow region of opposite conductivity type. Electrical connections are made to each of the three regions.

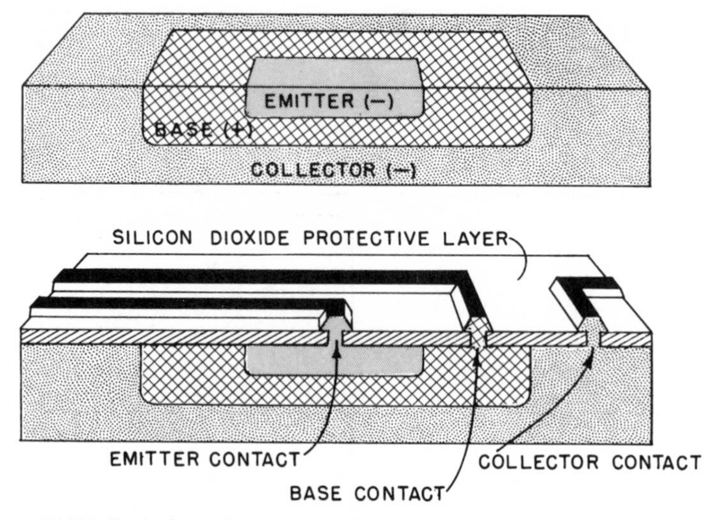

FIGURE 10 Simplified diagram of a planar transistor. The P- and N-regions are formed by selective diffusion of appropriate impurities.

be made by batch processing in which many transistors are made at the same time on a slice of silicon, as illustrated in Fig. 11. Integrated circuits, which include both transistors and associated circuitry, can be fabricated in the same way. A portion of a silicon slice with identical integrated circuits is shown in Fig. 12. The slice is diced and connections

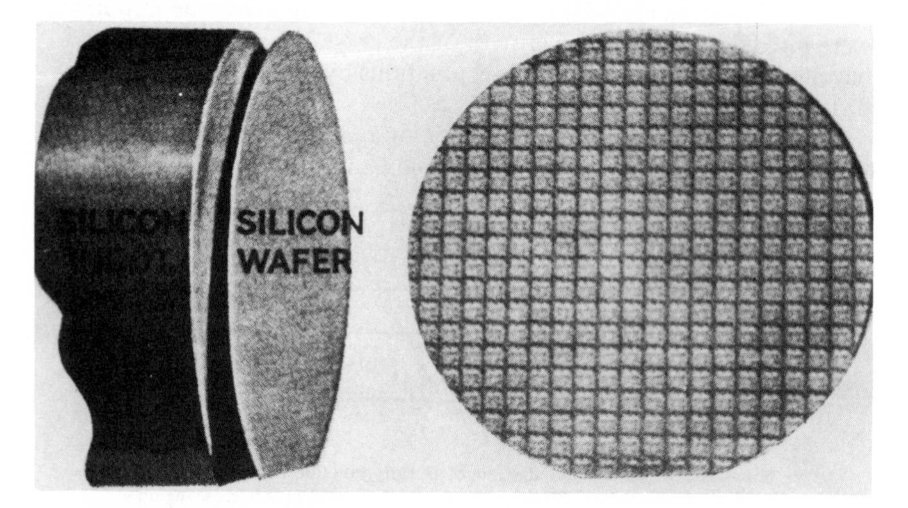

FIGURE 11 Transistors and integrated circuits are formed by batch processing of silicon wafers in which many identical units are formed at the same time.

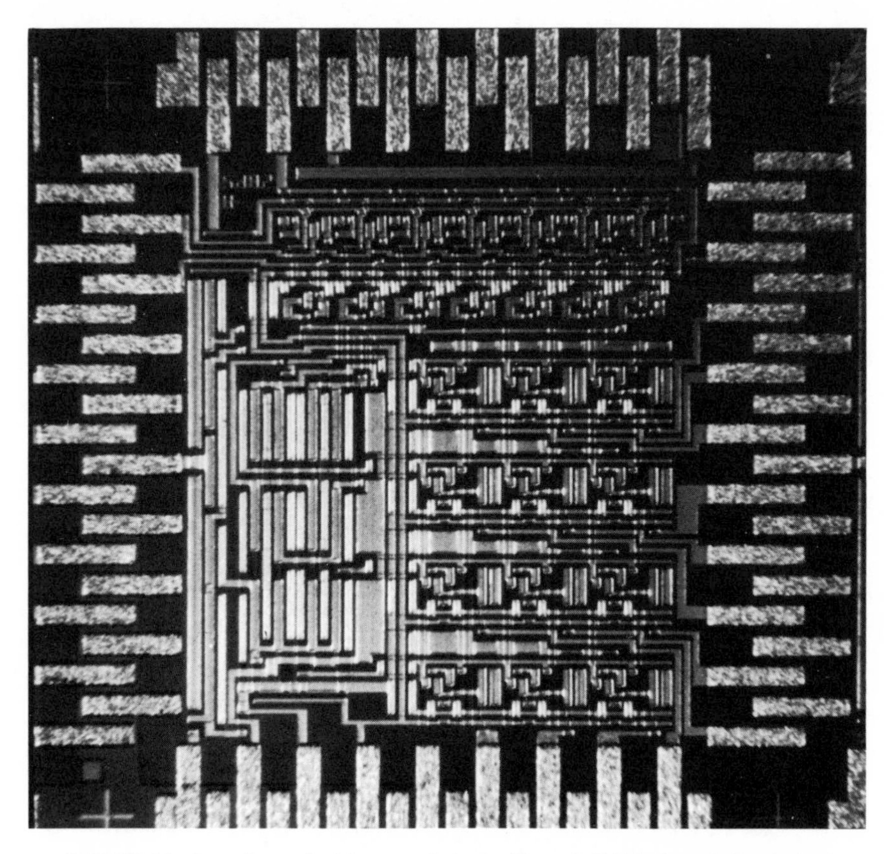

FIGURE 12 A moderate-sized integrated circuit. [From Bell Telephone Laboratories.]

made to form the individual circuits. The building blocks of electronics are now integrated circuits rather than discrete components. Large-scale integrated circuits may contain thousands of transistors. A memory element for Illiac IV, a large computer designed at the University of Illinois, is shown in Fig. 13. There are over 1000 transistors on a single crystal of silicon. Such large-scale integrate circuits can be made and sold for the order of one tenth of a cent per transistor. The technology of large-scale integration is still advancing at a rapid rate. The pocket computer is just one of the many projects dependent on large-scale integration.

Aside from electronic circuits, there are many other applications of semiconductors, including (1) junction-power rectifiers; (2) junction luminescence, including lasers; (3) solar batteries; (4) radiation

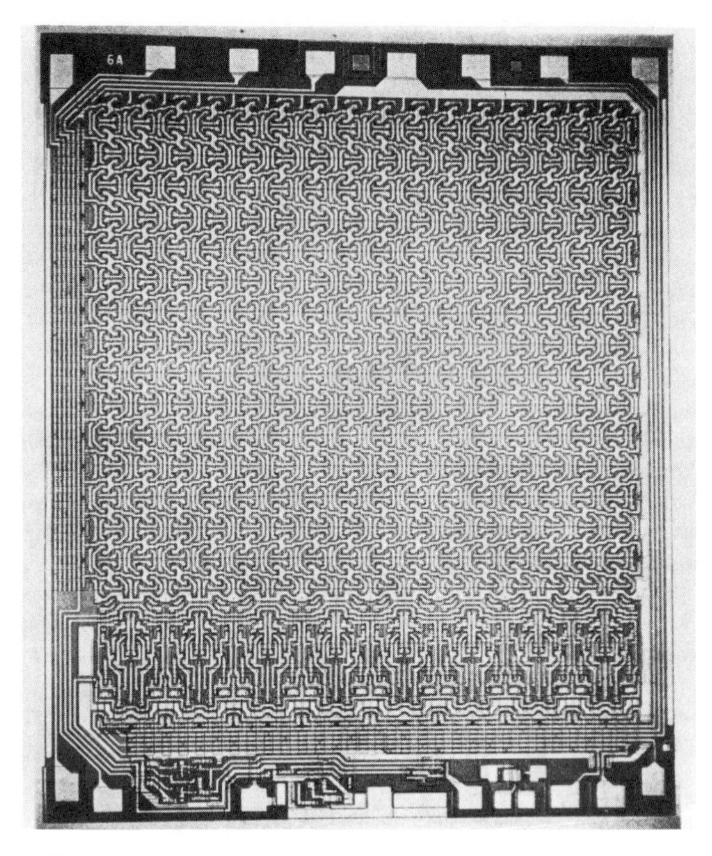

FIGURE 13 Large-scale integrated circuit for the memory element of Illiac IV, a large parallel processing computer designed at the University of Illinois.

detectors; (5) microwave oscillators; and (6) charge-coupled devices for computer memories.

Silicon P–N junctions are widely used as power rectifiers and have many applications in the electric power industry. P–N junctions made of III–V compounds such as GaAs and GaP can convert low-voltage direct current into light with reasonably high efficiency. They are just coming into widespread use for display devices. Such junctions also can be made to emit laser light. They can also do the reverse—convert light into electric power, as in solar batteries. Many detectors for high-energy radiation make use of semiconductors. With use of Gunn oscillators and other methods, one can generate microwave radiation by passing a large current through a semiconductor. A recent innovation of the Bell

Laboratories is that of charge coupled devices for computer memories, displays, and other purposes.

Light-emitting diodes (LED's) are just coming into widespread use, after ten years of development. One important application is for displays in pocket computers. Diodes of GaAsP emit in the red, and those of GaP, with nitrogen as an impurity, in the green. Other materials, for example, InGaAs, emit in the yellow. The first diode to emit laser light was GaAs, which emits in the infrared. With further advances using sophisticated technology with ternary and even quadrinary compounds, it has been possible to extend the spectrum to the yellow and green. A leader in this work has been Zh. Alferöv of Leningrad. I will give just one example based on an experiment of N. Holonyak and his students at Illinois. It is not a P–N junction but a thin platelet of InGaAs about a micrometer thick, as shown in Fig. 14. When the platelet is excited by green light from a gas laser, it emits laser light in the yellow. We may expect to see much more widespread use of junction luminescence in the coming years.

Figure 15, from Bell Labs, shows a remarkable device used as the target in the camera in the Picturephone. It makes use of integrated-circuit technology and consists of an array of nearly 800,000 photodiodes on a chip of silicon a half a centimeter on a side.

Aside from semiconductors, there are many other applications of solid state, such as the following:

1. Nonconducting magnetic oxides (ferrites). Used for memory cores in computers, tape recorders, etc.
2. Composite materials for strength
3. Xerography (selenium photoconductor)
4. Solid-state lasers, nonlinear optics
5. Holography (image recording)
6. Liquid crystals

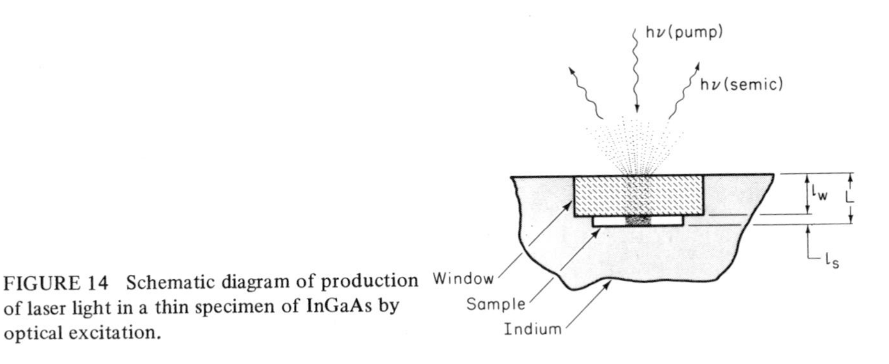

FIGURE 14 Schematic diagram of production of laser light in a thin specimen of InGaAs by optical excitation.

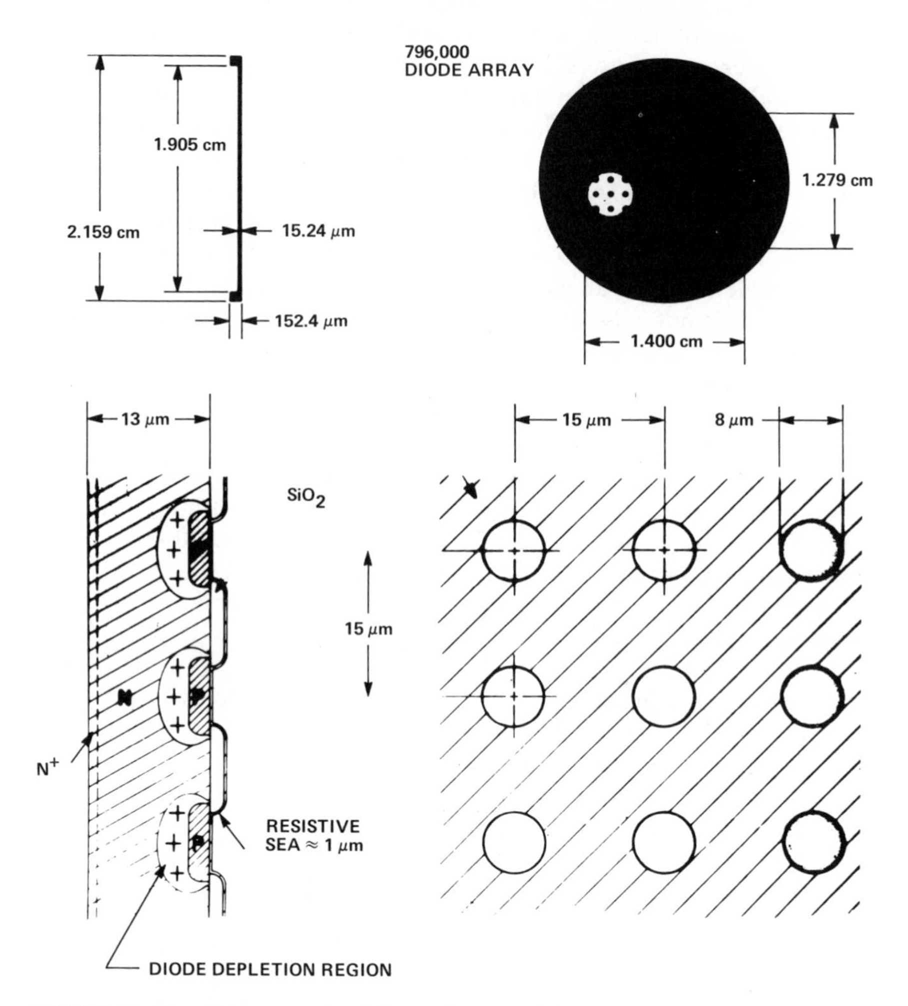

FIGURE 15 Target in camera tube of Picturephone consisting of an array of almost 800,000 photodiodes on a silicon slab about 2 cm in diameter. [From Bell Telephone Laboratories.]

7. Bubble memories (magnetic domains in garnet)
8. Superconducting electromagnets, motors, generators, sensitive detectors.

I will discuss only the last two items on the list, magnetic bubble memories and applications of superconductivity.

In Fig. 16 is shown an array of crystals of various substances used in the communications industry. At the top is quartz, used in crystal filters. The one on the left is a crystal of silicon. In the center is a synthetic ruby crystal. It is a ruby crystal that was used for the first solid-state laser. On the right is a garnet crystal, an oxide of a rare earth and iron. It is an orthoferrite, a nonconducting ferromagnetic material. The magnetic properties of this class of materials were discovered by Louis Néel and co-workers at Grenoble following the discovery and development of ferrite crystals, another class of nonconducting ferromagnets, by the Philips Laboratory at Eindhoven.

Garnets are used in magnetic bubble memories for computers under development by Bell Labs. The memory requires use of a perfect synthetic crystal of garnet in the form of a thin slab. Figure 17 shows in schematically ferromagnetic domains is a garnet crystal in zero magnetic field. The light-colored domains correspond to spin up, black to spin down. If a magnetic field is applied in the downward direction, the black domains will grow at the expense of the light until all that is left are small isolated light domains, or bubbles, about 100 μm in diameter. In a perfect crystal, these domains can be made to move by apply-

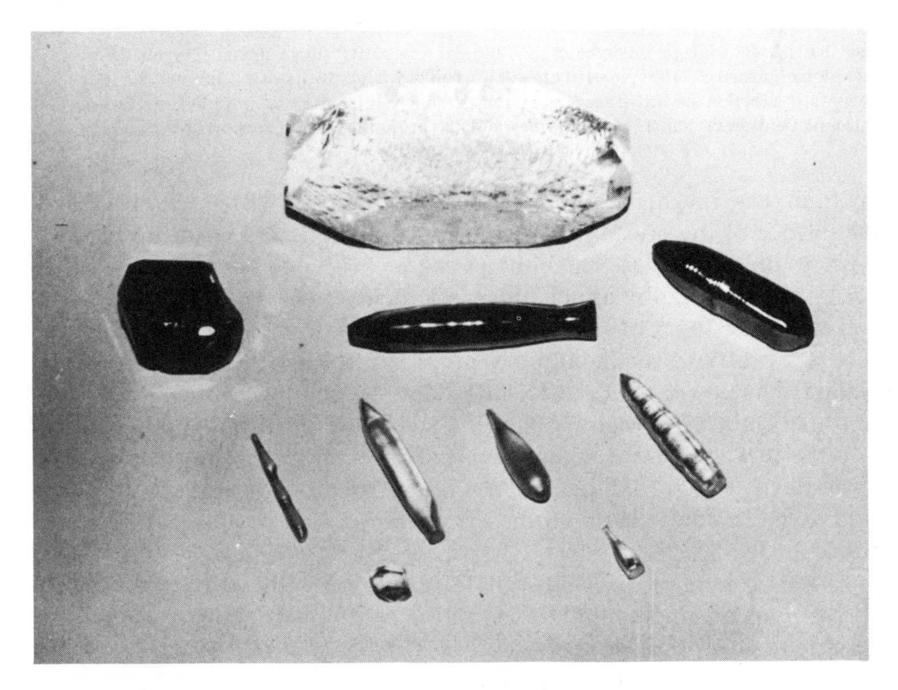

FIGURE 16 An array of crystals used in the communication industry. [From Bell Telephone Laboratories.]

FIGURE 17 Ferromagnetic domains in a thin slab of a garnet single crystal. The small domains that remain polarized in the upward direction when nearly all of the slab is magnetized in the downward direction are called magnetic bubbles. They can be made to move by applying magnetic fields in directions parallel with the slab. [From the Bell Telephone Laboratories.]

ing transverse magnetic fields. Permalloy strips (Fig. 18) are applied to the surface of the crystal to provide guides and pinning positions for the domain or bubbles. The bubbles can be made to shift from one position to another by applied magnetic fields. In this way, shift registers and other functions of a computer memory can be accomplished. It is very pretty to see the bubbles move in real life or in a motion picture. The power required to shift a bubble from one position to another is much less than required for switching with transistors. Under study at IBM and elsewhere is the use of Josephson junctions for memories in which flux quanta are shifted from one position to another with no more power required than that needed for the cells of the brain.

As a final example of solid-state technology, I will discuss potential applications of superconductors in the electric-power industry. This development is based on the discovery of superconductors that withstand very high magnetic fields. Figure 19, from John Hulm of the

Westinghouse Corporation, summarizes developments of superconducting compounds with relatively high transition temperatures and of applications. High-field superconductors are called type II; they permit flux to penetrate in the form of singly quantized vortex lines. At each Low-Temperature Conference, the Fritz London award is presented to a leading low-temperature physicist. This year's award went to A. A. Abrikosov, a former student of Landau, who developed a theory of type II superconductors and published it in 1957. It is his theory that really belongs on this chart in place of the BCS theory. Although Abrikosov unfortunately was not able to come to Boulder to receive his award in person, he did send an acceptance speech that was read for him and was one of the highlights of the meeting. He said that he had actually worked out the theory of type II superconductors in 1953, but his work was rather abstract and mathematical and Landau did not understand the physics behind it, so Abrikosov put it away in his desk drawer and did not publish it. It was not until several years later, when Feynman discussed quantized vortex lines in liquid helium, that Landau

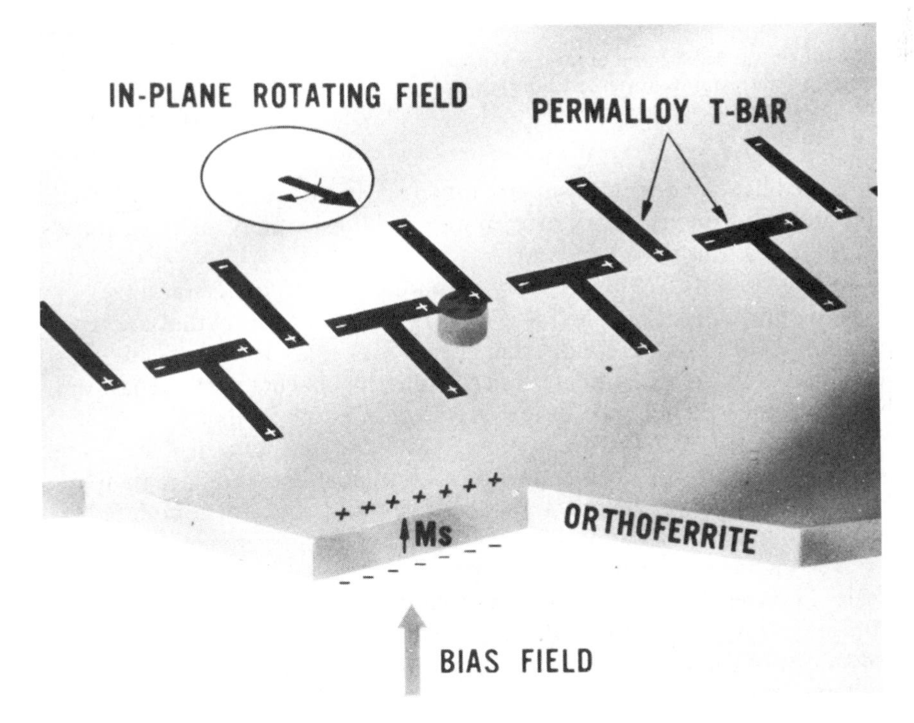

FIGURE 18 Principle of magnetic bubble memeory, in which bubbles are guided along permalloy strips to form a shift register. [From Bell Telephone Laboratories.]

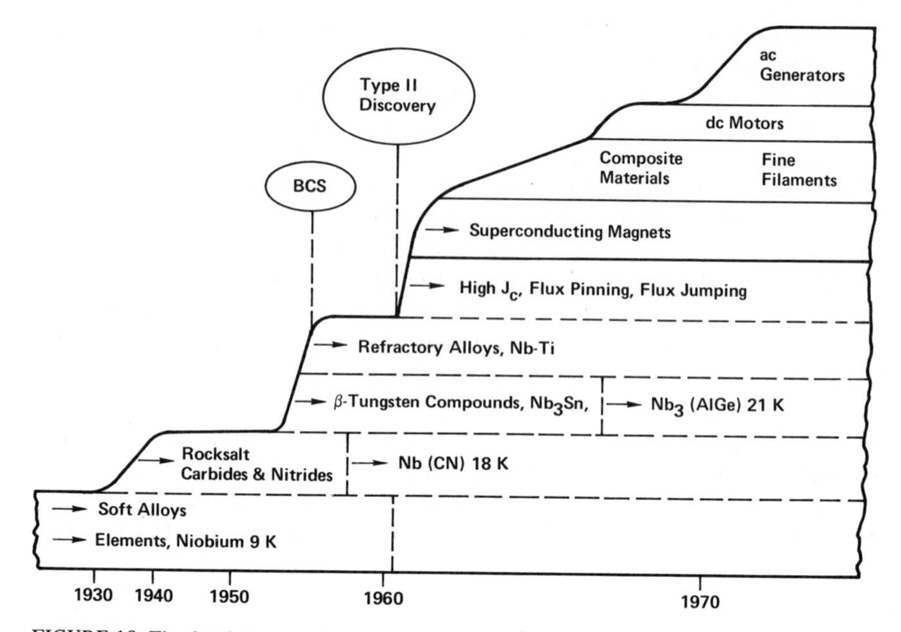

FIGURE 19 The development of superconducting materials and of superconducting technology. Abrikosov's theory of the vortex structure in type II superconductors was published in 1957, the same year as the BCS theory of superconductivity. [From J. K. Hulm, Westinghouse Research Laboratories.]

could see the analogy with superconductors and Abrikosov was allowed to publish his theory. Abrikosov in his speech told many amusing stories of his discussions with Landau.

In any case, Abrikosov's theory is fundamental to understanding type II superconductors, and it is type II superconductors that are used in high-field superconducting magnets. A great deal has been learned as to how to fabricate superconducting magnet wire used in the windings for stability and for low loss in alternating fields.

With use of high fields from superconducting magnets, electric generators and motors can be made much smaller and compact than those using iron-core magnets. Figure 20 shows a recently completed experimental 5-MV A experimental superconducting generator made by Westinghouse. The helium tank in the foreground indicates the scale. The Westinghouse workers believe that superconducting generators are entirely practical in the near future. Many other potential applications of superconductors are under investigation. These include underground power-transmission lines, magnetic levitation of high-speed trains, and electric motors. Superconductivity appears on the verge of taking off on large-scale applications.

These are just a few highlights of the remarkable development of solid-state science and technology of the past quarter century. What of the future? Solid-state physics is now emerging into a mature science. The very rapid growth of the recent past has leveled off, but we can expect a continued slow growth at least at the level of the economy as a whole. The field will continue to be of great importance to technology as well as to science. Performance of most products is limited by the properties of the materials used. Improved properties of materials is reflected directly into improved products. Thus we can expect that the field will continue to be important for many years to come.

Among the growing areas of solid-state physics at the present time are surface physics, use of laser beams for light-scattering experiments, the study of quantum crystals and fluids, superfluids, applications of superconductivity studies of amorphous materials, and the study of phase transitions. Studies are being extended to increasingly more complex materials.

Concepts of solid-state physics are being applied in other areas such as nuclear and particle physics, astrophysics, and biology. They help us

FIGURE 20 A superconducting 5-MV A generator developed by Westinghouse. The helium tank in the foreground illustrates the scale. [From Westinghouse Research Laboratories.]

to understand charge transport, optical excitation, and other properties of organic and biological systems, a field often called the organic solid state. It is thought that neutron matter in pulsars is probably a superfluid. These are aside from the widespread use of solid-state instrumentation in all branches of science.

We may compare the growth of solid-state physics with that of a fruit tree. The seeds of the field were planted in the early years of the century. The seeds sprouted, and the young plant began to grow vigorously following the discovery of quantum mechanics in 1926. It took some years, however, before it was mature enough to bear fruit, and the first real harvest did not occur until after World War II. The tree continued to grow rapidly, sending out many new branches; and each year brought an increasingly large harvest of ripe fruit. Now the tree is mature and is not growing as rapidly as in its youth, but it is still vigorous and healthy. Each year brings a new harvest of discoveries in science and applications. Some branches may be dying out, but others are sending out vigorous new shoots in various directions. We expect the tree to bear fruit for many years to come.

JACQUES FRIEDEL *heads the Solid State Physics Laboratory at the University of Paris. He is a past president of the French Physical Society and presently is serving on the council and the executive committee of the European Physical Society. He is a foreign honorary member of the American Academy of Arts and Sciences, and he has served as Secretary of the IUPAP Commission on Publications.*

7 *Dislocations in Solids and Liquids* / J. FRIEDEL

INTRODUCTION

Asked to talk about the plastic properties of materials, I could have given you a progress report on recent technical achievements in this field. This would certainly have included topics such as *ausforming* treatments of steels, which much increase their "toughness"–i.e., nonbrittle hardness–the use in turbines of *composite materials*, and the development of *fracture mechanics* in the aerospace industry.

In a few words, ausforming is a simultaneous use of straining and heat treatments; it probably produces very fine precipitates on a very dense network of dislocations–it has, for instance, allowed a doubling in the last year of the purely hydrostatic pressures available in the laboratory. Composite materials imbed hard but brittle fibers or elongated precipitates in a softer matrix, resulting in materials able to sustain large stresses in some directions. Fracture mechanics has seen the systematic use of a K factor, which essentially measures the amount of elastic energy of a crack under stress that is plastically relaxed at its tip and regulates its propagation; this knowledge of the K factor for the materials used in airplanes allows companies to use with a good conscience planes that are full of cracks.

But such technical progresses, important as they are, keep today within the limits of either chemical recipes or macroscopic and phenomenological concepts. Metallurgy is an old lady who never waited for our knowledge of modern physics to work; and indeed we are still barely grasping, from a fundamental point of view, the grossest features of the plastic properties of the surrounding world.

To complement Professor Bardeen's broad review on the solid state, I am therefore finding it more useful to concentrate on a single fundamental concept, that of the crystal defect called *dislocation*, which is responsible for the plasticity of these materials. This is far from a new field; but I want to stress some significant progress, which relates the brittleness of materials to the chemical nature and geometry of their bonding; I also want to place the problems of crystal dislocations in a broader context, which includes solid glasses, magnetism, liquid crystals, polymers, and biological materials. This paper might show how cross-fertilization can work on a broad front. It will be more in the nature of a progress report than a list of technical achievements.

VOLTERRA'S DISLOCATIONS

The concept of dislocation was introduced by Volterra, at the beginning of the century, to describe states of internal stresses in macroscopic classical *continuous solids*. It can be shown that the most general state with internal stresses that are continuous everywhere except along a line L can be obtained by the following recipe:

1. Cut the solid V, assumed without internal stresses, along a surface S limited by the line L (Fig. 1, a and b).
2. Move the two lips S′, S″ of the cut with respect to each other by a pure displacement, without change of form.

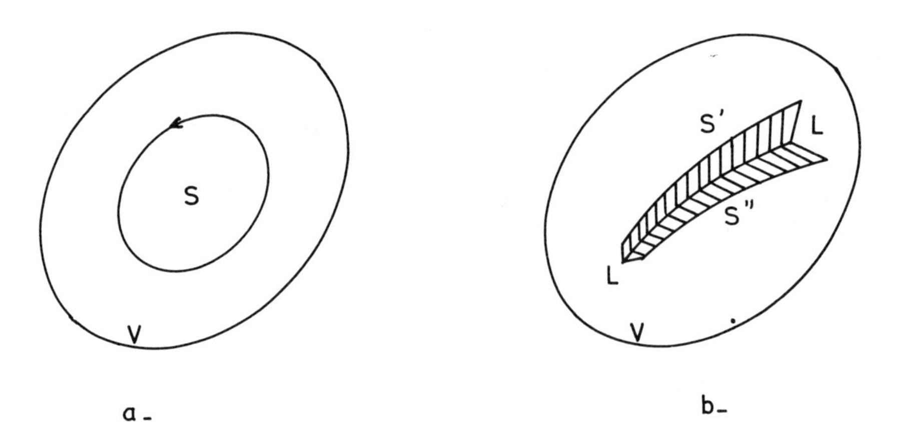

FIGURE 1 Creation of a Volterra dislocation: a, view in perspective; b, cross section of the solid.

3. Add matter in the empty space thus created (Fig. 1, b) or remove excess matter, if any.
4. Re-establish the interatomic bonds across the cuts, and relax the external stresses applied during these processes.

The result is of course a solid with internal stresses. Because the relative displacement of the lips S′, S″ does not involve distortions, it produces no discontinuity in the stresses along the surface S but only along the line L. The dislocation line thus produced is characterized by its geometrical *form* L and its strength, a displacement. This in turn can be considered as the sum of a *translation* **T** and a *rotation* **R**.

Volterra was interested in materials such as prestressed concrete, in which all internal stresses are small and often within the elastic range. Dislocations of infinitely small strength are then of interest; and it can be shown that any state of elastic internal stresses can be described by a *continuous distribution of such infinitesimal translation dislocations.* Rotation dislocations are not necessary, as they can be considered as a continuous distribution of translation ones. This is clear in Fig. 2, a and b, which shows that the effect of inserting a sector is equivalent to inserting a collection of staggered sheets of constant thickness.

Such continuous distributions of infinitesimal dislocations are much used to study elastic problems involving what can be treated as an external continuously varying stress: solids with precipitates, cracks, magnetic materials with nonuniform magnetization. They also appear in the low-temperature plasticity of glass. Under suitable high stresses, glass can slip without cracking, and the limit between a region that has slipped and the rest is obviously a translation dislocation of total strength **T** equal to the relative translation of the two parts that have slipped (Fig. 3). For slip to progress without separating the lips S′, S″ of the cut, the dislocation must obviously *glide parallel to the translation* **T**. There is, however, every reason to believe that slip does not start abruptly (Fig. 3, a) but progressively (Fig. 3, b) at the edge of such a slip line, i.e., the total translation **T** is split over a certain width $L'L''$ into a continuous distribution of infinitesimal dislocations each with a fraction of **T**. The splitting arises from the fact that the elastic distortion energy due to a dislocation is proportional to the square of its strain, thus of its translation vector; and the square of a sum is more than the sum of the squares. The only limit to this splitting is the fact that, along the slip plane, some relative positions of the two parts of the crystal favor locally the formation of stronger atomic bonds. But as such a progressive slipping progresses, and at least in amorphous solids where each atom or molecule has many and not strongly directed

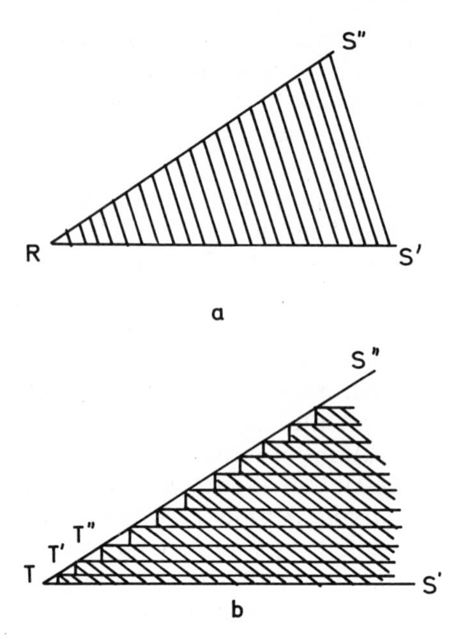

FIGURE 2 Equivalence between a, a rotation dislocation R, and b, a continuous distribution of infinitesimal translation dislocations T, T', T'' (in section).

bonds, one expects on the average practically as many bonds to be strengthened as to be weakened: the propagation of such a progressive slipping should therefore require a stress definitely smaller than an abrupt slipping (or a crack).

TRANSLATION DISLOCATIONS IN CRYSTALS

Slip has been, however, mostly studied in crystals. Here, to leave no permanent damage along the slip plane S, the translation **T** must be a *period of the Bravais lattice*, so that the two lips S′, S″ are left after slip

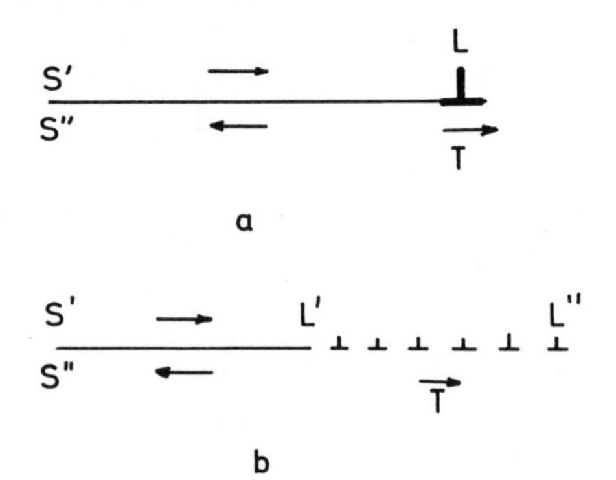

FIGURE 3 End of a slip line in a glass: a, abrupt slipping; b, progressive slipping.

in coherency and can reform their atomic bonds. By the same agrument, the edge of a slip line will split into a collection of elementary dislocations, each with a translation vector equal to an elementary period of the lattice.

Translation vectors are therefore finite. Stresses and strains are still small at large distances, in the range of linear elasticity. As they die out only slowly (in r^{-1}), most of their energy is in this range, and their main energy properties (line tension, energies of interaction between themselves and with applied stresses, etc.) that can be deduced are very much the same whatever the material. There is, however, a central region where the large strains are out of the elastic range. It is this core region that regulates details of the line tension that might induce dislocations to favor some crystallographic *slip planes*; it also plays the essential role in the *mobility* of dislocations. It is in the *dislocation core* that one must look for an understanding of the wide difference of plastic behavior in various materials.

A serious study of this core region involves a good knowledge of the nature of the bonding and of the geometrical arrangement of atoms. It is thus hardly surprising that general ideas are slow to emerge here.

Ionic Solids

The first reliable computer studies of dislocation cores have been made on ionic solids, where the nature of interatomic forces is best known: long-range Coulomb interactions and short-range exchange replusions. They have confirmed previous general inspired guesses.

1. At least in simple compact structures such as the *cubic rock salt*, dislocation cores have a tendency to take an elongated form along a close-packed plane. This close-packed plane contains the translation vector **T** of the dislocation, which can therefore slip in its elongated form (Fig. 4, a).
2. One can compute to a good approximation the spreading of the dislocation by balancing a misfit energy across $l_1 l_2$ with the elastic energy gained by splitting the whole dislocation **T** into a continuous distribution of infinitesimal dislocations $l_1 l_2$. This misfit energy can in turn be computed as if interatomic forces were very short range: the energy for a misfit x in M is practically the same per atom along the cut as for a uniform misfit x across a large area of cut.

In ionic solids, the misfits are always ions of the same sign; the energies involved are thus large and the splitting $l_1 l_2$ is small—of atomic

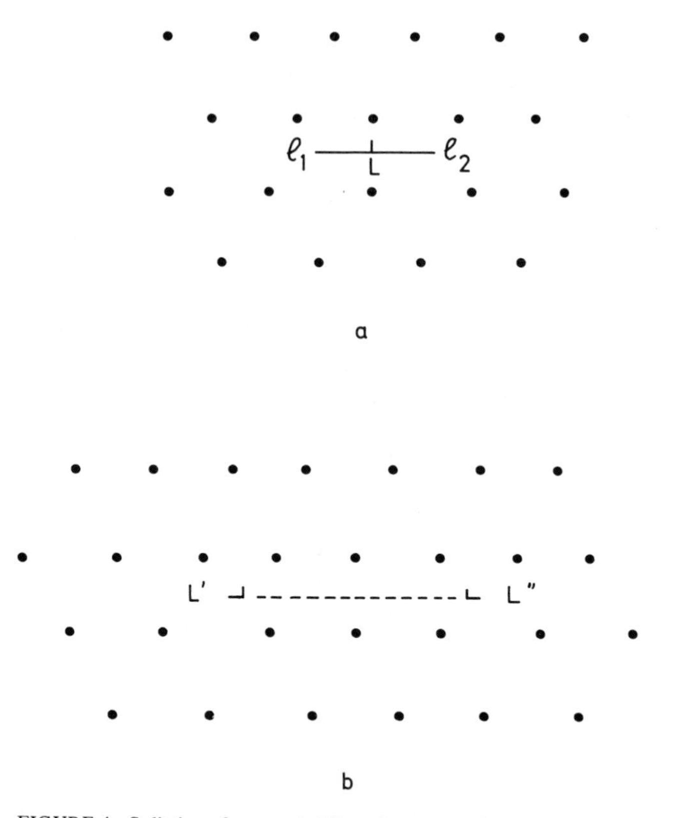

FIGURE 4 Splitting of a crystal dislocation: a, continuous splitting; b, splitting into two "partials" L', L'', separated by a "stacking fault" s of assumed low energy.

dimensions. It is, however, enough to induce dislocations to lie along these splitting planes. As in glasses, the slipping is sufficiently continuous for the dislocations to slide easily: the motion of the center of gravity L of the dislocation along the slip plane does not involve great changes in energy because the local order is not much disturbed, at least as far as average interionic distances are concerned.

If you take a piece of rock salt in your hand, you might think it is a hard and brittle material. However, this is due to a strong "pinning" of the existing dislocations by impurities or precipitates. It becomes very ductile if you free its dislocations by purifying or heating or prestressing. Thus it has been known for ages by miners, and later geologists, that layers or domes of salt are very ductile under the conditions of moderate temperature and pressure usually prevalent in the upper crust of the earth.

A study of *other rocks* shows clearly that not all ionic solids are so plastic. Thus the slightly less simple structure of calcite (CO_3Ca) has, in the same geological conditions of the upper crust, a finite but reduced plasticity. This is further limited by its possibility of twining, which will not be discussed here. Other types of rock, which develop in the deeper layers of the crust, have more complex structures; they are usually very hard and brittle as soon as they come out of the conditions of high pressures and temperatures where they were formed. Thus the slip of a dislocation in alumina (Al_2O_3) involves a very complex coordinated rotational motion of atoms in its core, which is practically impossible at low temperatures. Quartz (SiO_2) has a complex structure with strongly covalent bonding, which makes its plastic properties somewhat similar to those of diamond, discussed below; and this is the same for many complex silicoaluminates (feldspaths or lavae of various kinds). Hence the traditional reputation of granite. Only layered structures such as mica, a silicoaluminate with strong ionocovalent bonding within layers but weak bonding between layers, which slips easily along these layers, explains the part taken by the bending of deeper mica schist layers in the Alpine chains.

Metals

A larger spreading of the core is expected in metals, because atoms are neutral and the interatomic forces are not directional. Indeed, in most compact phases, there are close-packed planes along which the misfit energy is especially low near the center of the dislocation core. The dislocation L(T) the splits into two well-defined halves, L'(T′) and L'' (T″) such that T = T′ + T″, separated by a "stacking fault" of low energy s (Fig. 4, b). Depending on the ratio of s to the elastic energy released by splitting, the width of such a split dislocation can vary from angstroms to hundreds of angstroms and is often directly visible in thin-film electron microscopy.

The exact nature and amount of this splitting can vary from metal to metal and explains their different plastic behavior. We shall discuss the simplest (cubic) structures.

In a *face-centered cubic* (FCC) metal, each dislocation is split along a close-packed (III) plane, which contains the translation vectors of its two halves (Fig. 5, a). The stacking fault has necessarily a low energy, as the misfit does not involve first nearest neighbors. Slip is very easy, corresponding to a very low and temperature-independent elastic limit σ_E (Fig. 6, a). But, with increasing strain ϵ, dislocations of different (111) slip planes can meet and recombine into straight barriers that cannot slip in either of the slip planes and are therefore immobile, or

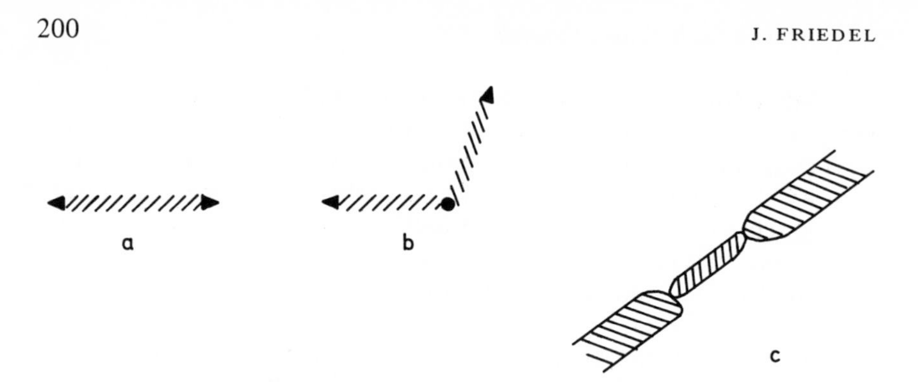

FIGURE 5 Dislocations in face-centered cubic metals: a, mobile split dislocation; b, sessile barrier (in section); c, cross-slipping dislocation (in perspective).

"sessile" (Fig. 5, b). The rapid increase in the number of barriers with strain produces a large increase in dislocation density and internal stresses, thus a strong hardening rate $d\sigma/d\epsilon$ (Fig. 6, a). This is partly released when dislocation loops can leave their initial slip planes by a thermally activated cross-slipping process, where under the applied stress it recombines in the initial plane and splits in another one (Fig.

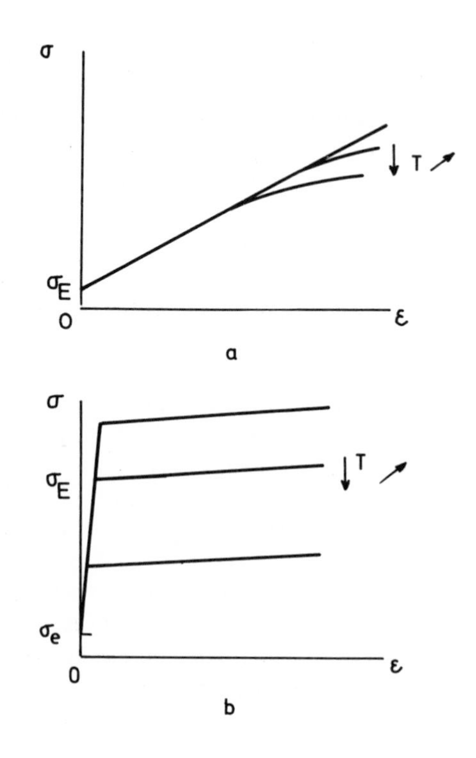

FIGURE 6 Stress–strain tensile curves $\sigma(\epsilon)$ in metals: a, face-centered cubic; b, body-centered cubic. The strain ϵ is the relative lengthening; the tensile stress σ is the applied force per unit section.

5, c). This softening, represented in Fig. 6, a, is less marked in strongly split dislocations, thus it occurs at higher temperature in metals or alloys of low stacking fault energy. One can explain in this way that hardening is less temperature-dependant in copper than in aluminium and even less in alloys such as FCC brasses (Cu, Zn) or bronzes (Cu, Sn), where the Fermi surface just touches a Brillouin zone boundary.

In body-centered cubic (BCC) metals, such as iron, dislocation loops are normally split along a close-packed plane (110 or 112) that contains their translation vectors (Fig. 7, a); they are thus again mobile at a low and temperature-independent stress, σ_e, with well-defined slip phases (Fig. 5, b). However, when such a loop reaches a position parallel to its translation vector (the "screw" position), it can split further into three "partials" bounded by stacking faults in three different planes. Such a split screw is obviously sessile. All mobile dislocation loops will rapidly transform into a pair of sessile screw dislocations bounded by a short length of mobile dislocation, which will disappear at the surface or be blocked: the initial easy straining is on a microscopic scale. The macroelastic limit σ_E observed after this microstrain stage corresponds to the motion of the screws. At 0 K, σ_E is very large, because it is the stress required to recombine two of the partials against their elastic repulsion (b → a, Fig. 7). At finite temperatures, a thermally activated motion of the screw is possible, by a local recombination of two partials somewhat similar to the cross-slipping of FCC metals. In most BCC metals, the splitting is not very large and screws glide easily under fairly low stresses, σ_E, near room temperature (Fig. 6, b).

The typical difference between the stress–strain curves of these two types explains their difference in ductility. In a tensile test, it is easy to see that straining is uniform (Fig. 8, a) as long as $(d\sigma / \sigma\, d\epsilon) > 1$, while

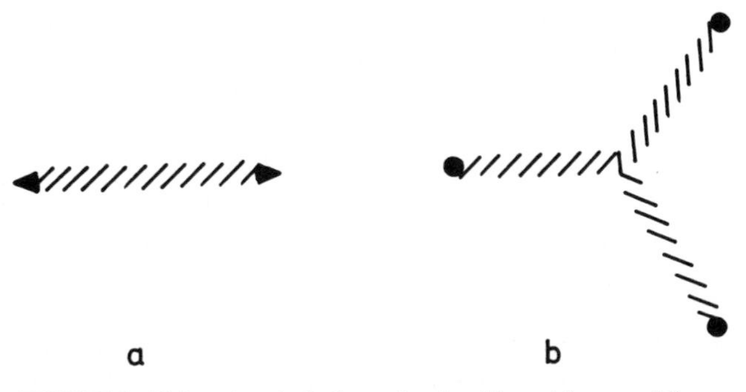

FIGURE 7 Dislocations in body-centered cubic metals: a, mobile split dislocations; b, sessile split "screw" dislocations (in section).

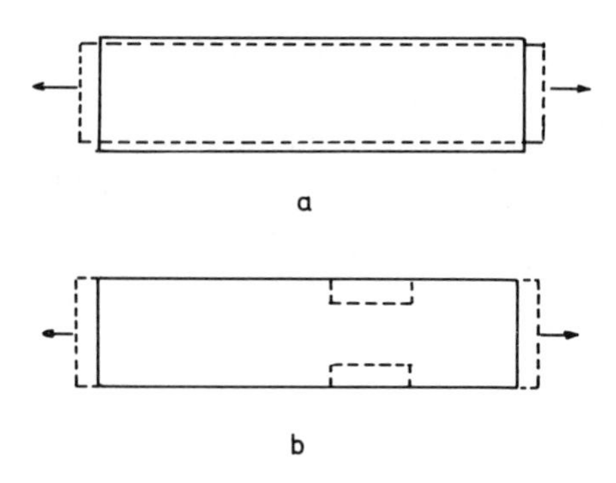

FIGURE 8 Tensile test: a, uniform strain; b, striction.

a "striction" appears otherwise (Fig 8, b), leading quickly to failure. In both types of metal, we can write approximately

$$\sigma \simeq \sigma_E + a\epsilon.$$

The striction condition is $\epsilon > (1-\sigma_E/a)$. The FCC metals, with $\sigma_E \ll a$, are ductile up to practically 100% deformation. The BCC metals, with a small and σ_E temperature-dependent, are completely brittle for $\sigma_E > a$, thus below a critical temperature that is around room temperature; even above this temperature, their ductility is less than FCC metals. The ductile–brittle transition of iron has much interested metallurgists; it was, for instance, the reason why soft steel tankers built by the United States during the World War II were apt to break in two on cold winter days. It was for a long time attributed to the pinning of dislocations by carbon impurities. It is now seen as an intrinsic limiting factor in the use of all BCC metals at low temperatures.

The striction condition just used is approximate. It neglects the possible variation of the hardness σ with *strain rate* $d\epsilon/dt$; such a factor must be included to explain the "superplastic" behavior of some materials, with no striction up to strains much larger than 100%. Instabilities can also originate from the heat released by straining, if the hardness σ is sensitive to *temperature*; such thermal instabilities occur more readily at low temperatures and, for FCC metals, at high strains. They are probably involved in the inhomogeneous straining of the crystalline crust of neutron stars, in preference to cracks, and possibly responsible for their sudden changes of form as measured by pulsar emissions.

We can note finally that metals with lower symmetries have not been much studied, except the *hexagonal* ones, because most have a more reduced plasticity. In hexagonal metals, the most stable dislocations have their translation vector usually perpendicular to the axis of hexagonal symmetry; and they split preferably parallel or perpendicular to that axis, depending on the ratios of the corresponding stacking faults. Other dislocations can probably split in a plane that does not contain their translation vector and become then sessile. Any way, easy slipping is restricted to a small number of systems, which do not allow the various grains of a polycrystal to deform coherently. Hence the low-temperature brittleness of these metals, tends to restrict the metallurgical use to metals such as zinc with not too high melting points. It is for instance one of the reasons why beryllium could not be used in nuclear-reactor technology.

Covalent

After these few comments on the bronze, iron, and nuclear age, a concluding remark on this topic should be about the age of silicon.

All tetravalent bonded covalent solids have a cubic structure somewhat akin to that of the FCC metals; and, as a result, their dislocations split along (111) planes in two halves widely separated by a stacking fault. This experimental observation agrees with the fact that the ensuing misfit again preserves the local bonding of each atom with its four neighbors; and, within the rough approximation, one then expects the stacking fault to have a low energy. Indeed, slipping occurs along well-defined (111) planes. It is, however, extremely difficult, except at very high temperatures: diamond is well known to top the hardness scale. The reason is surely in the covalent nature of cohesion, with little and strongly directional bonding: the cores of the two partials possess necessarily nonsaturated "dangling bonds," and moving these partials involves the breaking of more bonds. The glide of each partial T′ thus involves the expenditure of a large energy U, the energy to break a bond, when the dislocation moves over an area A of atomic dimensions. The work of the applied stress σ on the lips of the cut, when the slip progresses in this way is the force σA multiplied by the relative translation T', thus $\sigma A T' = \sigma v$, where v is a volume of atomic dimensions. Under thermal agitation, the average speed of each dislocation will therefore have an activation energy $U - \sigma v$, and one obtains

$$\frac{d\epsilon}{dt} = A \exp\left(-\frac{U - \sigma v}{k_B T}\right),$$

or

$$\sigma = \frac{U}{v} - \frac{k_B T}{v} \ln\left(\frac{d\epsilon}{A\,dt}\right).$$

Such laws are well followed experimentally with reasonable values of the parameters.

ROTATION DISLOCATION

Simple Crystals

We have not mentioned rotation dislocations in crystals so far. They play indeed very little role. The reason is twofold.

First, as pointed out earlier, the rotation **R** must, by itself or in conjunction with a translation **T**, be a symmetry operation of the crystal structure.* The angle of rotation is therefore large (60°, 90°, or 120°), and furthermore the long-range strains produced are so large that the resulting *energy* is very large indeed in large crystals.

Rotation dislocations are therefore mostly expected in small crystals. Indeed in crystalline *membranes* the long-range stresses are relaxed if the membrane takes a conical shape; and polyhedral membranes, which are frequent in nature, all contain regularly spaced rotational dislocations.

One also meets a number of rotation dislocations when studying the lattice structure *of vortex lines in supraconductors of the second kind.* The reason is here that these vortices undergo from the lattice defects a friction somewhat comparable in strength with their mutual interactions; they take therefore their most stable configuration only with difficulty if they are far apart, i.e., under small applied fields. Even there, these rotation dislocations usually appear as neighboring pairs of opposite signs, so as to compensate most of their long-range stresses. Inspection of Fig. 9 shows that such a pair of opposite rotation dislocation **R** and **R**′ is equivalent at long range to a pure translation dislocation T:

$$\mathbf{R}' + \mathbf{R} \simeq \mathbf{T}.$$

The strength of **T** is related to the strength and positions of **R** and **R**′ by an obvious geometrical relation.

Second, it is furthermore obvious from Volterra's recipe that the strain energy of a rotation dislocation increases very fast if its geometrical line is not along its axis of rotation. A rotation dislocation in a crystal is therefore necessarily *straight* and *immobile* ("sessile"). It cannot play a role in plasticity.

*A pure rotation dislocation (*R*) has been called a "disclination," a helical dislocation (**R** + **T**) has been called a "dispiration."

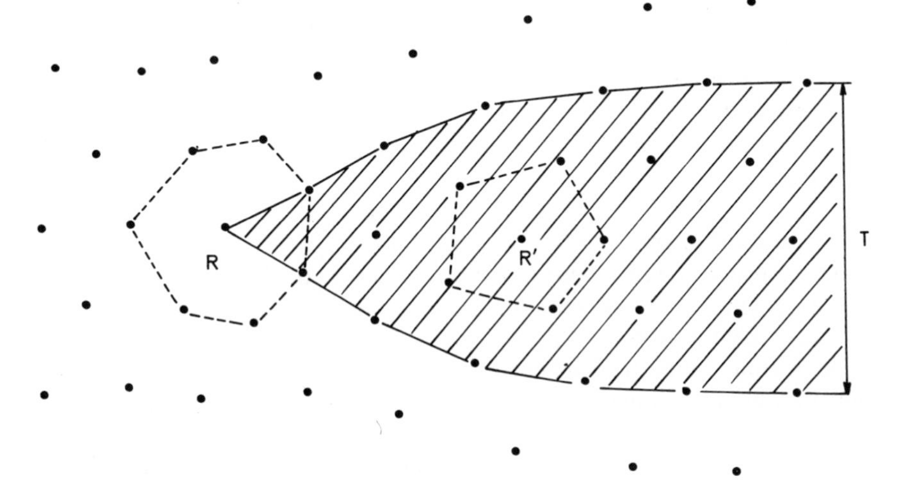

FIGURE 9 Equivalence to a translation dislocation T of a pair of parallel rotation dislocations R and R' of opposite strength.

Isotropic Liquids

One could imagine producing translation or rotation dislocations in an ordinary isotropic liquid by an instantaneous Volterra mechanism. But such a defect would quickly *relax* and disperse into an extended distribution of infinitesimal dislocations of practically no energy, in times comparable with transport times in the liquid. The driving force is the same as for glasses: the lowering of elastic energy by dispersion of the defect.

Intermediary dynamic stages in this relaxation involve a curly motion of the liquid. In normal liquids, one can analyze the vortices in terms of a continuous distribution of rotation dislocations in the *velocity* field, which disperses progressively. In quantum liquids, the irrotational nature of the motion is only broken along the vortex lines, which are necessarily quantized: one finds here for the velocity field the same difference as when going from a glass to a crystal for the position field.

These topics of hydrodynamics are well known to mechanicians and physicists, and I do not wish to dwell on them. I can only regret a lack of contact in this field with the metallurgists interested in dislocations and, perhaps, an insufficient interest in basic problems.

Mesophases

Liquid crystals are essentially liquids with anisotropic properties, due to the tendency of their constituent elongated molecules to show a strong directional ordering. Hydrodynamics of such liquids constitute a fascinating and perhaps technically useful topic and certainly involve the creation of vortex lines. But simple *static defects* are also of interest, and I will restrict myself to them.

The only partial order shown by the mesophases has interesting consequences.

1. In *solid* phases that would have the same symmetry properties as the corresponding mesophases, one can create two types of dislocations, with a strength (**T** or **R**) that either can vary continuously, as in a glass, or is quantized as in a crystal.

The simple "nematic" phase, for instance, is made of elongated molecules that are all parallel but possess no long-range order in their positions (Fig. 10, a). Rotations **R** perpendicular to the axis of the molecules are quantized ($R_\perp = n\pi$, n an integer) (Fig. 11); rotations $\mathbf{R}_{\parallel}$ parallel to the axis and all translations **T** are not quantized. A similar difference arises in the other (cholesteric and nematic) phases, which can actually be prepared more easily in a solid state (Fig. 10, b and c).

2. In *liquid* mesophases, the quick diffusion produces a series of effects.

(a) All dislocations with continuously varying **R** and **T**'s are relaxed, as in a normal liquid. The dislocations with *quantized* **R** or **T**'s only subsist in a static state. Dislocations with $R_\perp = n\pi$ only exist in the nematic phase, for instance.

(b) Even for such dislocations, a large part of the strains is relaxed. Strains corresponding to a spatial variation only subsist in the orientation of molecules. The dislocations are lines of discontinuity in the

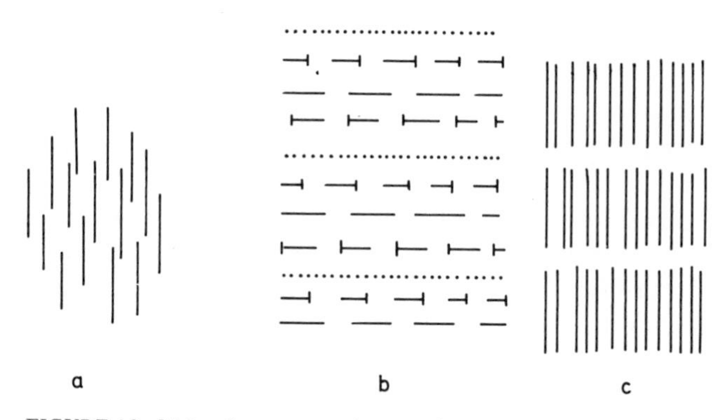

FIGURE 10 Molecular structure in mesophases: a, nematic; b, cholesteric; c, smectic.

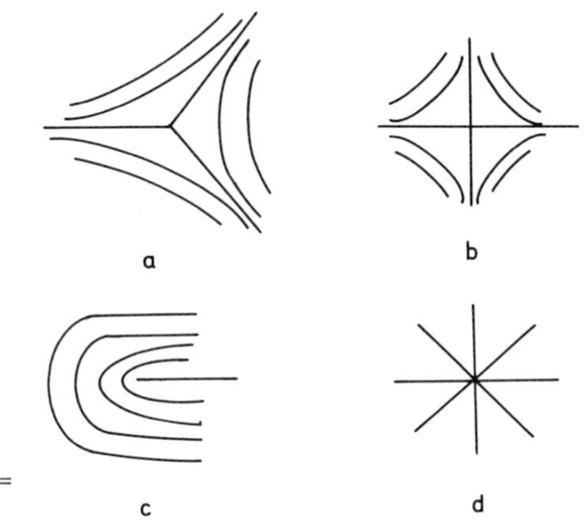

FIGURE 11 Symmetries of rotation dislocations $R_\perp = n\pi$ in a nematic mesophase (Frank unrelaxed configuration): a, $n = -\pi$; *b*, $n = -2\pi$; c, $n = \pi$; *d*, $n = 2\pi$.

orientation of molecules, and the elastic energy stored is now much less than in a solid and comparable with that of a translation dislocation.

(c) The liquid relaxation also allows these dislocations to be much more *flexible and mobile* than in a solid. In the nematic phase, a rotation dislocation can bend easily if it can locally carry the axis of rotation $\mathbf{R}'$ with it. This is possible because, as pointed out in Fig. 9, this means adding or subtracting to $\mathbf{R}$ a translation $\mathbf{T}$, which is easily produced or relaxed by the mesophase. Similarly, less general motions are possible in other mesophases.

As a result, few dislocations subsist in liquid crystals, even if they are stirred very hard: most disappear quickly at the surface or annihilate each other. The few that remain, often at a defect, are clearly visible under the optical microscope, because these distorted anisotropic liquids give strong optical contrasts.

Indeed dislocations were observed in nematic and cholesteric phases before the end of last century. These "threads" have given their name to the *nematic* phase, and their structure, following pictures such as Fig. 11, were understood as early as 1920; they were, therefore, by far the first dislocations observed as such in nature.

(d) Some care must, however, be taken about these dislocations. Rotation dislocations $R_\perp = n\pi$ look much broader and fuzzier for n even than for n odd. This is due to a further relaxation of the core, which is possible in such cases. Figure 14 pictures, for instance, the simple case of a nematic cylinder enclosed in a tube with boundary conditions such that molecules should arrive normally to the surface of the tube. The nematic must therefore necessarily possess a dislocation $R_\perp = 2\pi$. However, contrary to what was assumed in Fig. 11, d, one

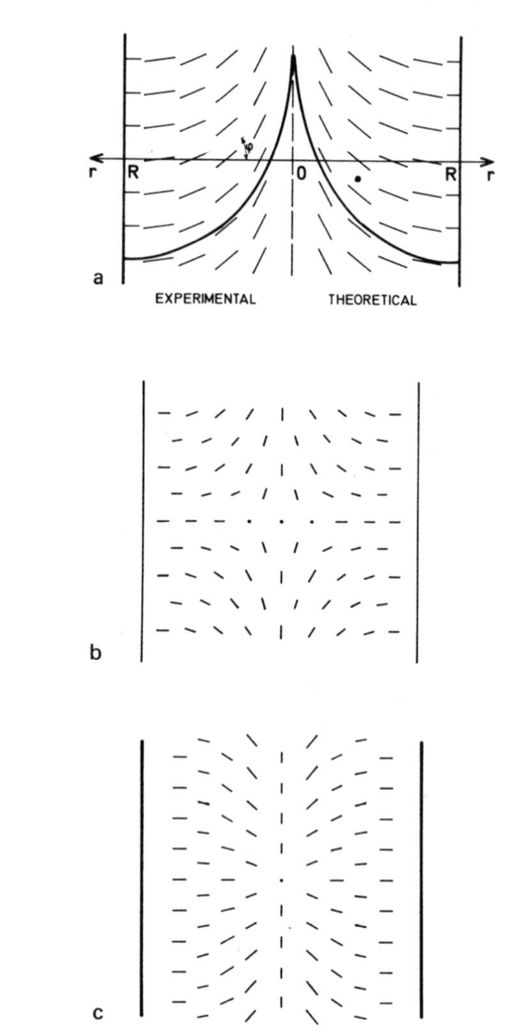

FIGURE 12 Nematic in a tube: a, simple relaxed core; b and c, singular points of opposite signs [schematic orientation of molecules, deduced from optical studies by C. Williams, P. Pieranski, and P. E. Cladis, *Phys. Rev. Lett. 29*, 90 (1972)].

sees in Fig. 12, a that the molecules do not lie in a plane: they bend out of it, so that the dislocation has a *continuous core*. Indeed the dislocation of Fig. 12, a can be said to end in two *singular points* at both ends of the tube. Such singular points can also be seen inside the tube (Fig. 12, b and c). These effects were predicted before they were observed, and the curvatures agree with what is known of the ratios of the various elastic constants of the liquid.

(e) Another case understood early was the *Grandjean walls*, obtained in the cholesteric structure. This is a structure where the molecules lie again parallel in a plane, but their direction changes in a helical way when the plane moves parallel to itself (Fig. 10, b). Let us put such a liquid between two glass plates suitably rubbed with greasy fingers, so as to give a preferred orientation to the surface molecules. When the two plates are not parallel, the cholesteric structure is favored at regular intervals, where the distance between the plates is a multiple of the half-pitch p of the helix.

There are necessarily regularly space zones of mismatch, which are clearly visible–these are the Grandjean walls. It was, however, understood before 1930 that matching was continuous along these walls too, except along lines. Because these correspond to a vernier effect–more half-pitches on the right than on the left–they clearly are translation dislocation lines. Recent studies have shown that these lines have the minimum possible period, i.e., one half-pitch, when the thickness of the sample is comparable with the pitch. The molecules then tend to lie all nearly parallel to the plates even in the core of the dislocation. This "de Gennes" configuration is pictured on the left of Fig. 13. For thicker samples, these elementary dislocations tend to regroup into pairs, giving dislocations with a translation vector equal to the pitch. These are actually *split* into two rotation dislocations of opposite signs (**R**, **R**′, Fig. 13). Such a splitting, similar in a way to that of Fig. 9, is especially stable because, as has been shown conclusively, both rotation dislocations have nonsingular cores.

This special stability of "double" translation dislocations and their tendency to split when roughly perpendicular to the cholesteric axis explain the zigzag form taken by such dislocations in the bulk of cholesteric phases, with split "edge" parts perpendicular to cholesteric

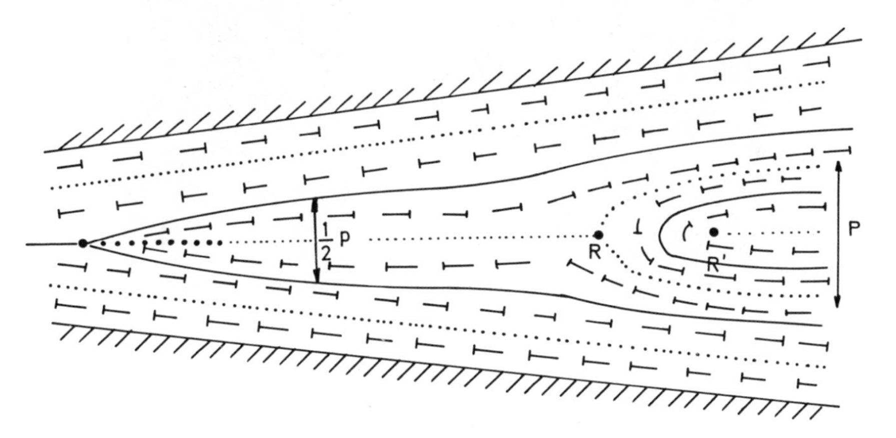

FIGURE 13 Grandjean walls.

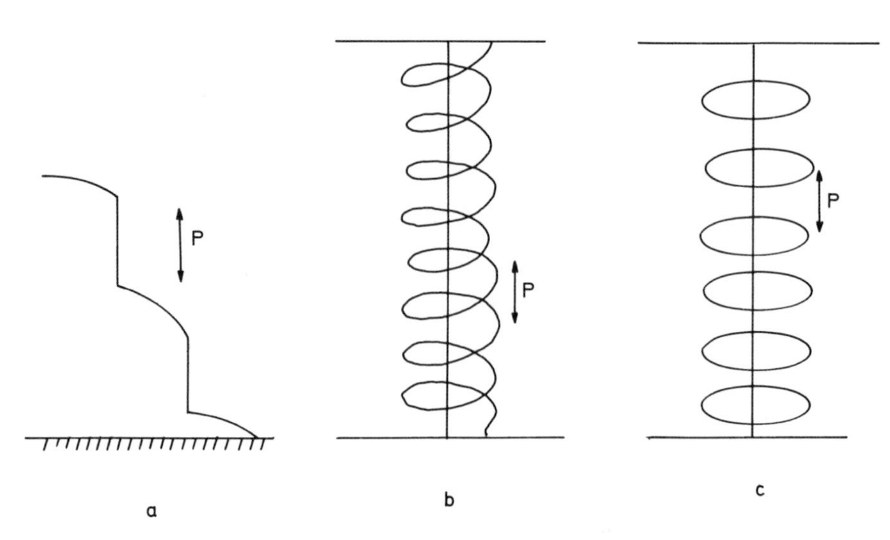

FIGURE 14 $R_\perp = \pm 2\pi$ dislocations in a cholesteric: a, edge and screw zigzags; b and c, two types of association (after J. Rault, Thèse, Orsay, 1972).

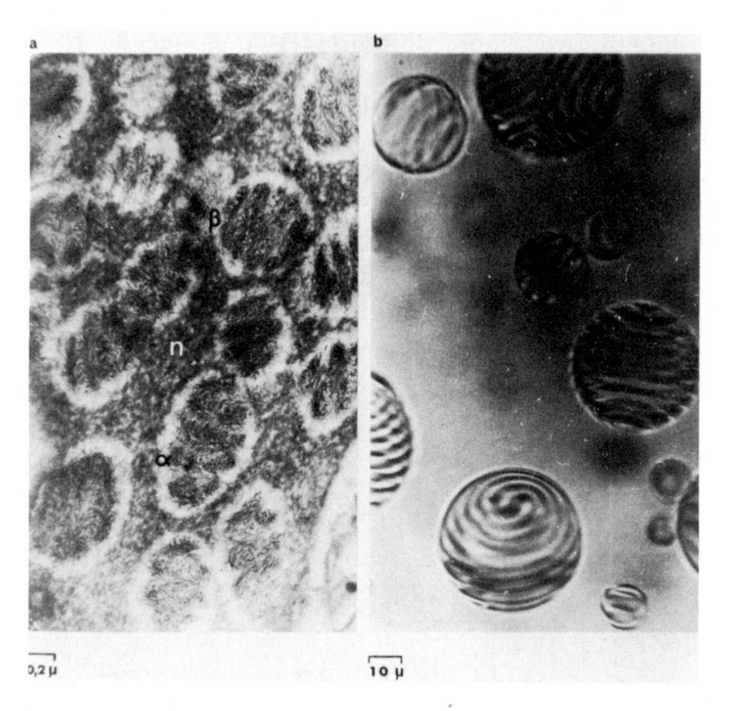

FIGURE 15 Chromosomes: a, compared to cholesteric droplets; b, after Y. Bouligand, *Tissue and Cells 4*, 189 (1972).

axis alternating with nonsplit "screw" parts parallel to it (Fig. 14, a). The tendency for such translation dislocations to go in pairs of opposite signs, one straight screw surrounded by a nearly edge helix (or by a collection of edge rings), can also be related to more complex splitting processes of the core (Fig. 14, b and c). Singular points that are also observed are very similar to those in nematics and have the same origin.*

Biological Structures and Polymers

Let me end by discussing two biological extensions. The first concerns the cholesteric arrangement of DNA molecules in some *chromosomes*. Figure 15 shows that in their normal state such chromosomes look like simple cholesteric droplets, the period in the layered structure giving the antiprojection of the half-pitch in the plane of cut and the curved contrast coming from the progressive turning of the molecules in successive planes. Figure 16 shows clearly that such chromosomes split by driving through a rotation dislocation of $(-\pi)$.

The second case is that of the *skin* of a number of crabs, where the cells have again a cholesteric structure, which, in simple cases, looks like Fig. 17 in the cut. Again the contrast comes from the fact that the cut is at an angle with the cholesteric axis. These layered structures show all the translation dislocations that one expects in this case: "screw" dislocations, parallel to the helical axis, which connect successive layers with each other and thus allow perhaps a better flow of fluids; "edge" dislocations perpendicular to the helical axis, which showed for the first time on a macroscopic scale a splitting equivalent to that of Fig. 15.

There is a last type of line that one observes in these skins, which is pictured Fig. 18. This is *not* a dislocation line but a "focal line."

Such lines are well known in all layered structures, i.e., structures made up of surfaces that must lie more or less parallel to each other. These surfaces have then common normals and to be able to fill the whole space with the structure, the envelope of these normals must reduce to a line, the "focal line" l (Fig. 19).

In structures where the layers are rigidly parallel, such as the smectic mesophase, with its parallel molecules (Fig. 10, c) or equivalent polymer structures, the focal lines are conics going by pairs. In cholesterics, where the helical pitch can change slightly by a small elastic distortion, these lines can take slightly different forms. In all cases, there is a tendency for these lines to be associated in rather complex but

*A layer of cholesteric structure, a dislocation line, and a singular point on the line bear a close analogy to a Bloch wall, a Néel line, and a Bloch point in magnetism.

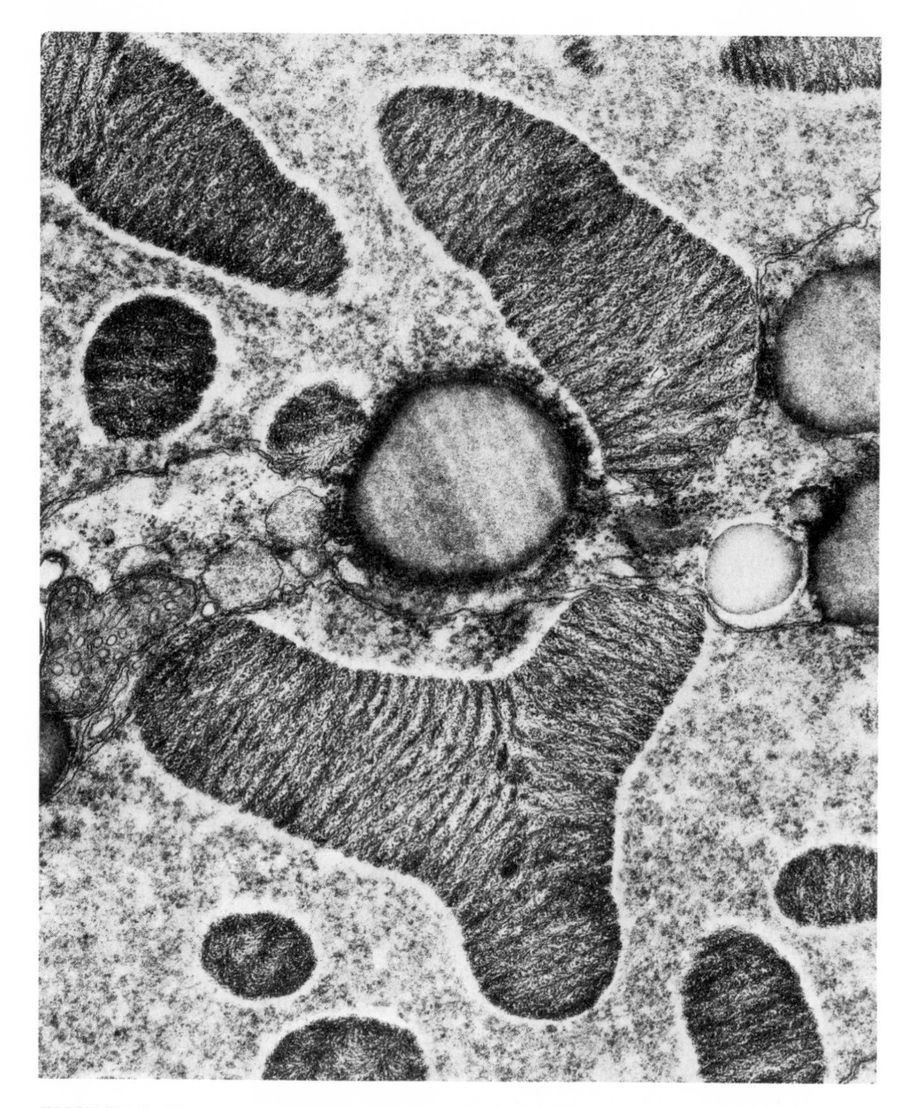

FIGURE 16 Chromosome in the process of splitting (*Plastodimin navicula* ×28,000, after M. O. Soyer, Laboratoire Arago, 66 Banyuls sur Mer, France).

regular patterns, corresponding to a series of interpenetrating protuberances. This type of arrangement explains most of the textures observed in all these layered structures.

The presence in these partly ordered structures of singular points and focal lines clearly shows an increasing complexity in the description of inner strains when one leaves the simpler field of crystals.

CONCLUSION

By the choice of examples I have taken, I have tried to show you the nature of the work in the dislocation field: most of the description and most of the effects are macroscopic in nature; but there is an essential core that regulates many properties and that involves the details of the atomic or molecular bonding. I have tried to show you various aspects that a single elementary concept can take, and the usefulness of connecting these aspects, even if they bear on very different applications. I have tried to show that this type of research is both alive and useful today.

FIGURE 17 Cholesteric structure of the skin of some crabs (cut at an angle to the cholesteric axis; after Bouligand, unpublished).

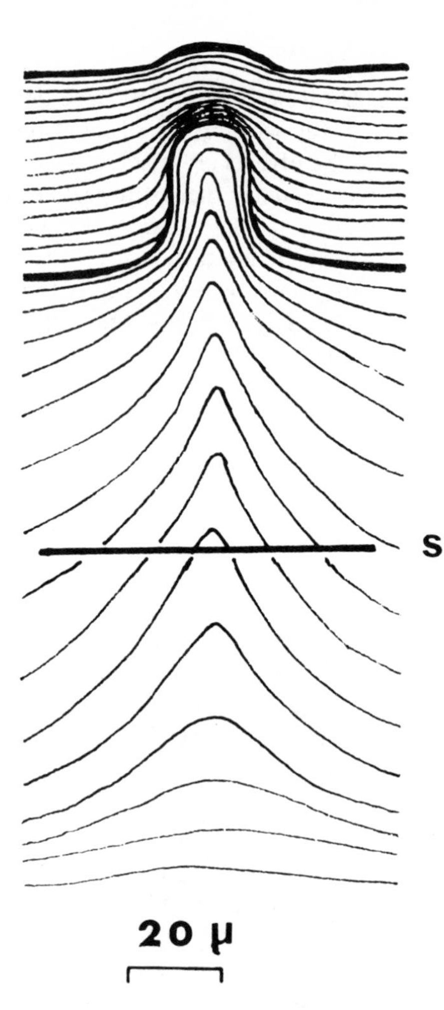

FIGURE 18 Focal line in the skin of a crab. The cholesteric layers are conical surfaces of axis vertical [schematic, after P. Drach, *Ann. Inst. Oceanogr. 19*, 103 (1939)].

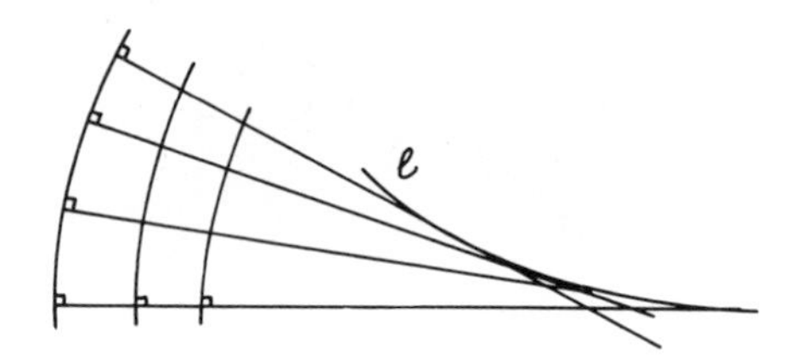

FIGURE 19 Definition of the focal line in a layered structure.

ROY W. GOULD *is on leave from the California Institute of Technology to serve as the director of the Division of Controlled Thermonuclear Research of the U.S. Atomic Energy Commission. At the California Institute of Technology, Dr. Gould is a professor of Electrical Engineering and Physics.*

8 *Plasma Physics: The Latest Infant of the Classics* / ROY W. GOULD

Plasma physics as a discipline is about 50 years old, the same age as the International Union of Pure and Applied Physics. It is probably fair to say that the work in plasma physics began seriously in about 1922 with Irving Langmuir when he began his studies and coined the term "plasma." Although gas discharges were studied in the 1800's by Faraday and many others, Langmuir was one of the first to apply systematically the principles of atomic and statistical physics to ionized gases. He also developed many of the important early ideas about collective oscillations of the plasma electrons, and these oscillations bear his name as a result. In this paper I will sketch briefly for you the history of the subject and then turn to an application of great interest, namely, controlled thermonuclear fusion for power generation. This may well turn out to be the most important application of the science of plasma physics.

In the early 1930's, there was a development in plasma physics of great significance to radio communication. This was the Appleton-Hartree theory of radio propagation in the ionosphere, the theory of refraction and absorption of radio waves in ionized gases. Later, in the 1940's, Alfvén made a big advance when he recognized the importance of the behavior of plasmas in astrophysics, particularly their interaction with magnetic fields. He developed the basic principles of magnetohydrodynamics, the dynamical interaction of plasmas with magnetic fields, and then applied them to natural phenomena in astrophysics and space science.

During the past 25 years or so, there have been many varied developments in plasma physics. At first, these developments were stimulated mainly by needs and interest in understanding space and astrophysical phenomena. The applications to these fields probably dominated the development of plasma physics as a science during the period between 1945 and 1960. The second major application, controlled thermonuclear fusion, has assumed a dominant influence in determining the developments in plasma physics as a science during the past 10 or 15 years.

Figure 1 will remind you of one of the most important of many beautiful accomplishments in space science that involves plasma phenomena—the observation of the interaction between the solar wind and the earth's magnetic field. This work culminated in the experimental demonstration of the existence of a standing bow-shock as the solar wind impinges on the earth's magnetic field. The main features of this phenomenon can be reasonably well understood in terms of the fluid theory of plasmas. The need to understand this and other phenomena in space physics and astrophysics in detail has stimulated the development of plasma physics greatly.

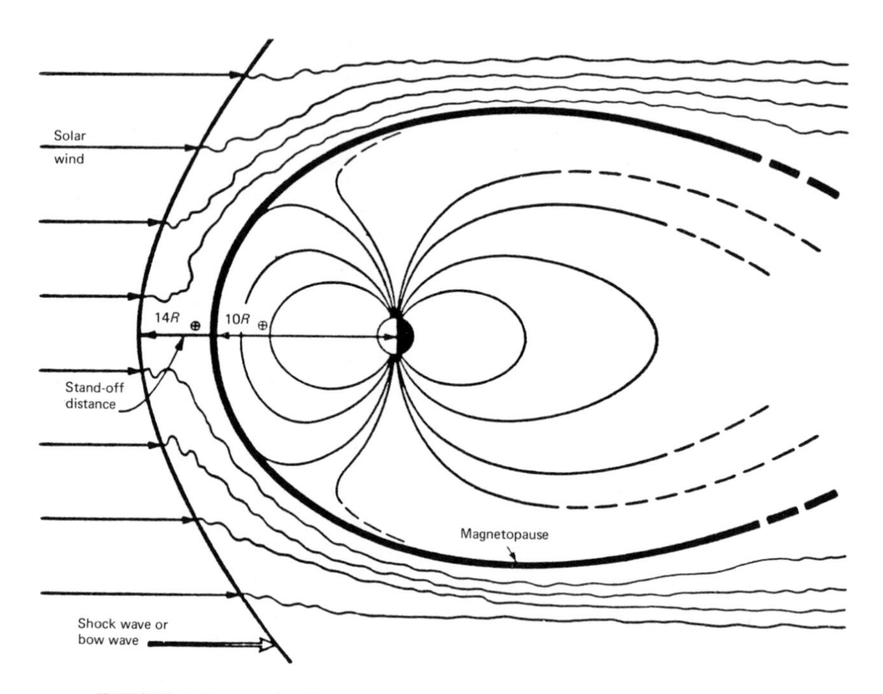

FIGURE 1 Interaction between the solar wind and the earth's magnetic field.

FLUID (MHD) EQUATIONS

$$\frac{\partial \rho_m}{\partial t} + \nabla \cdot (\rho_m \mathbf{v}) = 0$$

$$\frac{\partial}{\partial t} (\rho_m \mathbf{v}) + \nabla \cdot (\rho_m \mathbf{vv} + \psi) = \rho \mathbf{E} + \mathbf{J} \times \mathbf{B}$$

(Heat flux equation)

$$\eta \mathbf{J} = (\mathbf{E} + \mathbf{v} \times \mathbf{B}) + \ldots$$

BOLTZMANN-VLASOV EQUATION

$$\frac{\partial f}{\partial t} + u \cdot \nabla f + \frac{q}{m} (\mathbf{E} + \mathbf{u} \times \mathbf{B}) = \left(\frac{\partial f}{\partial t}\right)_{\text{coll}}$$

$$\mathbf{J} = \sum_{\text{species}} q \int \mathbf{u}\, f\, d\mathbf{u}$$

FIGURE 2 The two principal theoretical formulations of the dynamical equations for plasmas.

Before turning to application to thermonuclear fusion, some remarks about the foundations of plasma physics are appropriate. There is a curious dichotomy that pervades all of plasma physics, the *fluid* versus the *individual particle* point of view. Figure 2 outlines what might be called the two principal theoretical formulations of the dynamical equations for plasmas. The fluid point of view treats the behavior of plasma in terms of the usual macroscopic variables used in ordinary fluids—the mass density, the mean velocity, the stress tensor—together with the all-important electromagnetic field. These equations, together with others, such as the heat-flux equation to account for the flow of heat and a generalization of Ohm's law, which relates the current induced in the plasma to the electromagnetic field, provide the simplest formulation of the electrodynamics of a conducting fluid. One can understand many important phenomena from the fluid formulation. In fact, it is remarkable just how much one can understand, even though the fluid equations cannot always be justified rigorously, particularly in view of the long collision mean free paths that generally occur both in astrophysical plasmas and in the laboratory plasmas being studied for controlled thermonuclear fusion. Nevertheless, many important phenomena are reasonably well described in these terms.

There are several important distinctions between ordinary fluids and plasmas. I have already mentioned the long mean free path. Among other things, this means that there is no mechanism that rapidly establishes thermodynamic equilibrium. Thus the concept of local thermodynamic equilibrium, so useful in the treatment of the dynamical behavior of ordinary fluids, is no longer applicable. Second, there is the importance of the electromagnetic forces on the motion of the plasma, as well as the generation of electromagnetic fields by the currents and charges that are induced in the plasma. A third important distinction is that many, if not most, plasmas are not in thermodynamic equilibrium, because of the manner in which they are produced and because collisions are not frequent enough to cause the plasma to reach an equilibrium state. As a consequence of the nonequilibrium state there are various sources of free energy that can drive instabilities: spatial gradients in composition, temperature, etc.; anisotropy of the velocity distribution; and others. These give rise to a large class of new phenomena that do not normally occur in ordinary fluids.

During the past 20 years, there has been very extensive work on the foundations of the kinetic equations for a plasma, i.e., providing a rigorous justification for the hierarchy of kinetic equations of which the Boltzmann-Vlasov equation in Fig. 2 is the lowest (collisionless) order. This formulation of plasma dynamics retains the main features of the multivelocity character of individual particles of the plasma, and it is appropriate to the situation in which collisions are infrequent. The coupled Boltzmann-Vlasov and Maxwell equations have turned out to be very rich in physical phenomena, particularly when the consequences of the nonequilibrium initial states and of the strong magnetic fields are included. A considerable effort has been expended in the study of small perturbations from various initial states, and many new dynamical phenomena have been discovered. The linear theory of waves, the stability of small perturbations in plasmas, the normal modes of oscillation, all have been intensively studied in idealized situations, but this subject is by no means fully explored. Contact between theory and experiment is somewhat limited but improving rapidly. Although it is not possible to dwell on all the facets of the instabilities that have been found, one important one is that instabilities in magnetically confined plasmas are sometimes responsible for the anomalous transport of plasma. There now seems to be a reasonably good qualitative, if not quantitative, understanding of the many instabilities that can arise in plasmas and of ways to eliminate these or to minimize their effect on transport, through the shaping of the magnetic-field configuration.

Nonlinear phenomena, always a difficult subject in physics, is now one of the principal areas in plasma physics in which there is extensive theoretical effort. One can expect interesting developments in this area in the years ahead. Nonlinear behavior of plasmas is probably even more important than the nonlinear behavior of ordinary fluids.

An important manifestation of the individual particle behavior in collisionless plasmas is the notion of collisionless damping of small perturbations. It was predicted in 1949 by Landau that the plasma oscillations, or Langmuir oscillations, in the electron gas of a plasma should be damped, despite the fact that there were no collisions. This is a basic idea, and the phenomenon of collisionless damping pervades much of plasma theory and is crucial to the understanding of a wide variety of important phenomena. It is interesting that collisionless or Landau damping was not verified experimentally until 1964, or 15 years after it was predicted. Figure 3 illustrates the important experimental results of Malmberg and Wharton. It shows the spatial decay of a wave whose frequency is several hundred megahertz. Both the phase of the wave, through an interferometer trace, and the exponential decay of its amplitude are shown. Most of the main features of collisionless damping were verified in this experiment, and rather good agreement with Landau's theory was obtained.

An important feature of the collisionless damping of perturbations in plasmas is that it is, in principle, reversible. A few years after the experimental demonstration of damping, it was shown theoretically that

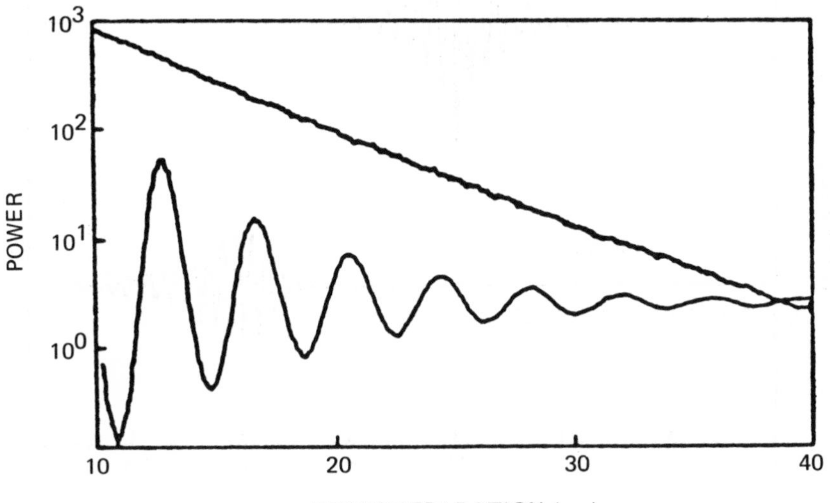

FIGURE 3 Spatial decay of a wave whose frequency is several hundred megahertz. Upper curve shows the exponential decay of its amplitude. Lower curve plots the phase of the wave through an interferometer trace.

the application of a second perturbation could be used to recover the information contained within the electron velocity distribution from the first perturbation, even though the macroscopic variables associated with the initial perturbation, such as the charge density and the electric field, have damped away. There is an illustration of reversibility—the plasma wave echo (see Fig. 4). This figure again shows the decay in both directions away from the source of a launched wave at 120 MHz together with the application of a second perturbation at a frequency

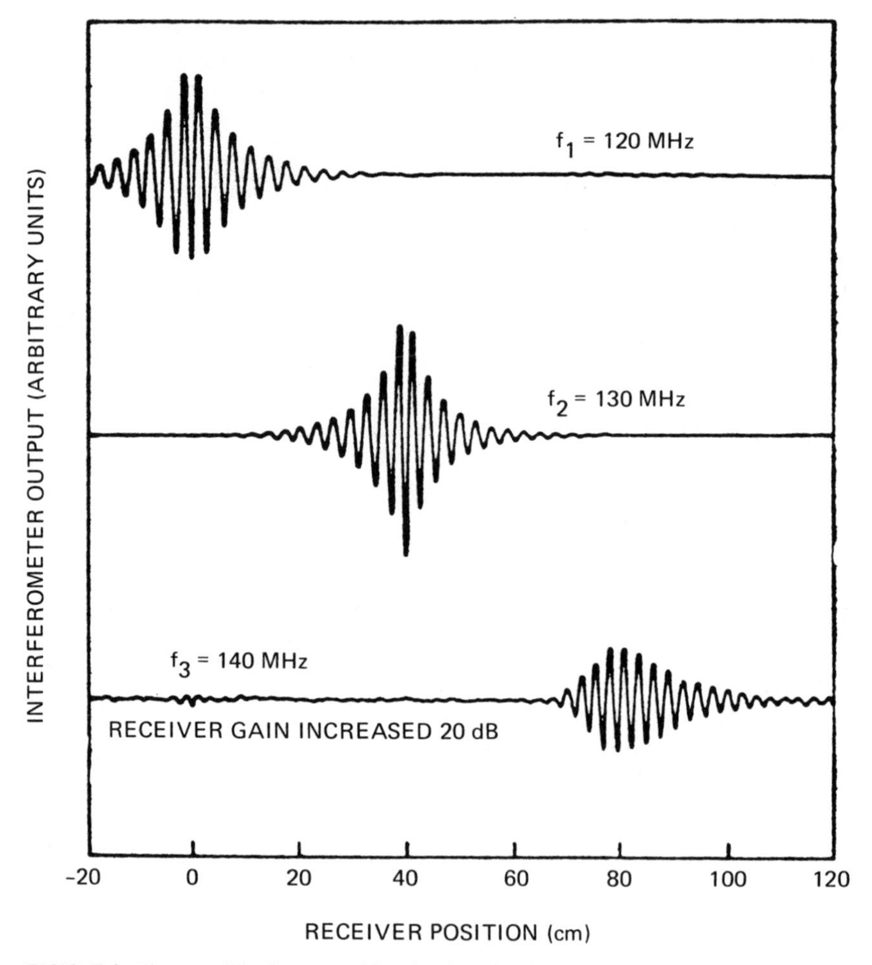

FIGURE 4 The reversible character of Landau damping. The transmitter probes are at 0 and 40 cm. Upper curve, receiver tuned to f_1; middle curve, receiver tuned to f_2; lower curve, receiver tuned to f_2 and gain increased 20 dB.

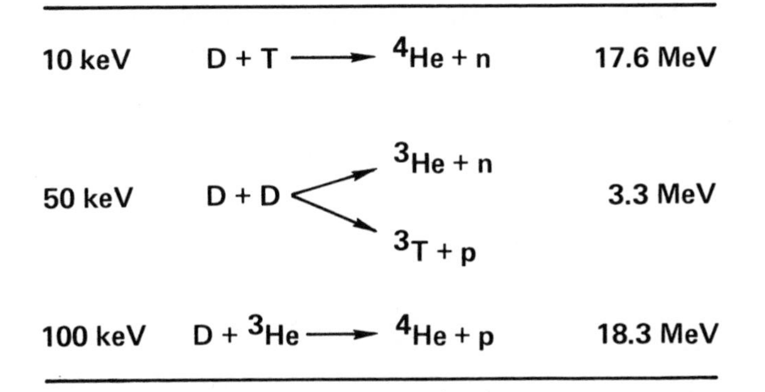

FIGURE 5 The principal fusion reactions.

of 130 Hz, which also decays away from the source. Through the nonlinear interaction with the second wave, the information from the first wave is regenerated at a spatially separated point and with a frequency of $2f_2 - f_1$. This conclusively demonstrated the reversible character of Landau damping, a concept that had been believed by theorists for many years and incorporated into many theories but only demonstrated in the laboratory about five years ago.

The remainder of this paper is devoted to what could be the most important application of plasma physics. First, I will describe the concepts involved in controlled thermonuclear fusion, and then I will indicate where we stand and where we are headed. Of course, the motivation for this work is the availability of a large amount of energy from the deuterium in the oceans. By "burning" this deuterium using nuclear fusion reactions, it would be possible to generate a great deal of energy. Furthermore, this form of energy generation could be environmentally attractive because of the reduced radioactivity and increased efficiency. The principal fusion reactions of interest are shown in Fig. 5. The first-generation fusion reactors would probably employ the deuterium–tritium (D–T) reaction because of its lower reaction energy. Since tritium is not a naturally occurring substance, the use of such a reaction would also require the production, or breeding, of tritium from lithium. Lithium would in effect be consumed as well as deuterium. Eventually, it is hoped that the D–D reaction could be used to consume only the deuterium that is available in immense quantities and at very low cost from the ocean. The energies obtained are tens of millions of electron volts per fusion reaction, and the energy required to make these reactions take place are only tens of thousands of electron volts. Thus there is a very handsome energy gain to be obtained.

TABLE 1 Requirements for Achieving Useful Power

1. Heat fusion fuel to nuclear-reaction temperatures.
2. Contain the reacting mixture away from material and walls and free from impurities long enough for a substantial fraction to react.
3. Extract the energy released and convert it to a useful form.

However, the requirements for achieving useful energy from these reactions, summarized in Table 1, are severe.

The fact that the temperature of the fuel must be greater than a hundred million degrees means that it is in the plasma state, and this makes possible its confinement with a magnetic field. Figure 6 illustrates what is perhaps the simplest magnetic confinement concept. Since charged particles are confined to move in small helical orbits *along* magnetic-field lines, generally speaking, closing the field lines on themselves forms a toroidal field configuration and provides a way of confining the plasma. The toroidal field is produced by the current-carrying windings in the form of a toroidal solenoid. The field lines differ slightly from the usual toroidal magnetic fields in that a line does not close exactly on itself after one traverse around the torus. A rotational transform, which causes a line to make its way systematically around the minor cross section of the torus, must also be provided. The line may close on itself after four or five times around. This magnetic configuration is the basis for Tokamak and Stellarator devices. Toroidal plasma confinement is the subject of intensive research throughout the

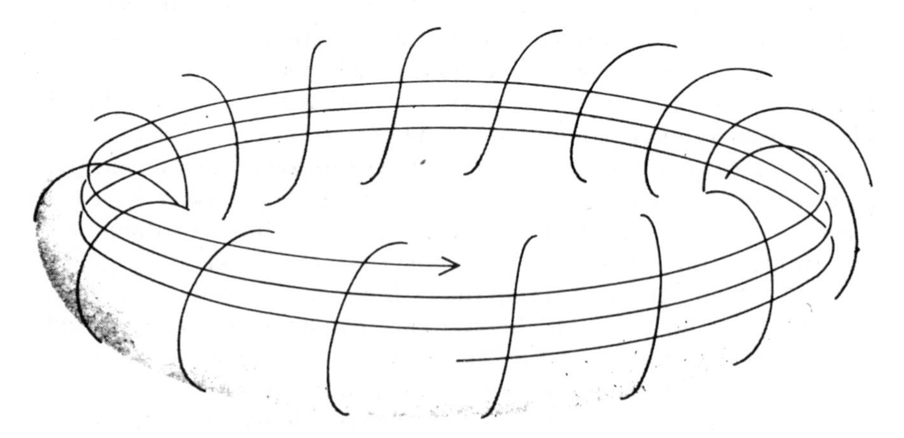

FIGURE 6 The simplest magnetic confinement concept.

world. Because a comprehensive review of other methods that are being studied for the magnetic confinement of plasmas will not be included in this paper, I will necessarily omit some important work on the use of theta pinches and magnetic mirrors to confine plasmas magnetically. Instead I want to give you some flavor of what is occurring, some highlights of the past several years, and some idea of what you may expect in the next few years.

Charged particles cannot be contained indefinitely, even in a closed-line or toroidal magnetic configuration because of collisions between charged particles, as rare as they may be at these elevated temperatures. Figure 7 compares the rate with which particles are scattered by Coulomb collisions with the rate at which they undergo fusion reactions. Scattering is considerably more likely than fusion, and this means that the particles must be confined for many collision times. This figure also

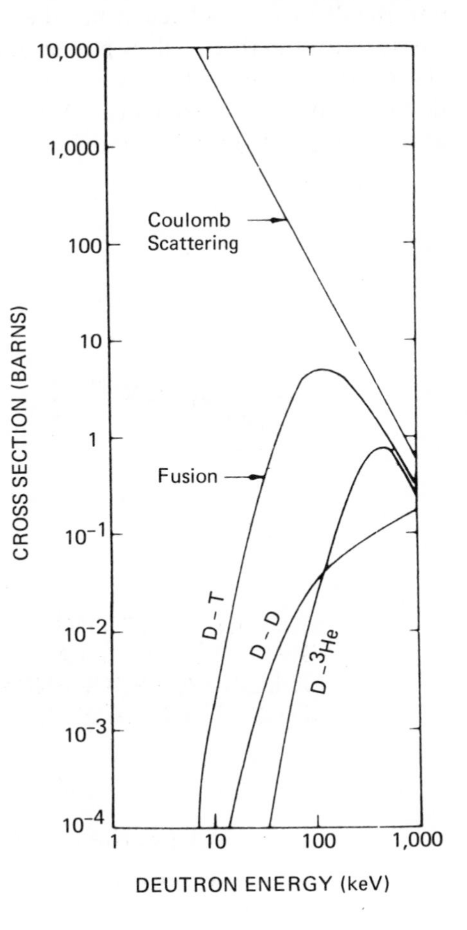

FIGURE 7 Comparison of the rate with which particles are scattered by Coulomb collisions with the rate at which they undergo fusion reactions.

shows that the reaction cross section for D–T is significantly higher than for D–D. For this reason, the D–T reaction appears most attractive for early reactors.

Coulomb collisions allow the particles to escape from the magnetic field in the manner illustrated in Fig. 8. A particle that undergoes a collision executes a new helical orbit whose center is displaced from the original. Through each collision the center particle's orbit can be displaced in a random direction across the field line by as much as the radius of gyration itself. A series of such collisions, which take place with a frequency v, each with displacement about equal to the cyclotron radius, can give rise to a cross-field *diffusion* coefficient whose magnitude is of the order of the collision frequency times the square of the step size. Thus there can be a diffusion loss of particles across the magnetic-field line. The frequency of Coulomb collisions that enters into the diffusion coefficient has the important property that it decreases with the three halves power of the temperature so that the plasma loss rate should decrease with increasing temperature if diffusion is the principal mechanism. For a straight cylinder of radius a, the time for a typical particle to diffuse out of the cylinder is given by

$$\tau \sim a^2 B^2 T^{1/2}/n.$$

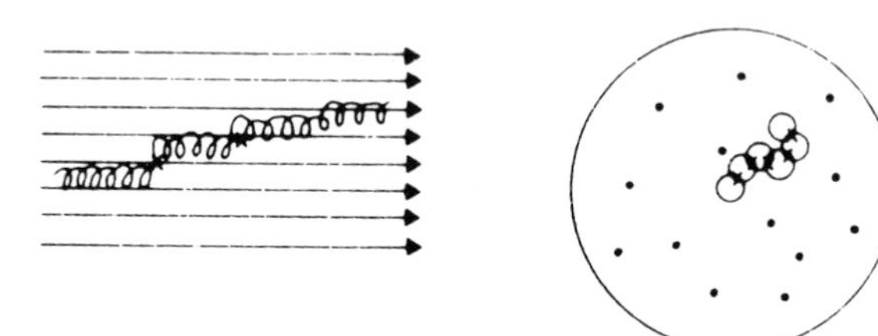

CONFINEMENT TIME	$\tau \sim a^2/D$
DIFFUSION CONSTANT	$D \sim \nu r_o^2$
COLLISION FREQUENCY	$\nu \;\alpha\; n/T^{3/2}$
CYCLOTRON RADIUS	$r_o \sim mkT/e^2B^2$

$$\tau \;\alpha\; a^2B^2T^{1/2}/n$$

(CLASSICAL CONFINEMENT)

FIGURE 8 Cross-field collisional diffusion.

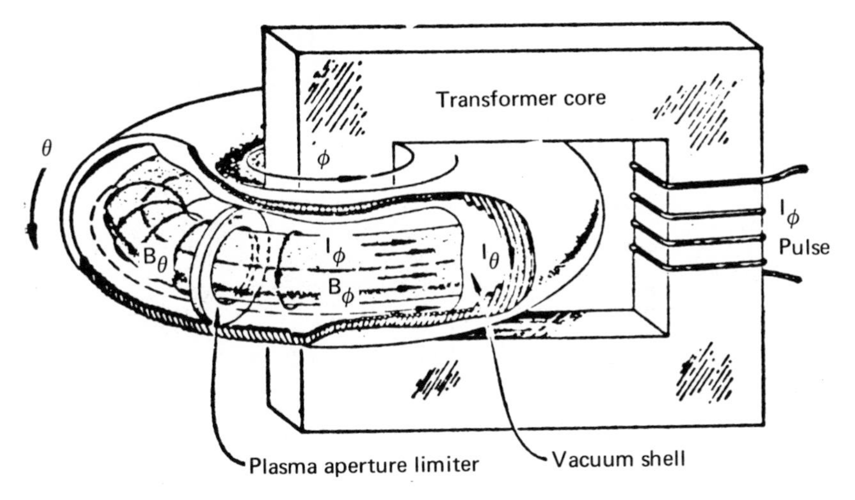

FIGURE 9 The Tokamak concept.

This represents an upper limit on confinement time as a consequence of Coulomb collisions. One of the surprising developments of the last few years is the discovery that the effect of toroidal curvature on this simple diffusion formula for a straight cylinder is substantial.

The most important applications of toroidal confinement are to the Tokamak devices and to the Stellarator devices. Tokamaks are toroidal plasma-confinement configurations, which were pioneered by Dr. Artsimovitch and his co-workers in the Soviet Union. That line of investigation was taken up in the United States nearly three years ago and now in Europe as well, as a result of achievements in the Soviet Union. The Tokamak concept is illustrated in Fig. 9. In addition to the strong toroidal field produced by the toroidal field winding, a poloidal field is produced by inducing a plasma current to flow around the torus. The poloidal field is proportional to the plasma current. An iron-core transformer in which the plasma acts as a one-turn secondary is used to induce that current. The role of the induced plasma current is twofold. It gives the rotational transform, or twist, of the magnetic field, and it heats the plasma by collisional dissipation of that current so as to increase the electron temperature. The energy is given mainly to the electrons, because they are affected by the induced electric field. Some of their energy is transferred to the ions by Coulomb collisions.

This flow of energy is depicted in Fig. 10. Energy is added to electrons by the induced electric field; this energy is lost by radiation, by thermal transport to the wall, by particle transport to the wall, and by collisional transfer of energy to the ions. The ions, in turn, lose their

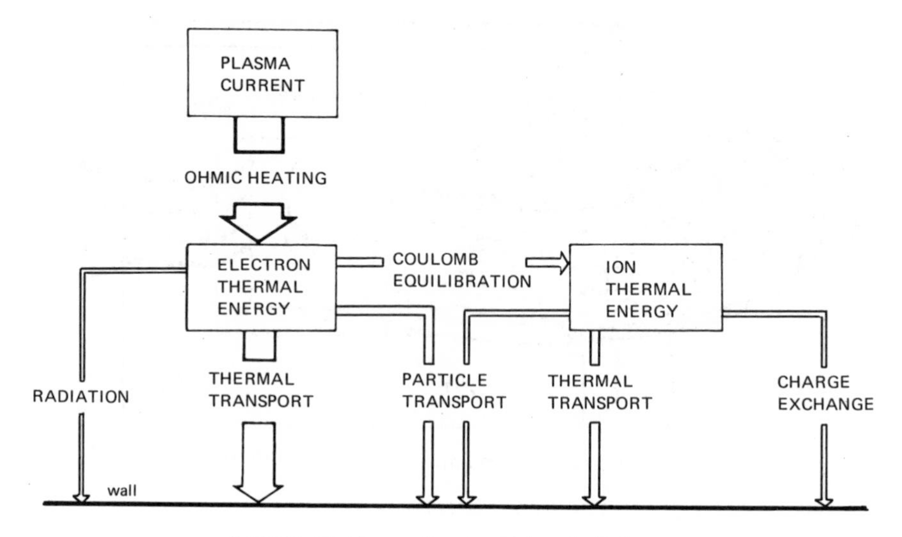

FIGURE 10 Energy flow in a Tokamak discharge.

energy by particle transport to the wall, by thermal transport to the wall, and by charge exchange with neutral atoms. This method of heating plasma fails at higher temperatures, because the collision frequency between charged particles decreases the conductivity, and hence the dissipation of energy decrease. In essence, the hotter the plasma the higher its conductivity and the less energy of the toroidal current is dissipated in heating the plasma. It is recognized that this method of heating can probably only bring such a toroidal plasma to within a factor of 2 to 4 of the temperatures required for reactors, and, consequently, methods of auxiliary heating are being explored.

Figure 11 shows the Princeton Tokamak. The toroidal field windings and the iron core of the transformer can be seen. Not seen in this figure is the flywheel motor–generator set, which can deliver 200 MW for 2 sec, used for producing the toroidal field and also the plasma current. Of course, an immense amount of auxiliary diagnostic apparatus must be used to determine what is actually happening within the plasma, which has a minor radius of about 15 cm and a major radius of a little over a meter. Incidentally, the vacuum, magnet, heating, and other technologies used in the production of fusion plasmas have been developed more or less simultaneously with the development of the basic science of plasma physics. The Princeton facility was originally built as a Stellarator by Lyman Spitzer about 12 years ago, and in many respects the Stellarator concept is similar to the Tokamak. The main difference between a Stellarator and a Tokamak is that the twist or rotational transform of the magnetic field is produced by helical windings that are located inside the main toroidal field coils (but outside of the plasma). As a consequence, considerable space that would otherwise be available for the plasma is required. When the Princeton Stellarator was converted to a Tokamak about three years ago, the plasma parameters improved dramatically, comparable with those that had been obtained by Artsimovitch and his co-workers in the Soviet Union. The elimination of the helical windings within the toroidal field coil made it possible to enlarge the plasma radius from about 5 cm to about 15 cm. According to the diffusion formula, the plasma is confined for a longer time, and, consequently, the loss of energy decreases. For the same energy input, the temperature of the plasma also rises, giving a further improvement of confinement. It is generally believed that the increased performance arises out of the fact that the plasma is simply larger, and that is an important point when contemplating reactors, which must be still larger.

Now I would like to describe some specific results that have been obtained on the Princeton Tokamak during the past year. Figure 12

FIGURE 11 Photograph of the Princeton Tokamak.

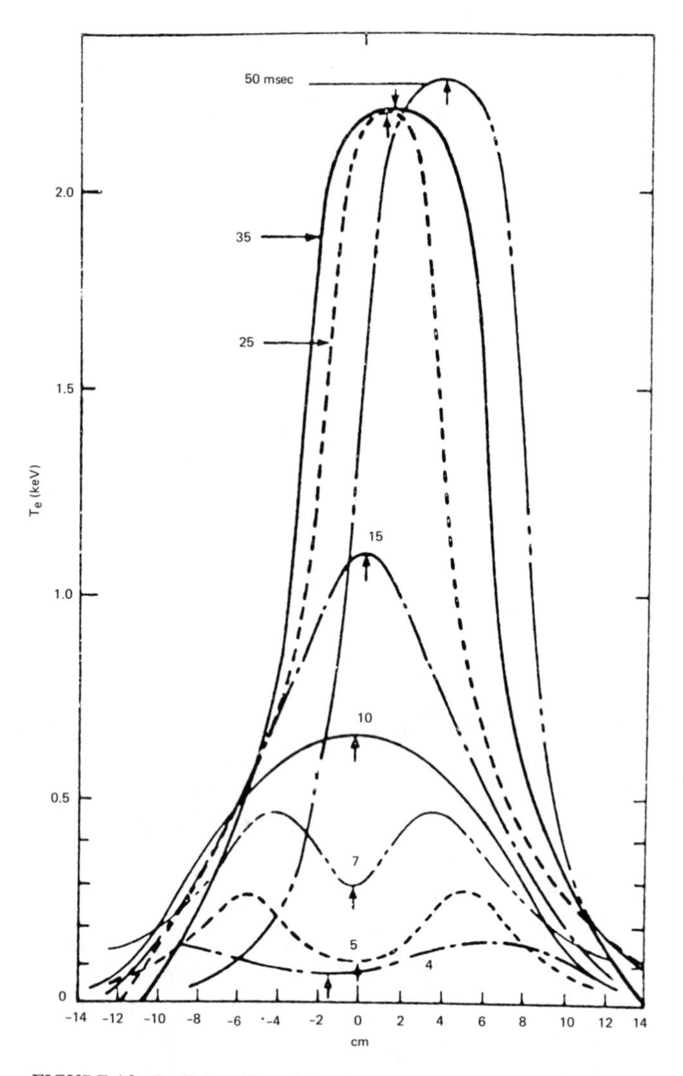

FIGURE 12 Radial profile of the electron temperature in the discharge of the Princeton Tokamak.

shows a radial profile of electron temperature in the discharge. These profiles are based on data obtained from Thompson scattering of laser light by the free electrons of the plasma. The line width of the scattered radiation gives directly the electron temperature or, more precisely, one component of the velocity distribution. The time evolution in a pulsed discharge is shown. Radial temperature profiles are shown at 4, 5, 7, 10, and 30 msec. The discharge is created originally with a rather low temperature. The temperature rises to above 2 keV, and, while it is not shown, the ion temperature rises also, although it lags behind the electron temperature and only reaches about 600 eV. This is nevertheless substantial. There is a tendency for the plasma to be hotter initially on the outside in an annular shell, rather than in the center. This is thought to be a manifestation of the classical skin effect exhibited by good conductors, as the plasma is. As a current is initiated in a good conductor, it first flows on the surface of the conductor; and only after a period of time, governed by the resistivity of the conductor, does the current penetrate into the central core.

The penetration of the current is reflected in the temperature profiles of the electrons, which are heated most by the current where it is highest. The hollow temperature profile tends to fill in with the passage of time, and this is accompanied by an overall increase in the temperature of the electrons. The actual rate at which the current appears to penetrate the plasma in this experiment seems to be three to four times faster than expected theoretically, and this is yet unexplained.

Another empirical relationship, which has been well documented by data from both the Soviet and Princeton Tokamaks, is shown in Fig. 13.

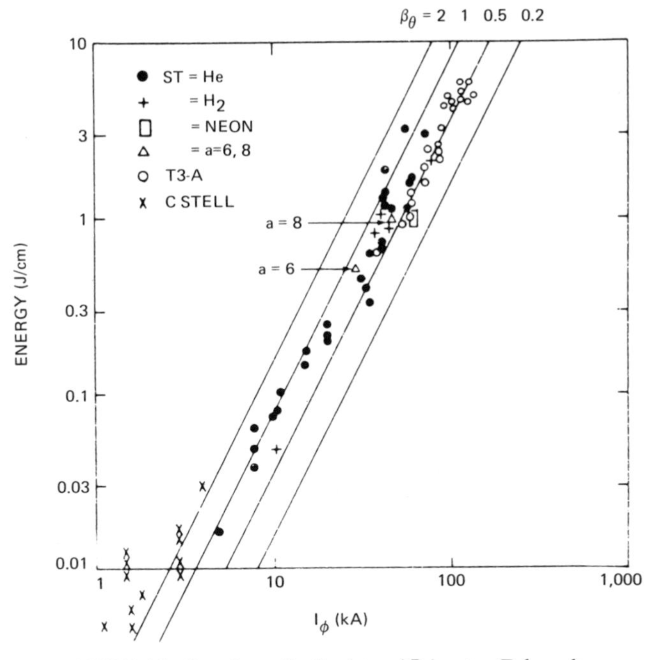

FIGURE 13 Data from the Soviet and Princeton Tokamaks.

This is a relationship between the thermal energy per unit length of the plasma column and the induced toroidal plasma current. The energy per unit length of the plasma increases as the square of the induced current over three orders of magnitude in the current. This relationship is such that the plasma energy content $\frac{3}{2} Nk\ (T_e + T_i)$ (N is the number density of electrons per unit length and T_e and T_i are the electron and ion temperatures, respectively) is approximately equal to the magnetic energy density associated with the *poloidal* magnetic field. Alternatively, one can say that the pressure of the hot plasma is balanced by the pressure of that part of the magnetic field that is produced by the induced current. The experimental data are consistent with this kind of relationship over a large range of plasma parameters. This gives rise to the notion that in fact the plasma is confined not by the toroidal field but by the poloidal field, which is weaker by a factor of 10 to 30 than the former. The toroidal field plays an important role mainly in stabilizing the discharge, making it relatively free from instabilities. Insofar as confinement is concerned, however, it appears as though the poloidal field is the important one.

Another empirical result, shown in Fig. 14, is that the confinement time is found to *increase* with temperature. This figure shows the confinement time in milliseconds as a function of electron temperature.

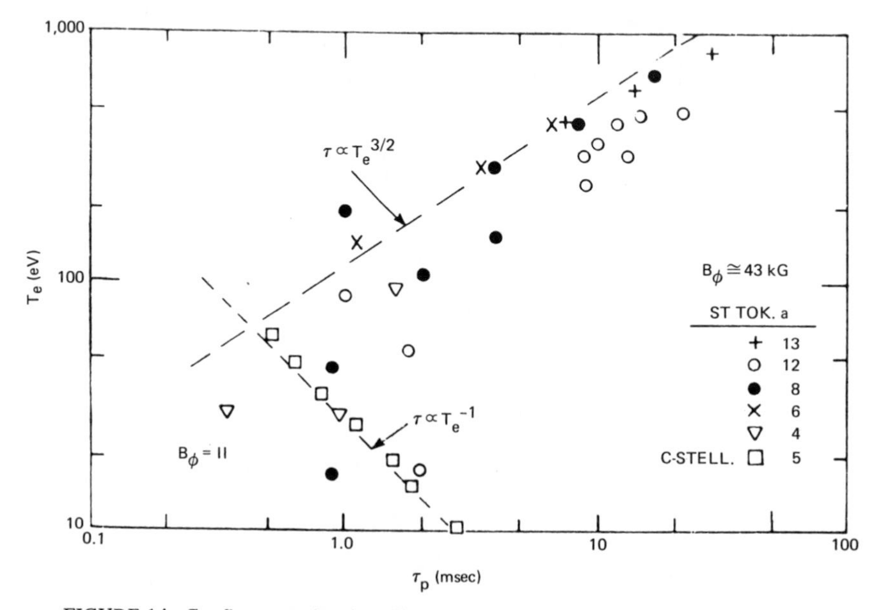

FIGURE 14 Confinement time in milliseconds as a function of electron temperature.

These data were obtained by varying the size of the plasma column and also by examining different times in the discharge, so that temperatures are different. One of the very encouraging results is that above about 100 eV, which Tokamaks readily achieve, the confinement time appears to increase as the three-halves power of the electron temperature, whereas in some of the earlier Stellarator experiments, and many other devices, confinement times were found to *decrease* with increasing temperature. Although the precise reasons for this different behavior are not yet understood in detail, there are some clues, and it is important that the expected and hoped for increase in confinement times with increasing temperature is now being observed. Incidentally, one might ask why confinement time is observed to increase as temperature to the three-halves power, whereas the simple arguments given earlier say it should increase only as the one-half power of temperature. When allowance for the first relationship, that the product of density and temperature nT is constant, is made in the earlier confinement time relationship, an extra factor of T is introduced.

The question of scaling to devices with larger size and higher temperatures is indeed a crucial one for this particular class of closed-line systems (Tokamaks and Stellarators). There is now a gap of about two orders of magnitude between the confinement times that have been achieved on the experimental devices to date and that which is required for a plasma that could produce as much thermonuclear energy from fusion reactions as is invested in creating and sustaining the plasma. Thus it is important to verify whether, in fact, the confinement time continues to increase with the parameters as indicated. There is only limited evidence on that question so far, but even that limited evidence is encouraging.

Favorable results to date have led to the initiation of still other experiments. Figure 15 gives a summary of the important features of the various Tokamaks that are now operating or are under construction. The one that I have been discussing is the Princeton ST. The TM3, T4, and T6 are Soviet Tokamaks. The minor cross section of the torus as well as the magnetic-field strength and the maximum plasma current for stability are shown. The Ormak at Oak Ridge has a minor plasma diameter of 23 cm and a magnetic field of 25 kG. There are additional Tokamaks under construction in Europe, and a large Tokamak, similar in size to the Princeton Large Torus (PLT), in the Soviet Union. This series of devices should fill out some of these important ideas: the details of scaling to higher temperatures and higher magnetic fields. These should be rather well elucidated during the next five years. At the top of this figure is shown a new large device now under construc-

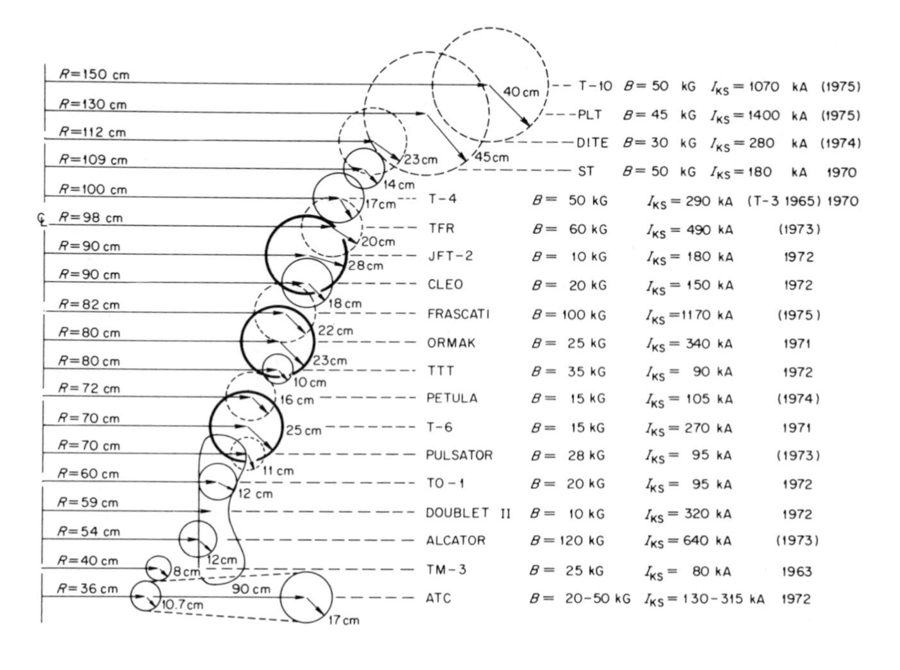

OPERATING BEING DESIGNED OR CONSTRUCTED

Comparison of Some Tokamak Experiments
(I_{KS} Calculated on the Basis of $q = 2.5$).

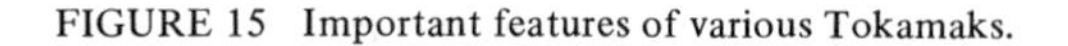

FIGURE 15 Important features of various Tokamaks.

tion at Princeton. It will have a plasma diameter three times larger than the existing Princeton Tokamak and a minor diameter of nearly 1 m. It should be capable of a plasma current of about 1.5 MA. The PLT facility and a comparable facility, the TIO, under construction in the Soviet Union will make possible crucial experiments insofar as the scaling to larger sizes and higher temperatures is concerned. If these experiments go as hoped, based on the limited experimental evidence now available, then these devices should come close to producing plasma conditions that would be capable of break-even thermonuclear energy equal to energy invested in the plasma.

None of these devices can actually reach the temperatures needed for break-even by ohmic heating alone. For example, it is predicted that the PLT may reach an electron temperature of 3 or 4 kV after heating for perhaps ½ sec. Some supplementary means of heating the plasma further to the 10 or 20 kV required will eventually be necessary. Thus there is a great deal of necessary study of supplementary heating methods, such as the injection of neutral atomic beams with great power and heating of the plasma by the absorption of radio-frequency energy in the form of waves at frequencies that resonate with the natural frequencies in the plasma (the ion cyclotron frequency or the hybrid frequency). These supplementary heating methods will be explored during the next few years.

One of the interesting scientific developments of the last several years has been in transport processes, such as diffusion and thermal conductivity in toroidal systems. It was discovered a few years ago by Russian scientists Galeev and Sagdeev that the curvature of toroidal confinement geometry has a major influence on plasma confinement. They found that the curvature of the toroidal field, and the variation of the field strength that goes with toroidal geometry, can cause a significant increase in the diffusion rates of particles and energy across the magnetic field. In effect, the previous formula for τ is modified rather significantly. As a very rough approximation, the magnetic-field strength that appears in the formula is not the strong toroidal magnetic field but rather the weaker poloidal field. Therefore, the confinement times to be expected are less than one would have expected by several orders of magnitude. This is, in a sense, another manifestation of nearly collisionless or long mean-free-path aspect of the particle motion. Figure 16 illustrates how the diffusion of particles across the magnetic field can be enhanced by the curvature effect. This figure shows a *projection* of magnetic-field lines and several orbits in a cross section of the torus at constant azimuth. As a consequence of the rotational transform of the magnetic field, each field line slowly progresses around the minor axis as with increasing azimuth, so that the projection of the field line in a plane of constant azimuth is a circle. As a first approximation, individual particles follow field lines, but two important additional effects arise because the field in a torus is stronger near the major axis. First, the spatial gradient in B causes the particles to drift away from the field line that they are on, and in this illustration that drift is *upward*. After encircling the minor axis, this drift returns the particles to their original (projected) position, shown in this figure as passing particle orbits. This is precisely the purpose of the rotational transform. Some particles as they approach closer to major axis of the torus find the small increase in magnetic field to be too large and are reflected by the magnetic mirror effect, producing the *banana* orbits (so-called because of their shape when projected to constant azimuth). These particles are referred to as "trapped particles."

The existence of these trapped particles makes possible an enhanced diffusion of particles across the magnetic field in the following way. It

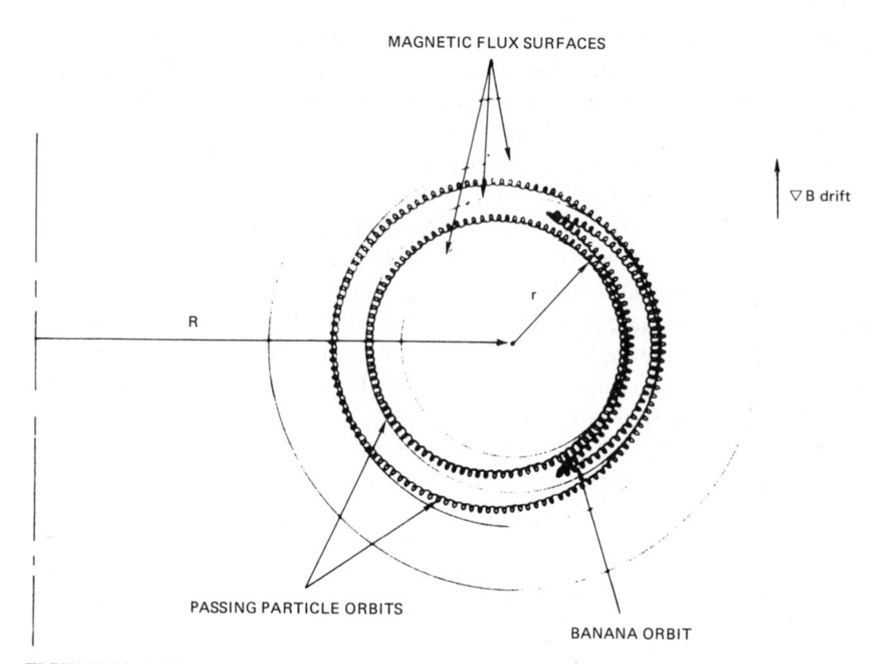

FIGURE 16 Diffusion of particles across the magnetic field enhanced by the curvature of toroidal confinement.

is possible for a particle that is initially a passing particle near the mirror axis to make a collision and causes it to take up a banana orbit. It executes a banana orbit for a while, and then, at some point further from the minor axis, it makes another collision, which puts it back into one of the passing orbits. In effect, this particle has made a step away from the magnetic axis of the system, which is a step much larger than the small cyclotron radius of the particle in the magnetic field and characterizes the step size in the usual diffusion process. In effect, the existence of these banana orbits allows the particle to take large steps across the magnetic field instead of small steps, thus greatly increasing the diffusion rate. Calculations show that the poloidal magnetic field strength is the relevant field in this banana diffusion process rather than the toroidal field. Some other factors also enter, but this is the major change. This was an important discovery, and in part it accounts for the fact that some of the earlier devices did not live up to our expectations, because our expectations were too high. Now a number of experiments are in general agreement with these ideas, but many details remain to be sorted out rather carefully and accurately. Since the larger banana diffusion rates scale with temperature in accord with the earlier classical formula, it is referred to as neoclassical.

There has been, in the last two years, a very nice verification of neoclassical diffusion by Ohkawa and his co-workers at Gulf General Atomic. They have measured carefully the decay rate of a low-temperature, low-density plasma in toroidal geometry, as seen in Fig. 17. In this experiment, the poloidal magnetic field is produced, not by induced plasma currents as in the Tokamak but by currents in four supported current-carrying rings. This produces an octupole magnetic field, which is somewhat more complicated than that of the Tokamak, so that an extension of the neoclassical theory of diffusion in toroidal magnetic fields was required.

Figure 18 shows the effective reciprocal lifetime of the plasma that was measured as a function of the characteristic size of the device divided by the collision mean free path. As the plasma decays, and becomes less dense, collisions become less frequent and the mean free path becomes longer and longer. Thus this parameter decreases during the decay of the plasma, and one can measure what, for practical purposes, amounts to the diffusion coefficient as a function of mean free path. The curve agrees rather well, quantitatively, with the predictions

FIGURE 17 A poloidal magnetic field is produced by currents in four supported current-carrying rings.

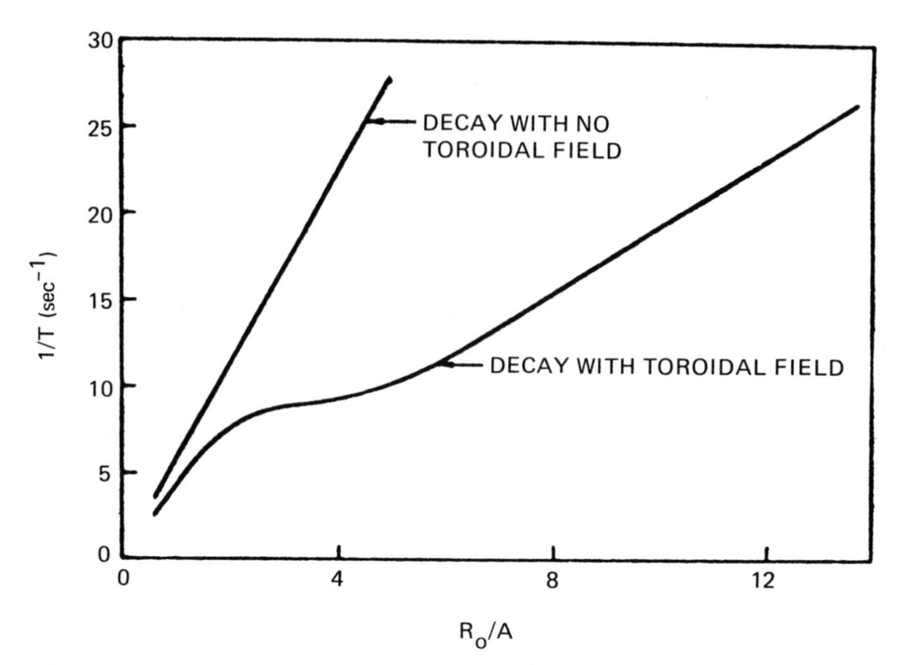

FIGURE 18 Effective reciprocal lifetimes of a plasma measured as a function of the characteristic size of the device divided by the collision mean free path. Lower curve, behavior in the combined toroidal and poloidal fields; upper curve, results with poloidal field only.

of the neoclassical theory. Insofar as this behavior continues to hold at much higher temperatures (this experiment was done in a 1-eV plasma), it should be pointed out that reactors will operate in this long mean-free-path regime, and in this experiment the long mean-free-path feature is obtained by operating at densities of 10^9 particles per cubic centimeter.

Now I would like to describe a more speculative concept for thermonuclear fusion, which has evolved and received a great deal of attention in the past several years, although it was proposed somewhat earlier. This is the idea of achieving controlled fusion through the use of lasers. It is a speculative idea in the sense that it is essentially a prediction based on computer calculations, and important substantive experiments are yet to be performed. The idea is based on the fact that laser energy can be focused to very small dimensions and produced in short pulses, thus achieving enormous power density. Imagine, as in Fig. 19, a small pellet of fuel that is irradiated by a short pulse of highly focused laser energy whose role is to heat the target to ignition temperature so that thermonuclear reactions will take place. The target must be heated to thermonuclear energies and the reactions must take place in a time comparable with, or shorter than, the time required for the heated pellet to fly apart. With particle energies of 10 keV the pellet will expand rapidly, in the order of a nanosecond or less for targets of a millimeter in radius, and you must get all of your useful return in energy before it flies apart. These pellets would be small, in order to keep the total ignition energy supplied by the laser small. Thus this would be a microexplosion. The energy in the blast must be captured and converted to useful form, some of which must be fed back to the laser to keep it running. This is the general concept and there have been some interesting developments in this area. A summary of the rudimentary energetics of this concept is given in Fig. 20. The capsule or pellet will fly apart in a time that is the order of its radius divided by the sound speed of the particles that are at an elevated temperature. The inertial confinement time turns out to be rather short for millimeter and submillimeter targets–fractions of a nanosecond. The energy that is released is proportional to the volume or the cube of the radius and to the reaction rate, which goes as the square of the particle density and also to the time over which the reactions can take place (the duration that the target holds together). I call your attention to an important feature of the energetics: the fraction of material burned increases linearly with the time that the target remains together, linearly with the radius, so that you can increase the fraction of the fuel that is burned by increasing its size. On the other hand, by increasing its radius, the

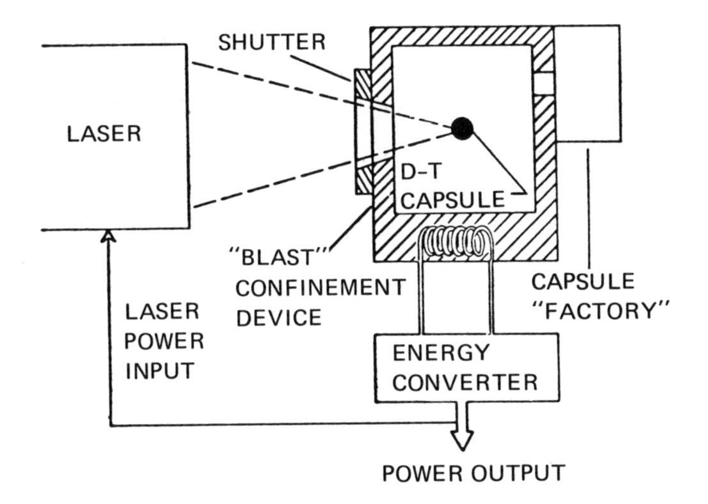

FIGURE 19 Speculative concept for a laser-fusion power generator.

volume of the pellet increases as the cube of the radius, as does the energy required to raise the pellet to ignition temperature. Thus to increase the "gain" of this process, you must pay a strong penalty in the form of greatly increased laser energy. The laser energy required to ignite such a solid target increases as the cube of the ratio of the fusion energy released to the laser energy. This simple scaling means that you pay a strong penalty for obtaining higher gain, and when you bear in mind that current lasers are rather inefficient and that thermoconversion efficiencies are at most 40 percent, then the gain must be rather high for useful production of energy. The laser energy required could be as high as a thousand megajoules, and, so far, energies of neodymium glass lasers produce pulses of only a few hundred joules. From Fig. 20 you can see that the reaction rate increases as the square of the density n, and if the matter is greatly compressed before the reaction takes place, the laser energy required to ignite can be greatly reduced by essentially the square of a factor by which the matter is compressed. This is an idea that was presented in March 1972 at the Quantum Electronics Conference in Montreal. This concept puts a new light on the problem of ignited fusion reactors, since it should substantially reduce the energy required for ignition.

The manner in which a very high compression could be obtained is illustrated in Fig. 21 when the target is irradiated from all sides with laser energy; the outer shell of the solid target is blown off, and the reaction force of the ablating material pushes inward on the remaining mass of material and compresses it. By carefully choosing the time history of the energy deposition in the outer layer of the pellet, a series of converging spherical shock waves can be initiated which raise the density in the center by a very large factor. Computer calculations show that this factor may be as high as 10^3 or 10^4. This would reduce the

LASER-FUSION BREAK-EVEN

$$E_{\text{fusion}} = \frac{4}{3} \pi R^3 \frac{n^2}{4} \langle \sigma v \rangle E_f \tau$$

$$\tau \sim R/v_s$$

$$\eta_{\text{abs}} E_{\text{laser}} = \frac{4}{3} \pi R^3 \cdot (3nkT)$$

$$E_{\text{laser}} = 2(n_s/n)^2 \frac{1}{\eta_{\text{abs}}^4} (E_{\text{fusion}}/E_{\text{laser}})^3 \text{ MJ}$$

FIGURE 20 Rudimentary energetics of the laser-fusion concept.

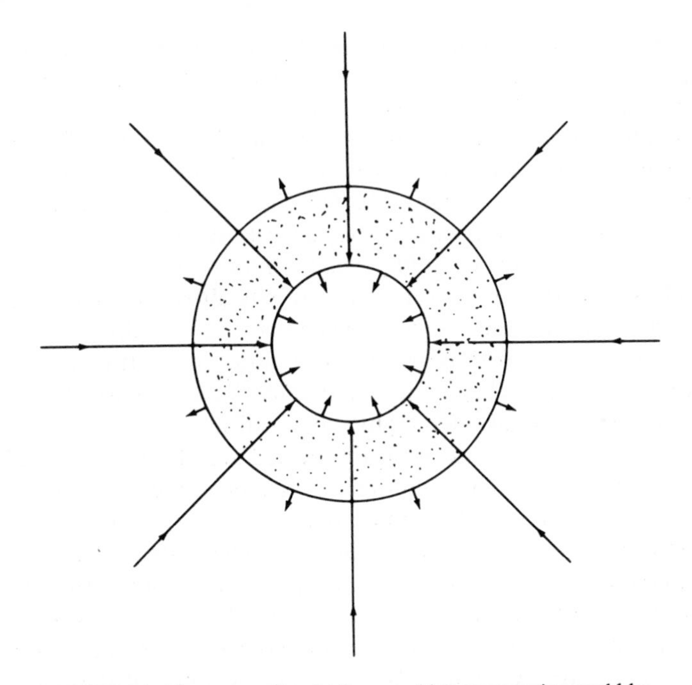

FIGURE 21 The manner in which a very high compression could be obtained in a laser-fusion reactor.

ignition energy requirements on the laser by 10^6 or 10^8, provided of course that these theoretical ideas bear up under experimental investigation. Simply, the physics experiments with condensed matters at these densities could be very interesting. If they are as large as predicted, then this could have an important impact on the controlled thermonuclear power production. It is somewhat speculative, however, because the ideas are still evolving and experimental work is just beginning.

I would like to conclude by summarizing some of the points that I have made. First, the science of plasma physics is about 50 years old. It began with Langmuir and has greatly matured, particularly in this period of most intensive development of the past 20 years. It is a science that is closely coupled and motivated by applications. These applications are to space and astrophysics, of which I have said little in this talk, and to controlled thermonuclear fusion, which is still in the research phase. As a science, plasma physics is a subject that is rich in physical phenomena, not only because of the strong coupling with the electromagnetic field but also because of the long mean-free-path or collisionless character of most plasmas. Individual particle effects survive in many of the important physical phenomena in plasmas, and the precise details of the velocity distributions of the plasma particles are frequently crucial to the microscopic behavior of the plasma.

Plasma physics is now recognized as an established scientific discipline within physics. One measure of this recognition is that the American Physical Society has had a Division of Plasma Physics for nearly 12 years, and it was one of the earliest divisions and also one of the larger ones. The European Physical Society now has a Division of Plasma Physics, and, although it is new, it is healthy and vigorous. The Soviet participation in that Society has been very useful. In the controlled thermonuclear fusion research program, international cooperation is very good; it has been cultivated since the declassification of the program in 1958 and has been highly successful.

As a science, plasma physics has developed greatly in the past 20 years; many important new phenomena have been identified and outlined and theories proposed. Contact between theory and experiment is still somewhat sparse but rapidly improving. There are gaps in the theory, particularly in the nonlinear regime, and I suppose that on Dr. Bardeen's scale, this field might be coming to the close of its *second* golden era, where many of the physical ideas are clearly identified. It is not yet a fully matured subject. We have reaped some of the fruits of the scientific developments, particularly in astrophysics and now in nuclear fusion. Possibly the fact that it is not yet a completely mature subject is why Dr. Artsimovitch chose to refer to the field as "the latest infant of the classics." If progress is maintained in the nuclear fusion area, we may very well be able to make this transition to Dr. Bardeen's "third golden era" in which we see a commercial application of this science that will be of great importance to mankind.

AAGE BOHR *and* BEN R. MOTTELSON *were asked to outline the present state of nuclear physics and to discuss what they felt were new perspectives opening up to increase our knowledge of the subject.*

AAGE BOHR *is a Professor of Physics at the Niels Bohr Institute of the University of Copenhagen, while Ben R. Mottelson is a Professor of the Nordic Institute for Theoretical Atomic Physics (*NORDITA*) in Copenhagen. They have worked closely together for over 20 years and are co-authors of a recent book on nuclear structure. Among many honors, they received the 1969 Atoms for Peace Award in recognition of outstanding contributions to nuclear physics in the service of mankind.*

The lecture at the symposium was delivered by Professor Bohr.

9 *The Many Facets of Nuclear Structure*/

AAGE BOHR *and* BEN R. MOTTELSON

EARLY DEVELOPMENTS

Fifty years ago, the nucleus of the atom had only rather recently been discovered. Its very existence had provided a decisive clue to the unraveling of the structure of atoms, a development that was rapidly leading to the establishment of quantal mechanics, but the internal structure of the nucleus remained unexplored. However, already the radiation emanating from the radioactive substances had given the first indication of the richness of the nuclear phenomena. The distinction between the α, β, and γ rays established by the early pioneers was gradually to be recognized as a manifestation of the hierarchy of the strong, weak, and electromagnetic interactions, and the nucleus was to become a laboratory for exploring the symmetries and structure of the novel interactions that go beyond the classical world of gravity and electromagnetism.

As the nucleus was subjected to more and more deep-going probes, a wealth of different, often contrasting facets was revealed, and the resulting picture of the nuclear structure underwent profound developments from decade to decade. The discoveries of the 1930's gave a broad view of the basic nuclear processes and revealed the complexity and subtlety of the strong interactions. During the first decade after the war, the nuclear physicists found themselves exploring a quite novel type of many-body system, in which the motion of the nucleons gives rise to a shell structure as well as to vibrational and rotational modes.

The subsequent period brought into focus the great variety of elementary modes in the nucleus involving shape oscillations, pair correlations, isospin, etc., and decisive progress was made toward understanding the occurrence of these modes in terms of the interplay of the individual particles. Current developments involve an expansion of the field along many frontiers and continue to reveal entirely new facets of the nuclear structure.

THE NUCLEUS AS A QUANTAL MANY-BODY PROBLEM

The developing understanding of the nuclear structure has been part of the broad development of quantal concepts appropriate to the description of systems with many degrees of freedom. The early phases of quantal theory gave emphasis to phenomena that could be described in terms of a few simple degrees of freedom or, as in electrodynamics, in terms of a perturbation in the motion of free quanta. Such phenomena could be directly mastered by a solution of the basic equations of motion. However, in the study of the many-body systems, ranging from the macromolecules and matter in bulk to the strongly interacting elementary particles, the situation has been a different one. The structural possibilities and the variety of correlation effects that may occur in a system such as the nucleus are so vast that a frontal attack on the Schrödinger equation in multidimensional configuration space has provided only limited guidance. The crucial problem has been the identification of the appropriate concepts and degrees of freedom to describe the observed phenomena. Progress in this direction has been achieved by a combination of many different approaches, including theoretical studies of model systems and the establishment of general relations following from considerations of symmetry. Above all, the development has again and again been given entirely new directions by clues provided by startling experimental discoveries.

The many-faceted nature of the nuclear systems makes it impossible to give a one-dimensional presentation, but a few examples of phenomena encountered on some of the active frontiers of nuclear research may, perhaps, convey a feeling for the flavor and perspective of the development.

INTERPLAY OF SHELL STRUCTURE AND NUCLEAR DEFORMATIONS

In the exploration of nuclear dynamics, the efforts to achieve a proper balance between independent-particle and collective degrees of freedom has been a recurring and central theme. This may be true of all many-body systems, but, in the nucleus, because of the possibility of detailed studies of individual quantum states, this issue was encountered in an especially concrete form.

The dual aspects of nuclear structure manifest themselves in a very direct manner in the energy of the nucleus considered as a function of its shape. While the general features of this "potential energy function" can be described in terms of bulk properties of the nuclear matter such as surface tension and electrostatic energy, the specific geometry of the quantized orbits of the individual nucleons contribute important anisotropic effects; a striking consequence is the occurrence of nuclear equilibrium shapes deviating strongly from spherical symmetry.

The effect of the shell structure on the nuclear potential energy has come into new perspective as a result of the recent discovery of metastable states of the heavy nuclei that decay by spontaneous fission. The potential energy function of such a nucleus is illustrated schematically in Fig. 1. The dashed curve represents the estimate of this function as given by the liquid drop model; the occurrence of a maximum, the fission barrier, results, as is well known, from the competition between the surface tension and the electrostatic repulsion. The quantized energies of the individual particles have a more specific dependence on the shape and symmetry of the potential and give rise to rather large modifications of the potential energy function. In particular, the shell-

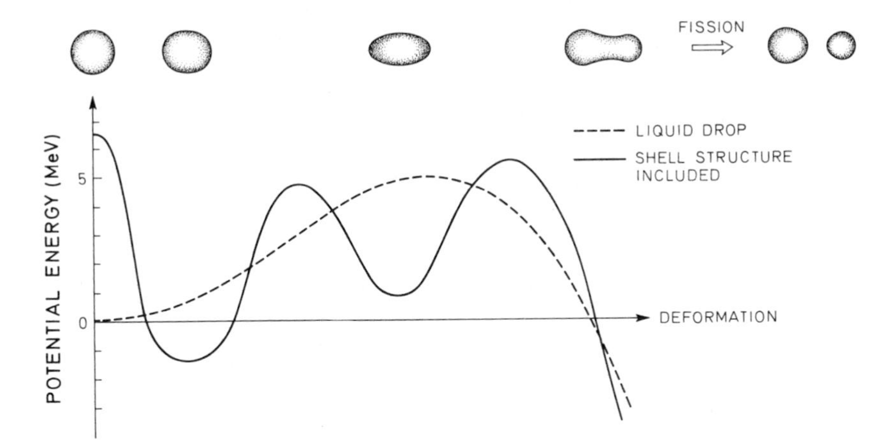

FIGURE 1 Potential energy function for deformations leading to fission. The schematic figure illustrates qualitative features in the structure of the fission barrier. The deformation parameter labels the path toward fission, as indicated by the shapes at the top of the figure. The shapes at the first and second minima correspond to those observed in the ground states and in the shape isomers in nuclei in the region of uranium.

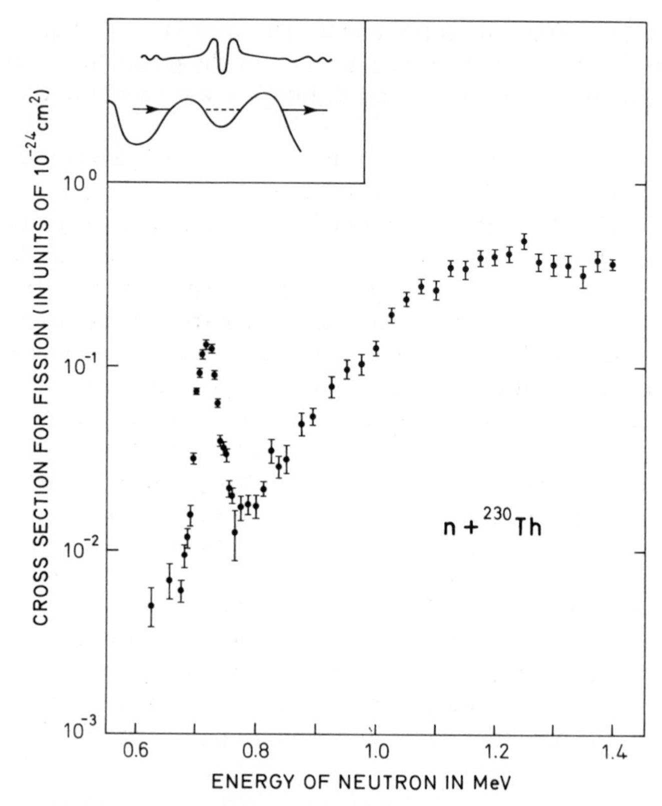

FIGURE 2 Resonance in threshold function for neutron-induced fission of ^{231}Th. The figure shows the cross section for fission as a function of the neutron energy (E. Lynn, preliminary data on Earwaker and James, quoted in *Physics and Chemistry of Fission*, p. 249, IAEA, Vienna, 1969). The insert illustrates the interpretation of the resonance phenomenon in terms of a semistationary state of vibration in the second minimum of the potential energy function.

structure effect implies a ground-state equilibrium with an eccentricity of about 20 percent and a second minimum at much larger deformations, in the region of the fission barrier. The occurrence of the second minimum is responsible for the existence of the isomeric states that decay by spontaneous fission with lifetimes shorter than those of the ground state by factors of 10^{20} or more. This feature of the potential energy function is also revealed by many striking phenomena in the fission process. An example is shown in Fig. 2, which exhibits a resonance structure in the cross section for fast neutron fission in ^{230}Th. The resonance occurring in the threshold region can be interpreted in terms of standing waves of vibrational motion in the second minimum, not unlike the resonance transmission in a Fabry-Perot interferometer. A wide range of ingenious approaches is currently being brought to bear in the developing spectroscopy of excitations in this new phase of the nuclear systems.

The occurrence of a highly deformed metastable state of the nucleus, referred to as a shape isomer, reflects a special stability associated with the shell structure. The nature of the new shells can be understood in a simple manner by reference to one-particle motion in a spheroidally deformed harmonic oscillator potential. As illustrated in Fig. 3, the degeneracies of the isotropic oscillator are removed by the deformation, but new major shells (degeneracies) reappear, when the oscillator frequencies in the different directions have rational ratios. Especially large effects occur for a deformation with the frequency ratio $\omega_{\perp} : \omega_3 = 2:1$, and the associated nucleon numbers for closed shells are $N = \ldots 110, 140, \ldots$. The nuclear potential differs from the harmonic oscillator in radial dependence as well as in the occurrence of a rather large spin-orbit coupling; as shown in Fig. 4, the inclusion of these effects leaves intact the main features of the oscillator shell structure in the 2:1 potential but modifies the closed-shell numbers to $N = \ldots 116, 148, \ldots$. The number $N = 148$ corresponds with the region of neutron numbers for which the shape isomers are observed to be especially stable.

The discovery of the shape isomers has given new scope to the concept of shell structure in quantal many-body systems and has raised the question of the general conditions on the potential that are necessary for the occurrence of significant deviations from uniformity in the eigenvalue spectrum. In particular, the development has focused attention on the intimate connection between the occurrence of shell structure in the quantal spectrum and the occurrence of degenerate families of periodic orbits in the corresponding classical motion.

INTERACTION OF NUCLEAR MATTER IN BULK (HEAVY-ION REACTIONS)

The study of the fission isomers is part of a much broader program exploring the stability of nuclear matter as a function of neutron and proton numbers as well as of the deformation parameters. One of the exciting perspectives is the possible occurrence of islands of metastable nuclei with mass numbers much greater than those hitherto encountered or with very different ratios between the numbers of neutrons and protons.

The part of the multidimensional particle-number and deformation space accessible to experimental study is being greatly expanded by the

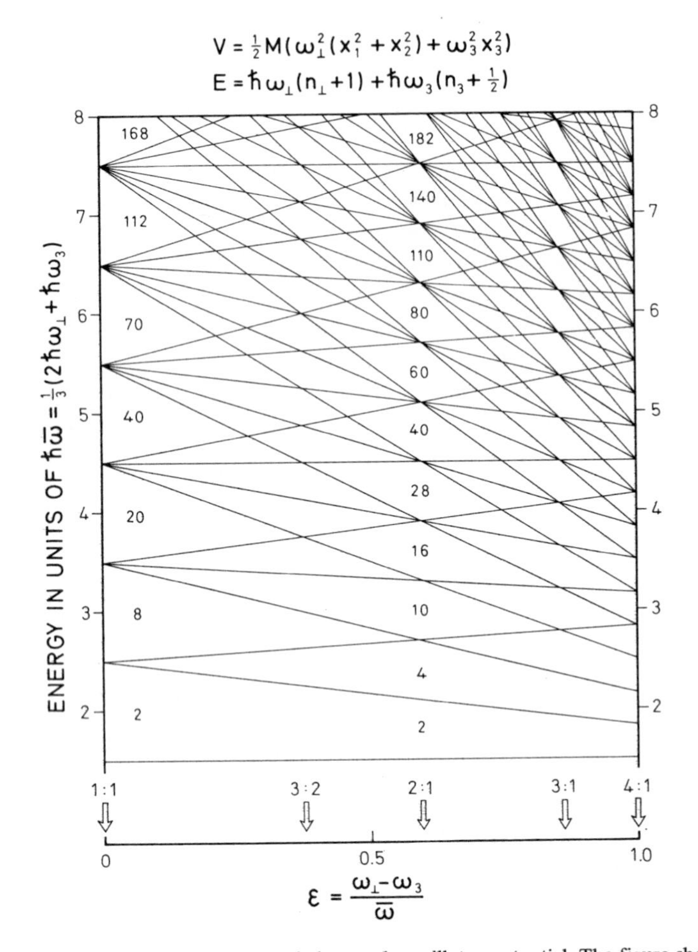

FIGURE 3 Shell structure in anisotropic harmonic oscillator potential. The figure shows the single-particle energy levels, as a function of deformation, in a prolate axially symmetric oscillator potential. The frequencies ω_3 and $\omega_\perp$ refer to motion parallel and perpendicular to the symmetry axis, while $\bar{\omega}$ is the mean frequency. The single-particle states can be specified by the number of quanta n_3 and $n_\perp$, and each energy level has a degeneracy $2(n_\perp + 1)$, due to the spin and the degeneracy in the motion perpendicular to the axis. Additional degeneracies leading to the formation of major shells may occur when the ratio of the frequencies $\omega_\perp : \omega_3$ is equal to the ratio between integers. The deformations corresponding to the most prominent shell-structure effects are indicated by the arrows labeled by the corresponding frequency ratio. For the shells with frequency ratio 1:1 (spherical shape) and 2:1, the figure gives the particle numbers for closed-shell configurations.

possibility of studying reactions initiated by the impact of heavy ions. As already indicated by pioneering experiments like those shown in Fig. 5, such reactions lead with high probability to the transfer of many particles between the interacting nuclei. The current discussion is concerned with the physical conditions that determine the numbers of nucleons transferred, the nature of the flow involved, and the extent to which a statistical equilibrium is established during the reaction.

Reactions between heavy nuclei are also found to provide new possibilities for exploring such features of nuclear matter as the texture of the surface. The example in Fig. 6 shows the elastic cross section as well as the inelastic excitation of a shape oscillation, considered as a function of incident energy. The semiclassical nature of the collision process makes it possible to relate the incident energy to the distance of closest approach, and the oscillations in the cross sections are a sensitive measure of the interactions that come into play in a grazing collision.

Studies have already been initiated with projectiles much heavier than those in the above examples. It is apparent that we are on the threshold of a whole new dimension of nuclear studies involving nuclear matter under pressure and strains that may only be rivaled by the extreme phases in cosmic evolution, such as in the formation of the neutron stars.

ELEMENTARY MODES OF EXCITATION; UNIFIED DESCRIPTION OF NUCLEAR DYNAMICS

The great sophistication that has been attained in the experimental study of nuclear spectra is making possible a penetrating analysis of the concept of the elementary modes of excitation. Examples of elementary excitations based on the closed-shell configuration of ^{208}Pb are shown in Fig. 7 (upper part). The single-particle and single-hole states are especially identified in reaction processes in which a nucleon is added to or removed from the ^{208}Pb ground state, and the figure shows the observed proton and proton–hole states in $^{209}_{83}$Bi ($\Delta Z = +1$) and $^{207}_{81}$Tl ($\Delta Z = -1$), respectively. The states are labeled by the conventional notation inherited from the early atomic spectroscopists. Collective modes corresponding to shape oscillations in ^{208}Pb have been identified by their large excitation probabilities in electromagnetic processes and inelastic scattering. While the shape oscillations can be resolved into particle–hole excitations, another type of collective excitation involves the addition or removal of correlated pairs of particles. The figure shows identified quanta of this type involving pairs of protons. [See, for example, the 0^+ states in $^{206}_{80}$Hg ($\Delta Z = -2$) and $^{210}_{84}$Po ($\Delta Z = +2$).]

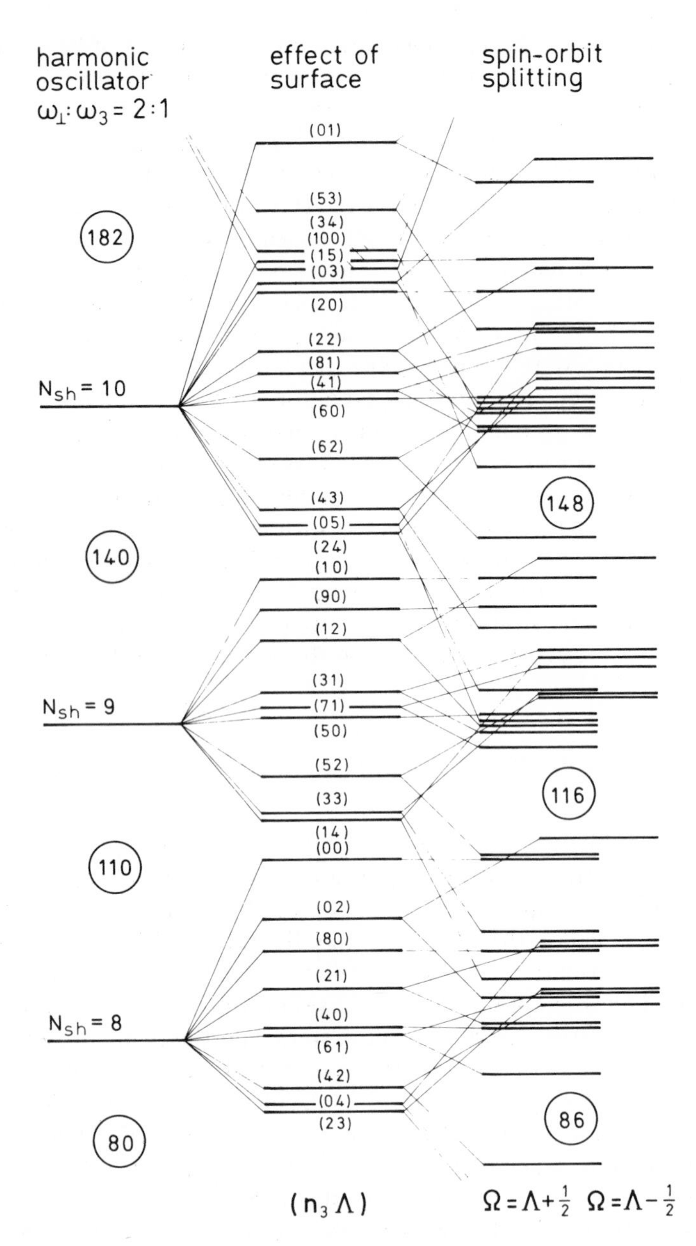

FIGURE 4 Shell structure in nuclear potential with 2:1 symmetry. The single-particle spectrum to the left is identical with that of Fig. 3, for the frequency ratio $\omega_\perp : \omega_3 = 2:1$. The shells are labeled by the shell quantum number $N_{sh} = n_3 + 2n_\perp$. The spectrum in the center shows the spreading of the shells that results from the deviation of the nuclear potential with its rather well-defined surface from that of a harmonic oscillator. The states are labeled by the

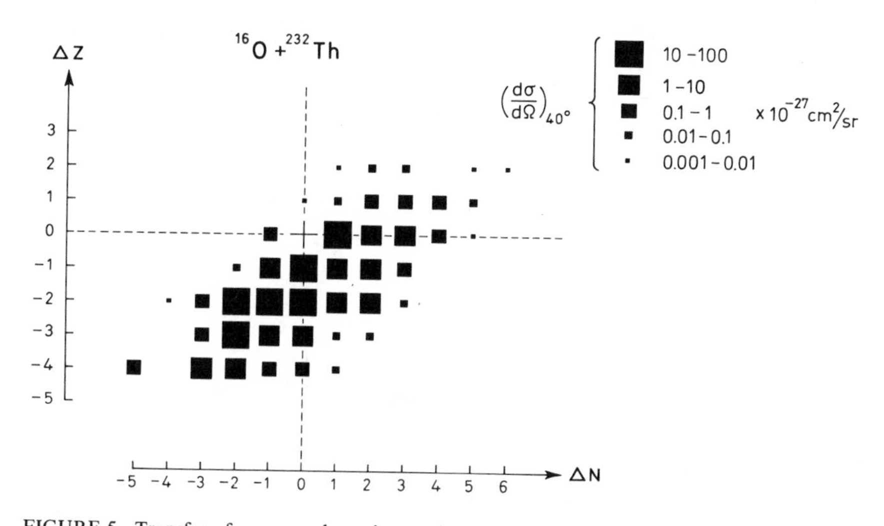

FIGURE 5 Transfer of many nucleons in reactions induced by oxygen ions. The figure shows that the intensities of the reactions involving the transfer of ΔN neutrons and ΔZ protons from the target nucleus to the impinging oxygen nucleus are shown. [A. G. Artukh, V. V. Avdeichikov, J. Erö, G. F. Gridnev, V. L. Mikheev, V. V. Volkov, and J. Wilczynski, *Nucl. Phys. A160*, 511 (1971).] The most probable transfers lead to an increase of the mass-to-charge ratio for the projectile, which can be understood in terms of a tendency toward a more uniform distribution of the neutron excess during the collision.

The elementary excitations provide the building blocks in terms of which one may attempt an analysis of the total excitation spectra. As an example, the spectrum of ^{209}Bi, in the lower part of Fig. 7, shows the occurrence of one-particle states as well as states corresponding to a single-particle or a single-hole combined with the boson excitations of ^{208}Pb.

The description in terms of independent elementary modes is an approximation that is limited by the interrelation of the different quanta.

---

component Λ of orbital angular momentum along the symmetry axis and by the quantum number n_3, which remains an approximate constant of the motion. Each level is fourfold degenerate, since the energy is independent of the sign of Λ and of the spin. The spectrum to the right includes the effect of the rather strong spin-orbit coupling in the nuclear potential, which splits the states with component of total angular momentum $\Omega = \Lambda \pm \frac{1}{2}$, in such a way as to favor parallel spin and orbit ($\Omega = \Lambda + \frac{1}{2}$). The resulting spectrum is seen to preserve the existence of the major shells obtained from the oscillator potential, though with a modification of the closed-shell numbers (given in circles) resulting from the fact that a few of the orbits with largest Λ and $\Omega = \Lambda + \frac{1}{2}$ have been shifted into the next lower shell. The effects illustrated in Fig. 4 are quite similar to those that govern the shell structure in spherical nuclei and lead to the closed shell numbers . . . , 82, 126 . . . , replacing the sequence . . . , 70, 112, . . . for the oscillator potential (see Fig. 3).

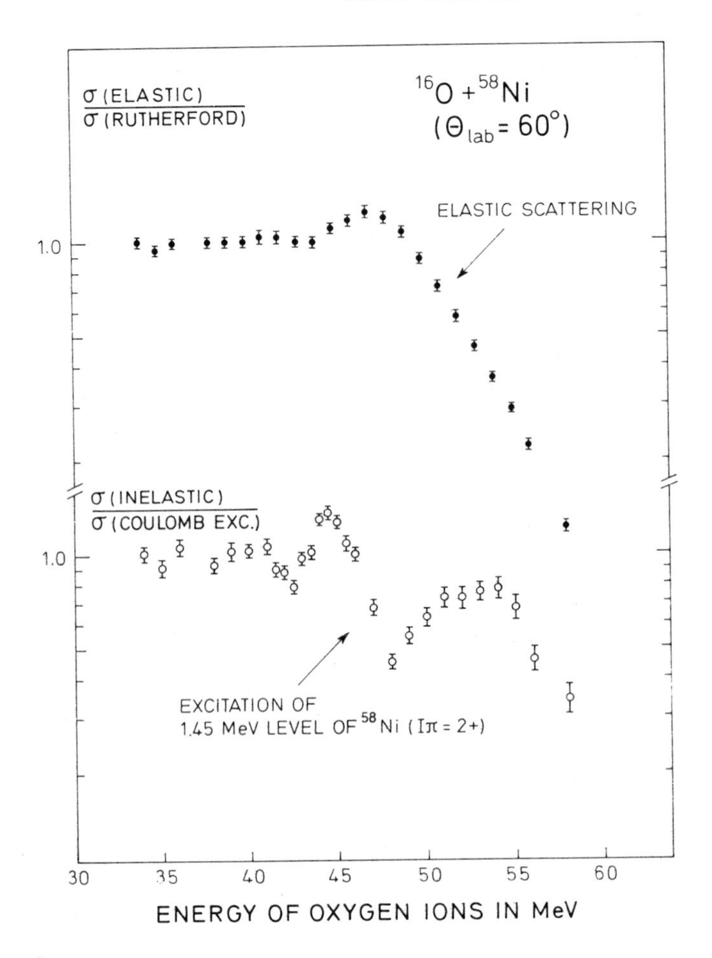

FIGURE 6 Elastic and inelastic scattering of ^{16}O on ^{58}Ni. The cross section for elastic scattering is given in units of the Rutherford cross section for scattering of two point charges. The inelastic scattering involves the first excited state in ^{58}Ni, which can be approximately described as a quadrupole surface oscillation; the cross section for this excitation process is given in units of the cross section for excitation by the electric field of the oxygen nucleus, assuming that the projectile does not penetrate into the target nucleus. The experimental data are from F. Videbaek, I. Chernov, P. R. Christensen, and E. E. Gross, *Phys. Rev. Lett. 28*, 1072 (1972). The cross sections in the figure refer to a fixed scattering angle of 60°, in the laboratory system, and are shown as a function of the energy of the incident ion. Because of the small wavelength of the heavy colliding particles, the process can be approximately described by considering the particles as moving along classical trajectories, and the incident energy and scattering angle therefore specify a rather sharply defined distance of closest approach. For energies below 45 MeV, this distance exceeds the range at which the nuclear interactions become effective, and the cross sections in the figure are close to unity, in the scale employed. The energy range from about 45 MeV to 50 MeV corresponds to collisions in which the nuclear interactions are beginning to have a significant role. There is still some ambiguity in the analysis of the observed oscillations, but a possible interpretation involves the interference between the attractive nuclear

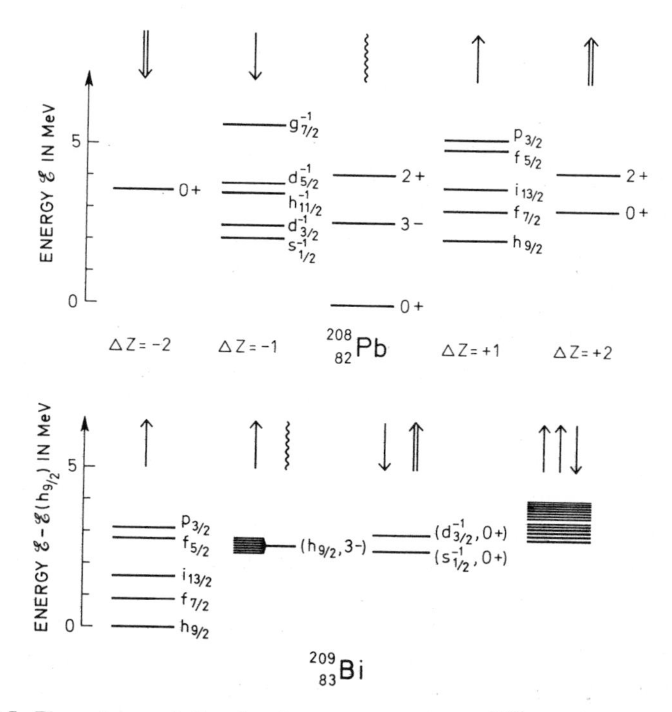

FIGURE 7 Elementary excitations based on the ground state of ^{208}Pb. The spectra of the nuclei around $^{208}_{82}$Pb can be described in terms of elementary excitations based on the ground state of ^{208}Pb, which involves closed shells of neutrons and protons and acts as the "vacuum state" for the excitations. The upper part of the figure shows fermion excitations involving the addition or removal of a single proton ($\Delta Z = +1$ or $\Delta Z = -1$) and boson excitations involving correlated pairs of protons ($\Delta Z = \pm 2$), as well as collective excitations in ^{208}Pb itself. The latter type of quanta can be expressed in terms of coherent particle–hole excitations with a resulting density oscillation approximately corresponding to that of a surface vibration. The energy scale employed in the figure involves a linear term in ΔZ, so chosen that the lowest one-particle and one-hole excitations ($h_{9/2}$ and $s^{-1}_{1/2}$) have the same ordinate. Additional elementary excitations not shown in the figure involve changes in neutron number ($\Delta N = \pm 1, \pm 2$). The lower part of the figure represents the low-energy spectrum of $^{209}_{83}$Bi. In addition to the one-particle excitations shown to the left, the spectrum exhibits excitations involving a single particle or a single hole combined with a collective excitation. The configuration ($h_{9/2}$, 3^-) gives rise to a multiplet of states with total angular momentum 3/2, 5/2, . . . 15/2 that have all been identified within an energy region of a few hundred keV. The configurations involving a hole and a pair quantum with $I\pi = 0^+$ each give rise to only a single state. At an excitation energy of about 3 MeV, a rather dense spectrum of two-particle, one-hole states sets in, as indicated to the right in the figure.

---

forces and the repulsive electrostatic interaction acting in a grazing collision. This interference leads to a decrease in deflection angle (resulting in an increase of the cross section per unit solid angle of scattering) and a decrease in the probability for excitation. For still smaller impact parameters, the strong nuclear interactions lead to more violent reactions, and the probability that an oxygen nucleus emerges intact decreases rapidly with increasing energy (for fixed scattering angle).

In the nucleus, the analysis of these interrelations can be based on the average potential fields that are generated by the collective motion; these dynamic fields represent a generalization of the familiar static self-consistent potential. The dynamic fields provide a coupling between the motion of the individual particles and the collective modes (see Fig. 8), which plays a similar role in the nucleus as the particle–phonon or particle–plasmon couplings in condensed media.

A systematic treatment of the particle–vibration coupling leads to a nuclear field theory, which is being actively explored as a basis for interpreting the rapidly accumulating evidence on the many interaction effects between the elementary modes including the anharmonicity in the collective motion and polarization effects leading to a renormalization of charge and moments of the particles.

Such a field theory also appears to provide a consistent treatment of the problem of the redundancy in the total degrees of freedom that has been felt as a paradoxical feature in the treatment of nuclear dynamics. Such a redundancy is inherent in the description of a system, like the nucleus, where the collective modes are manifestations of the same underlying degrees of freedom as the particle modes of excitation. As an illustration of this point, Fig. 9 shows the Feynman diagrams for the leading order interaction energy between a single particle and a collective shape oscillation, as in the $(h_{9/2}, 3^-)$ septuplet in ^{209}Bi (see Fig. 7). The fact that the particle configuration also appears as a component in the particle–hole expansion of the vibrational excitation finds expression in the exchange interaction represented by the last diagram.

It is seen that the four diagrams in Fig. 9 are completely equivalent to those describing the Compton scattering. Indeed, it is a remarkable feature of the spectrum of the nuclear many-body system that the vibrational quanta though ultimately composed of particle excitations appear just as elementary as the photon in electrodynamics.

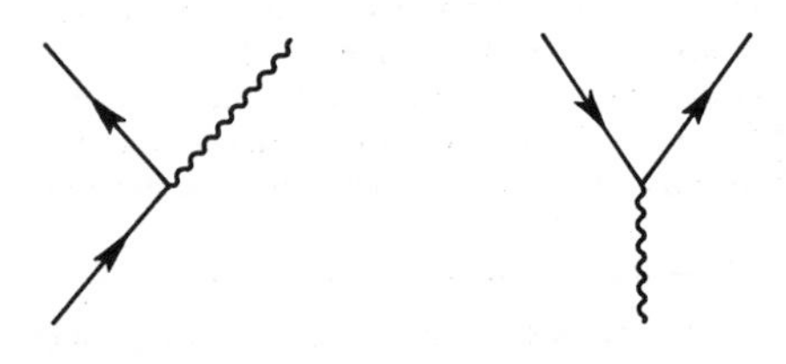

FIGURE 8 Particle–vibration coupling. The Feynman diagrams illustrate the basic coupling between particle and collective motion arising from the average one-particle potential generated by the collective vibrational motion. The first diagram represents the scattering of a particle with emission of a vibrational quantum (phonon); the second diagram represents the transition of a phonon into a particle–hole pair.

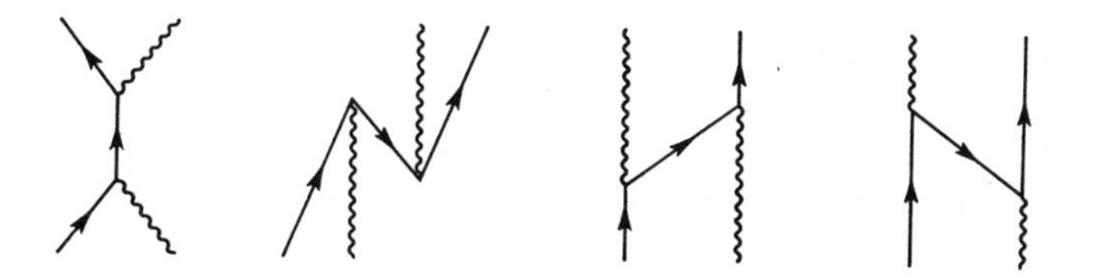

FIGURE 9 Interaction energy of a particle and a phonon. The basic particle–vibration coupling (see Fig. 8) acting in second order gives rise to an interaction energy between a particle and a vibrational quantum, as illustrated by the four Feynman diagrams in the figure. This interaction energy contributes, for example, to the splitting of the $(h_{9/2}, 3^-)$ multiplet in the spectrum of ^{209}Bi (see Fig. 7). The effects represented by the diagrams in Fig. 9 include the renormalization of the phonon in the presence of the particle, as well as the consequences of the identity of the particles and of the boson degrees of freedom.

FINE-GRAIN STRUCTURE OF NUCLEAR MATTER; MESONIC DEGREES OF FREEDOM

The interpretation of the equilibrium density and binding energy of nuclear matter in terms of the forces acting between the nucleons has been a challenge since the early days of nuclear physics. The problem has turned out to require a much more deep-going analysis than could have been anticipated, not only because of the complexity of the strong interactions but also because of the many subtle correlation effects that can be involved in collective properties of many-body systems.

A vast expansion of this field is resulting partly from the discovery of the great variety of collective modes generated by average fields of different symmetries and partly by the development of experimental probes capable of resolving the fine-grain structure of nuclear matter. Such resolution demands momentum transfers comparable with or larger than the Fermi momentum, as can be achieved by the scattering of high-energy particles from the nucleus.

The high-energy probes that are becoming available for the study of the nucleus can also excite the internal degrees of freedom of the nucleons, and thus a new field of strong interaction physics is opening up. Investigations on this frontier will illuminate the nuclear dynamics from new points of view and may reveal novel aspects in elementary quanta occurring as components in a strongly interacting system.

The question of elementarity appears in another guise, when one considers the nuclei in the broader context of the hadronic spectra. It is a provocative feature of the strong interactions that they lead to bound systems of multiple baryonic number, but with binding energies so small that it is useful to consider these systems as composed of a definite number of neutrons and protons. In the corresponding problem of quantum electrodynamics, the existence of atoms and condensed matter describable as nonrelativistic many-body systems reflects the smallness of the fine-structure constant, but one may ask: Where is there a small number in the structure that underlies the strong interactions?

COMPOUND NUCLEUS; STATISTICS OF QUANTAL STATES

Most of the available tools for probing the nuclear spectra can only resolve the individual levels in the low-energy region, where relatively few excitation quanta are involved. However, the whole development of nuclear physics has been decisively influenced by the existence of a small window, in the region of the neutron binding energy, within which the slow-neutron reactions provide a probe of enormously greater resolving power. Already the earliest experiments with slow neutrons revealed, very unexpectedly, a dense spectrum of resonances. This discovery led to the recognition of a strong coupling between the motion of the incident neutron and many degrees of freedom of the target. The coupling gives rise to the formation of a compound system with a lifetime very long compared with the one-particle periods.

The refinement that has been achieved in the study of the neutron resonance spectra is illustrated in Fig. 10, which shows the total cross section for neutrons incident on ^{232}Th, in the energy range up to about 200 eV. The information provided by data of this type has led to significant new developments in the characterization of statistical equilibrium and the formulation of the concept of randomness at the level of individual quantal states of a many-body system. A model that has especially been explored expresses the randomness in the couplings of the different degrees of freedom in terms of an ensemble of matrix elements that is invariant with respect to transformations of the chosen basis. This formulation leads to predictions concerning the distributions of eigenvalues and decay widths that have been quantitatively tested. Thus, the repulsion between neighboring levels of the same spin and parity gives rise to a low probability for finding very small level spacings, and this short-range order has been well established for over a decade. The theory also predicts a long-range order of a rather subtle type, which has been confirmed only very recently by experiments such as those illustrated in Fig. 10. The test of this long-range order is shown in Fig. 11, which gives the extent of the deviations of the observed level sequence from that of a completely ordered uniform distribution. While a level distribution with only short-range order implies mean-square

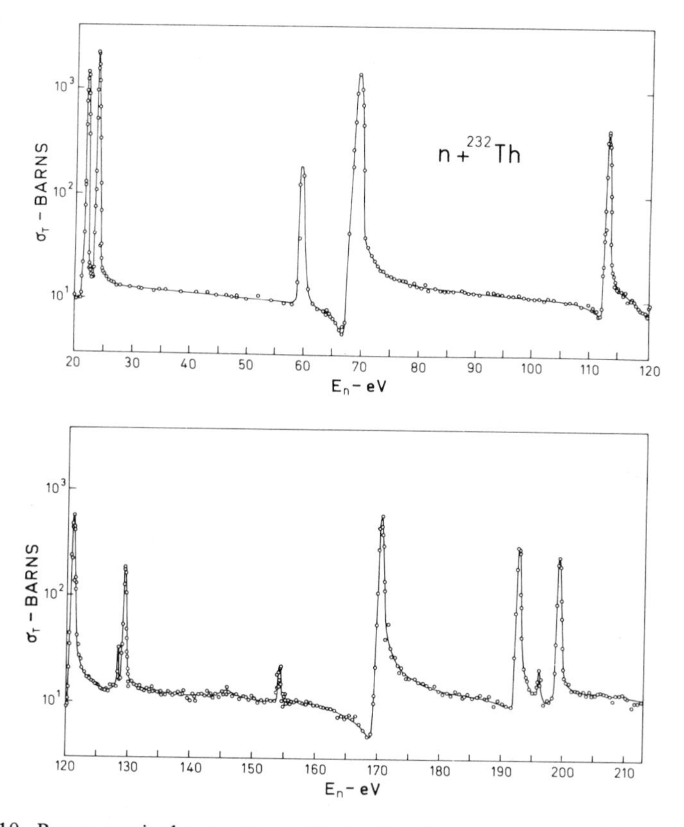

FIGURE 10 Resonances in slow neutron scattering. The figure shows the total cross section for neutrons incident on ^{232}Th; the resonances correspond to metastable states in the compound nucleus ^{233}Th that all have total angular momentum $I = \frac{1}{2}$ and positive parity. The experimental data are from the compilation Neutron Cross Sections, Sigma Center, Brookhaven National Laboratory, BNL 325, Suppl. 2, Brookhaven, N.Y., 1964. See also the more recent data by F. Rahn, H. S. Camarda, G. Hacken, W. W. Havens, Jr., H. I. Liou, J. Rainwater, M. Slagowitz, and S. Wynchank, *Phys. Rev. 6C*, 1854 (1972).

deviations increasing linearly with the number of levels, the inclusion of long-range interactions between levels, represented by random couplings, implies a much higher degree of ordering resulting in a mean-square deviation that is only logarithmic in the number of levels. Present efforts are directed toward identifying the limitations on the concept of randomness, as formulated above, that result from a more detailed analysis of the degrees of freedom involved in the nuclear spectra.

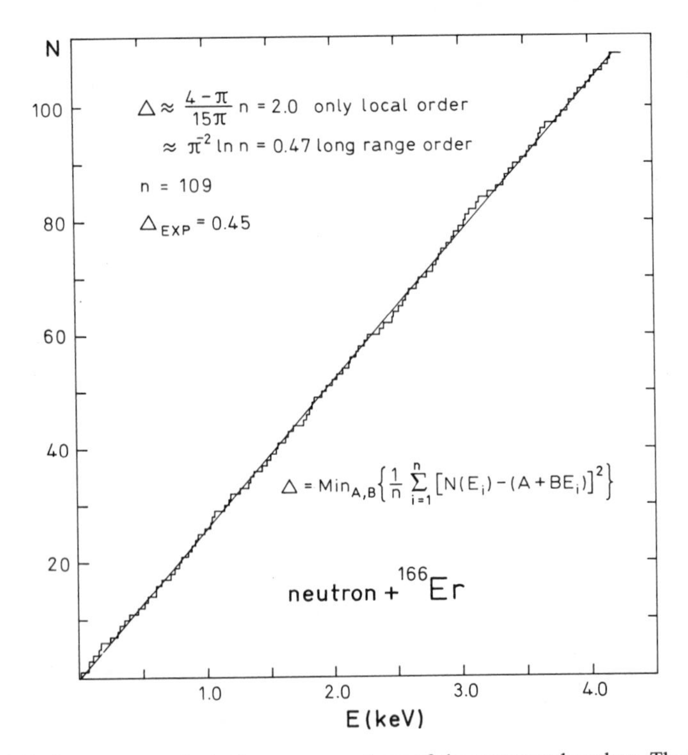

FIGURE 11 Long-range order in the energy spectrum of the compound nucleus. The figure shows the number of levels N of spin and parity $I\pi = 1/2+$ that have been observed for neutron energies up to a given value E [H. I. Liou, H. S. Camarda, S. Wynchank, M. Slagowitz, G. Hacken, F. Rahn, and J. Rainwater, *Phys. Rev. 5C*, 974 (1972)]. The mean-square deviation Δ from a uniform level spacing (corresponding to a straight line in the figure) is compared with the predictions based on a model (local order) that only includes the effect of repulsion between neighboring levels and a model (long-range order) that includes the repulsive effect of more distant levels as given by a random matrix obeying the time reversal and Hermiticity symmetries.

STRENGTH FUNCTION; DIRECT INTERACTIONS

The establishment of statistical equilibrium leading to the quantal randomness of the fine-structure resonances requires a time that is sufficiently long so that the couplings between the elementary modes of excitation can be effective. This time scale is revealed by the widths of the strength functions (or gross structure resonances), which show that the lifetimes τ_{coupl} of the elementary excitations correspond to energies that are typically of the order of MeV. Such a lifetime, though many orders of magnitude smaller than the periods τ_{comp} of the compound nucleus, is large compared with the time τ_{sp} for a nucleon to traverse the nucleus:

$$\frac{\tau_{sp}}{(\Delta E \sim 10 \text{ MeV})} \ll \frac{\tau_{coupl}}{(\Delta E \sim 1 \text{ MeV})} \ll \frac{\tau_{comp}}{(\Delta E \sim 10 \text{ eV})} .$$

The nuclear reactions, therefore, also exhibit important effects involving only a single or few degrees of freedom ("direct" interactions). The possibility of studying processes extending over six orders of magnitude in the time or energy dimension (varying from the single-particle to the compound periods) provides the opportunity of exploring in considerable detail the different levels of complexity and order of which the concepts of direct interactions and compound processes represent two extremes.

ISOBARIC ANALOGUE RESONANCE

Strength functions with especially long lifetime may result from symmetries or other dynamical features that lead to approximately conserved quantum numbers. Figure 12 shows the cross section for elastic scattering of protons on $^{58}_{26}Fe_{32}$. The remarkable resolution achieved in such experiments makes it possible to study the dense spectrum of resonances interfering with the Coulomb scattering. Superposed on the individual compound levels is a gross structure that can be attributed to the charge independence of the strong interactions, as expressed in the conservation of the total isospin. The compound levels have isospin $T = 5/2$, as in the ground state of $^{59}_{27}Co$ formed by adding a proton to $^{58}_{26}Fe$; but in the energy region studied, there is in addition a single state with $T = 7/2$, which has the same (relatively simple) internal structure as the ground state of $^{59}_{26}Fe$ (isobaric analogue state). If there were no interactions in the nucleus distinguishing between neutrons and protons, the isospin would be exactly conserved, and the $T = 7/2$ state would appear as a single sharp level in the spectrum of $^{59}_{27}Co$. However, the Coulomb forces that act between the protons give rise to a spreading of the properties of the $T = 7/2$ states into the adjacent compound states with $T = 5/2$. The lower part of Fig. 12 shows the cross section for radiative capture of protons on ^{58}Fe, a process that is even more sensitive than the proton scattering in revealing compound states in several angular momentum and parity channels but which does not respond specifically to the isobaric analogue state.

The distribution of the strength of the isobaric analogue state as revealed in the proton scattering has a width of only about 20 keV, which reflects the smallness of the electromagnetic interactions as compared with the strong interactions. The analysis of this strength function involves a symmetry breaking produced by a well-defined term in the

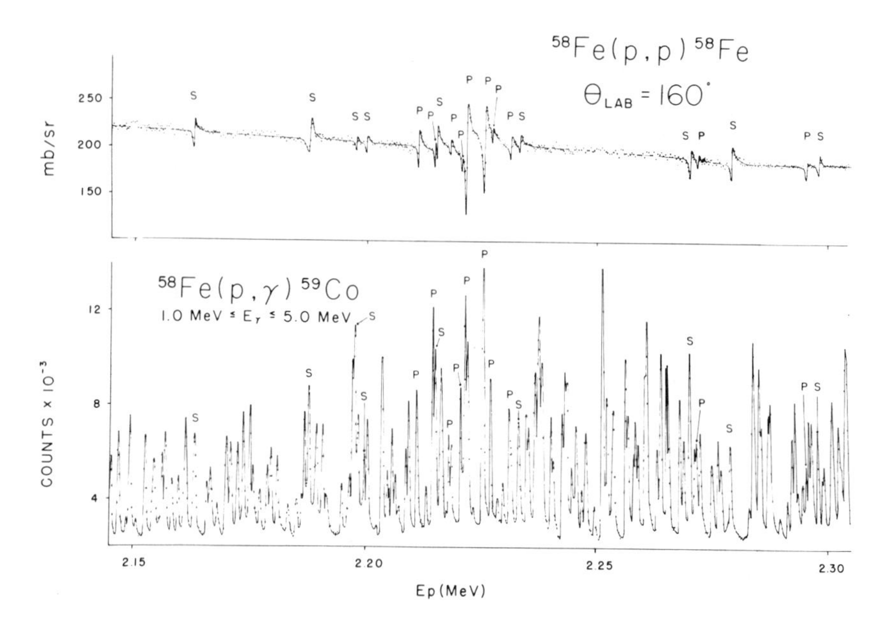

FIGURE 12 Gross structure in the resonance scattering of protons associated with the occurrence of isobaric analogue states. The upper part of the figure shows the elastic scattering cross section for protons on ${}^{58}_{26}$Fe as a function of the energy of the incident proton. The background represents nonresonant scattering primarily resulting from the Coulomb interaction between projectile and target nucleus; the resonances correspond to metastable states in the compound nucleus ${}^{59}_{27}$Co. For the proton energies considered, the resonances associated with s-wave scattering ($l = 0, I\pi = 1/2+$) are clearly resolved from the background. The p-wave scattering ($l = 1$, $I\pi = 1/2-, 3/2-$) is expected to give rise to three times as many resonances as the s-wave scattering, corresponding to the relative statistical weights, but the p-wave resonances are weakened by the reduced penetration through the Coulomb barrier that results from the centrifugal effect. Hence, most of the p-wave resonances remain undetected in this experiment. However, strongly enhanced p-wave resonances are found to occur in a narrow energy interval around $E_p = 2.22$ MeV; this gross structure effect can be associated with the approximate conservation of the isospin quantum number. The target nucleus ${}^{58}_{26}\mathrm{Fe}_{32}$ has total isospin $T = 3$ [the smallest value compatible with the isospin component $M_T = \frac{1}{2}(N - Z) = 3$]. Similarly, all the low-lying states in the compound nucleus ${}^{59}_{27}\mathrm{Co}_{32}$ have $T = 5/2$. The first $T = 7/2$ state belongs to the same T-multiplet as does the ground state of ${}^{59}_{26}\mathrm{Fe}_{33}$, which has $I\pi = 3/2-$; the energy of the corresponding state in ^{59}Co (isobaric analog state) can be obtained by adding the extra Coulomb energy ($\approx$8 MeV) associated with the transformation of a neutron into a proton inside the nucleus. The resulting energy just corresponds to the incident proton energy in Fig. 12, for which the strong p-wave resonances are observed. If the isospin quantum number were exactly conserved, the isobaric analogue state would give rise to a single $I\pi = 3/2-$ resonance, with a very large strength reflecting the simple internal structure of this state; in fact, the ground state of ^{59}Fe is known to be qualitatively described in terms of the ^{58}Fe core with the addition of a particle in a $p_{3/2}$ orbit. The proton width of the isobaric analogue state is therefore of a magnitude corresponding to a single particle p-wave resonance in a potential. In contrast, the $T = 5/2$ states have the complexity of a compound nucleus with many strongly coupled degrees of freedom resulting in a high-level density and a small decay width for particle emission. The violation of the isospin symmetry by the electromagnetic interactions (especially the electrostatic forces) implies that the properties of the $T = 7/2$ isobaric analogue state in the spectrum of ^{57}Co are shared among

Hamiltonian; the problem, however, is not a trivial one, since one is studying the effects of the perturbation in the actual physical states of the many-body system. Indeed, the total Coulomb potential in a heavy nucleus is not small compared with the nuclear potential, on account of the long range of the electric forces; and the weakness of its symmetry-breaking effect was only recognized after the discovery of the sharp isobaric analogue states. This unexpected discovery added a powerful new tool to the arsenal of reactions available to the nuclear spectroscopists.

COLLECTIVE MODES IN THE NUCLEAR PAIR FIELD (TWO-PARTICLE TRANSFER REACTIONS)

Collective modes associated with variations in the shape and density are familiar from classical systems. The more specifically quantal degrees of freedom associated with spin, isospin, and nucleon number give new dimensions to the nuclear dynamics. The exploration of the great variety of collective modes that can occur in these dimensions has only recently been initiated but appears to be a frontier of considerable perspective.

Collective motion with quanta-carrying nucleon number is related to the nuclear pairing effect, which was recognized as a systematic feature in nuclear properties at a very early stage of nuclear physics. The striking difference in the binding energies of nuclei with even and odd numbers of nucleons finds a rather dramatic expression in the different fissility of the even and odd isotopes of uranium. However, the collective significance of the pair-correlation effect and its far-reaching consequences for many nuclear properties were only gradually recognized.

A powerful tool for the study of the nuclear-pair correlations has become available in the two-nucleon transfer reactions. Figure 13 shows an example of a (t, p) process on ^{118}Sn, by which a pair of neutrons is added to the target nucleus. The spectrum is dominated by the transition

---

adjacent compound states with $T = 5/2$, each of which thereby acquires an enhanced p-wave resonance strength. The lower part of Fig. 12 shows the yield of the proton capture process with the emission of gamma rays. Because of the absence of a strong nonresonant background, this process is an even more sensitive detector of the compound resonances than the elastic scattering. The labels s and p exhibit the $l = 0$ and $l = 1$ resonances that are also detected in the elastic scattering process (upper part of figure); as expected, the capture process is not sensitive to the isobaric analogue structure. The total density of levels observed in the (p, γ) process corresponds approximately to that estimated for s– and p-wave resonances, on the basis of the observed spacing of the s-wave resonances in the (p, p) process. The experimental data in Fig. 12 are from W. C. Peters, G. E. Mitchell, and E. G. Bilpuch (to be published). We are indebted to Dr. Bilpuch for communications of these results prior to publication.

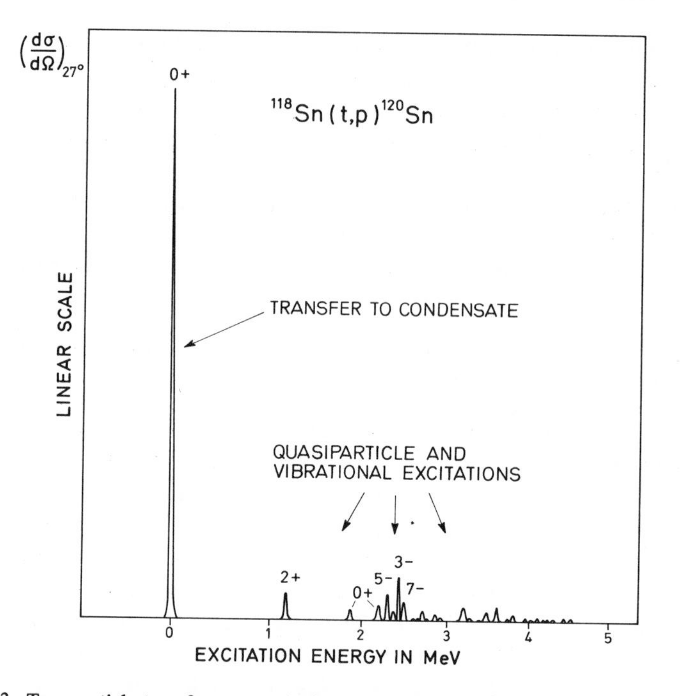

FIGURE 13 Two-particle transfer process in Sn nuclei. The figure shows the cross section for transfer of two neutrons from an incident triton ($t = {}^3$H) to a target of ${}^{118}_{50}$Sn [J. H. Bjerregaard, O. Hansen, O. Nathan, L. Vistisen, R. Chapman, and S. Hinds, *Nucl. Phys. A110*, 1 (1968)]. The transfer takes place as a "direct interaction" with an angular distribution characteristic of the transferred angular momentum. The cross sections in the figure refer to protons emerging at an angle of 27° from the direction of the incident beam; this angle corresponds to a maximum in the angular distribution for zero angular momentum transfer. The transition to the ground state involves a transfer of a pair of neutrons in a correlated state with angular momentum and parity 0^+. The process receives a further strong enhancement by the presence, in the target nucleus, of a rather large number of such correlated pairs formed out of neutrons in the partially filled shells. (Closed-shell configurations correspond to neutron numbers 50 and 82.) This enhancement is a quantal effect associated with the identity of the quanta (boson factor; see caption to Fig. 14). The occurrence of many bosons in a definite quantal state is referred to as a condensate, and the pair correlations in the nucleus have basic properties in common with the condensates of electron pairs in superconductors and of ^{4}He particles in superfluid helium. On account of the finite size of the nucleus, which is small compared with the correlation length for the pairs, supercurrents do not occur in nuclei (but may occur in superfluid nuclear matter in neutron stars).

to the ground state of ^{120}Sn, for which the cross section is found to be between one and two orders of magnitude greater than would correspond to the transfer of two neutrons each into a definite orbit in the nucleus. The enhancement reflects the fact that, in the Sn isotopes, the many neutrons in the partially filled shells form a condensate consisting of correlated neutron pairs, similar to the condensate of the paired electrons in a superconductor. In fact, the transfer process may be compared with the transmission of electron pairs between superconductors, as in the Josephson junction.

In a closed-shell nucleus (as in an insulator), there is no possibility of pair correlation and therefore no condensate. However, additional particles or holes can form correlated pairs that constitute elementary modes of excitation, as referred to above. The successive addition or removal of pairs leads to a vibrational-like spectrum, as illustrated in Fig. 14, which shows the neutron-pair vibrations with angular momentum zero, based on the closed-shell nucleus, ^{208}Pb. There are two distinct quanta in the spectrum corresponding to the addition or removal of a pair of particles with respect to the closed-shell configuration, and

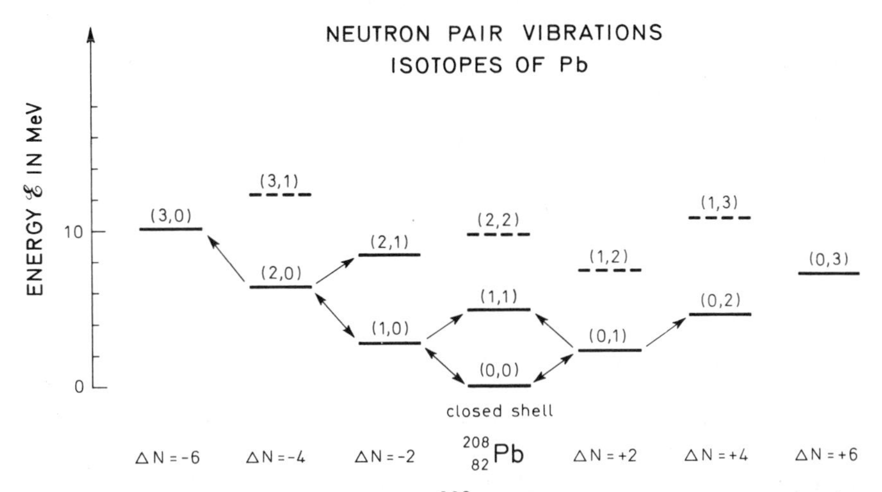

FIGURE 14 Neutron pair vibrations based on ^{208}Pb. The figure shows the spectrum of excitations involving quanta corresponding to the addition or removal of correlated pairs of neutrons with angular momentum and parity 0^+ (compare Fig. 7). The levels in the pair-vibrational spectrum are labeled by the number of pair-removal and pair-addition quanta. The levels $(n, 0)$ and $(0, n)$ correspond to the ground states of the even Pb isotopes. The energy scale includes a term linear in ΔN, which has been chosen in a similar way as for the excitations with different ΔZ in Fig. 7. The observed levels are indicated by solid lines. In the harmonic approximation (neglect of interaction between quanta), the energy would be a linear function of the number of quanta, but the observed level positions reveal interaction effects that can be approximately described in terms of a quadratic form in the number of quanta (pairwise interactions of the quanta). The dashed lines indicate predicted pair-vibrational excitations with energies that have been calculated with the inclusion of the interactions derived from the observed levels. The pair vibrations are characterized by large cross sections in the two-neutron transfer processes. The enhancement reflects the correlation of the neutrons in the individual quanta. When several identical quanta are present, the transition probability for the processes $n \leftrightarrow n-1$ involves an additional factor n (boson factor). The observed (t,p) and (p,t) transitions, indicated by the arrows in the figure, confirm the collective character of the pair-vibrational excitations.

the quantum numbers on the levels in Fig. 13 give the numbers of quanta of each kind. The arrows in the figure indicate experimentally observed strong transitions in two-neutron transfer processes populating ground states of the even Pb isotopes [$(n, 0)$ and $(0, n)$], as well as some of the excited states of the type $(n + 1, 1)$, occurring at excitation energies of 5–6 MeV.

The vibrational spectrum in Fig. 14 is an example of a collective family involving states in different nuclei. The vibrational motion is associated with the oscillation of a field—the pair field—that creates two nucleons. These oscillations occur not in the usual space but involve other dimensions including a so-called gauge space. The nucleon number operator, which appears as the angular momentum in gauge space, is usually assigned a rather passive role, as an overall constant of the motion, a superselection rule, that divides the phenomena into sharply separated compartments, each with a definite value of nucleon number. In the nuclear pair-correlation effects, however, we are dealing with phenomena that relate states with different nucleon number and that therefore involve operators, such as the orientation of the pair field in gauge space, that are complementary to the nucleon number. In the transfer processes, these operators are directly measured and the new dimensions are therefore experienced in a very real manner.

ROTATIONAL SPECTRA

The response to rotational motion is a basic property of physical systems that has played a prominent role in the development of dynamical concepts, ranging from celestial mechanics to the spectra of elementary particles. The question of whether nuclei possess rotational spectra was raised in the early days of nuclear spectroscopy. Quantized rotational motion was known from molecular spectra, but atoms provide examples of quantal systems that do not rotate collectively. The early discussion of this issue was, however, hampered by the expectation that rotational motion would either be a property of all nuclei or would be generally excluded and by the expectation that the moment of inertia would have the classical value, as for rigid rotation.

The establishment of the nuclear shell-structure provided a description in terms of single-particle motion that might seem to preclude the occurrence of collective rotation. However, a new situation was created by the recognition that the shell structure may lead to equilibrium shapes that deviate from spherical symmetry. It was evident that such a collective deformation, which defines an orientation of the system as a whole, would imply rotational degrees of freedom, but one was faced with the need for a generalized treatment of rotations applicable to quantal systems that do not have a rigid or semirigid structure like that of molecules.

An example of rotational band structure in a nuclear spectrum is shown in Fig. 15, which gives the two lowest bands observed in ^{166}Er. The angular momentum and parity quantum numbers of the states occurring in the rotational bands imply a deformation with axial symmetry and invariance with respect to space and time reflection. The energies can be represented by a power-series expansion in the angular momentum, which converges rather rapidly for the range of angular momentum values included in the figure.

Similar expansions can be given for the matrix elements of tensor operators characterizing electromagnetic transitions, β decay, particle transfer, etc. As an example, Fig. 16 shows the electric quadrupole (E2) matrix elements between the two bands of ^{166}Er in Fig. 15. The analysis exemplified by Figs. 15 and 16 is based only on the symmetry of the deformation, and it is seen that such an analysis provides an appropriate framework for interpreting the detailed body of evidence on the nuclear rotational spectra.

GENERALIZED ROTATIONAL MOTION

The recognition of the deformation, and its degree of symmetry breaking, as the central element in defining rotational degrees of freedom, opens new perspectives for generalized rotational spectra associated with deformations in many different dimensions including isospace and gauge spaces, as well as orbital space. The resulting rotational band structure may involve comprehensive families of states labeled by the different quantum numbers of the internally broken symmetries, and there may be relations between quantum numbers referring to different spaces.

The Regge trajectories that play such a prominent role in the current study of the structure of hadrons have features reminiscent of rotational spectra, but as yet there appears to be no definite evidence concerning the nature of the deformation that would define an orientation of the intrinsic structure of a hadron.

The condensates in superfluid Fermion systems involve a static deformation of the pair field, and the processes of addition or removal of pairs from the condensate constitute a rotational mode in gauge space. An example of such a rotational excitation is provided by the two-neutron transfer process linking the ground states of the even Sn iso-

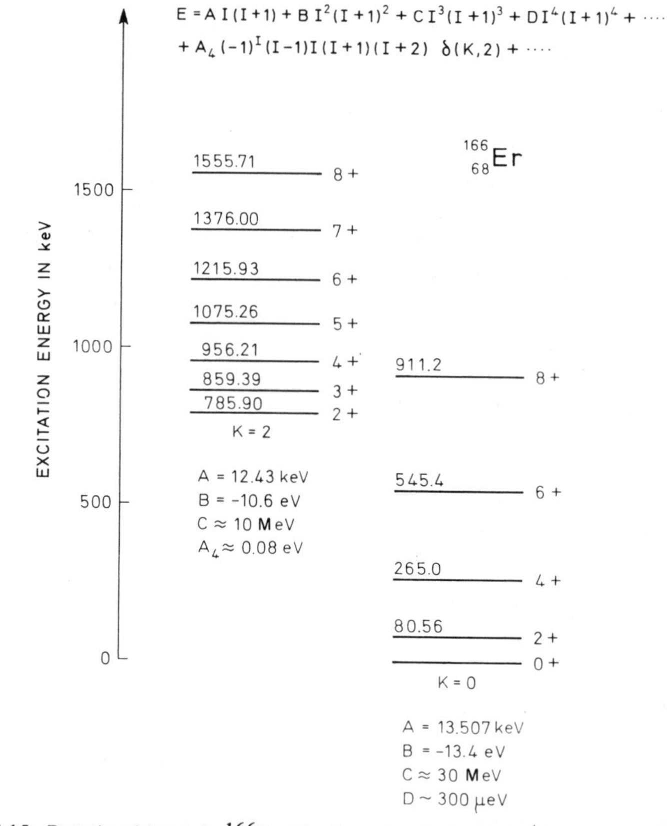

FIGURE 15 Rotational bands in ^{166}Er. The figure shows the observed energy levels belonging to the two lowest rotational bands in ^{166}Er, as studied in inelastic excitations and radioactive decay processes (C. W. Reich and J. E. Cline, *Nucl. Phys. A159*, 181 (1970); see also C. H. Lederer, J. M. Hollander, and I. Perlman, *Table of Isotopes*, 6th ed. (John Wiley & Sons, Inc., New York, 1967)]. Each energy level is labeled by the total angular momentum and parity $I\pi$, and the bands are further labeled by the quantum number K, which represents the component of the total angular momentum on the intrinsic symmetry axis. The collective deformation of the nucleus is directly revealed by the strong transitions between members of a band. For these transitions the electric quadrupole strength is several hundred times larger than for a single-particle transition. Moreover, the E2 transition probabilities for the various transitions within a band follow intensity relations with a simple geometrical basis (similar to those known from the band spectra of axially symmetric molecules). The nuclear shape is approximately spheroidal, with axial symmetry and reflection invariance. Such a deformation represents an only partially developed anisotropy (intrinsic breaking of rotational symmetry), and the rotational degrees of freedom are correspondingly reduced. Thus, the axial symmetry implies the absence of collective rotations with respect to the symmetry axis, and each rotational family of quantum states therefore involves only a single sequence with a fixed value of K. (The K = 2 band in ^{166}Er appears to represent a vibrational rather than a rotational mode of excitation.) The invariance of the nuclear shape with respect to a rotation of 180° about an axis perpendicular to the symmetry axis implies that this operation is a property of the intrinsic motion and leads to the selection rule I = 0, 2, 4 . . . , for the K = 0 band in ^{166}Er. The observed energy levels can be

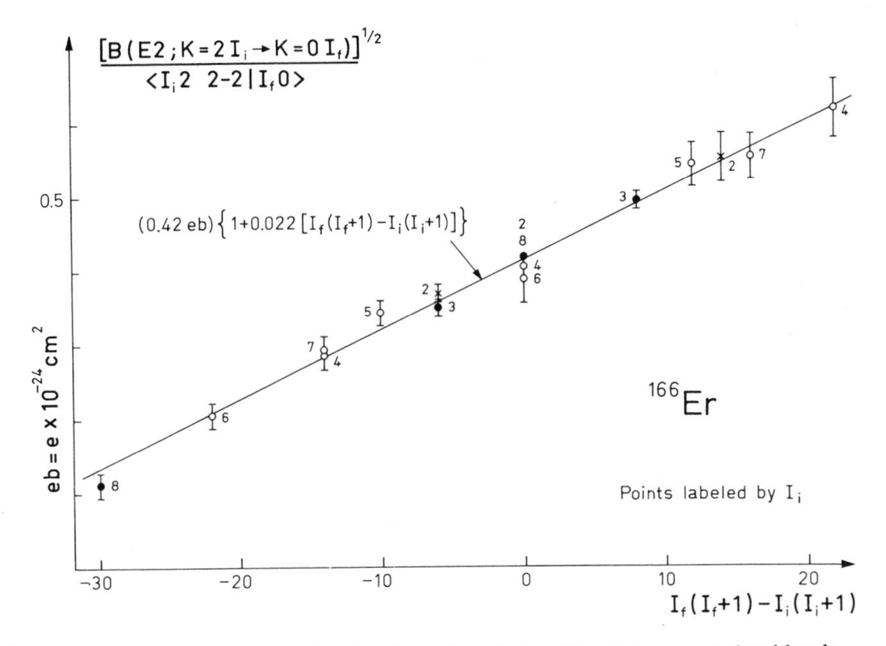

FIGURE 16 Intensity relation for electric quadrupole transitions between rotational bands. The figure shows the measured reduced electric quadrupole transition probabilities B(E2) for transitions between members of the $K = 2$ and $K = 0$ band in ^{166}Er shown in Fig. 15. [C. J. Gallagher, Jr., O. B. Nielsen, and A. W. Sunyar, *Phys. Lett. 16*, 298 (1965); C. Gunther and D. R. Parsignault, *Phys. Rev. 153*, 1297 (1967)]. An expansion similar to that of the energies in Fig. 15, but taking into account the tensor properties of the E2 operator, leads to an expression for $[B(E2)]^{1/2}$ that involves a Clebsch-Gordan coefficient (geometrical factor) multiplied by a power series in the angular momenta I_i and I_f of the initial and final states. The leading-order term in this expansion is a constant, and the next term is linear in $I_f(I_f+1) - I_i(I_i+1)$. The experimental data are seen to be rather well represented by these two terms. The angular-momentum-dependent terms in the expansion of energies and matrix elements can be interpreted in terms of the coupling between bands produced by the Coriolis and centrifugal forces. Thus, a term in the E2 matrix element propertional to $I_f(I_f+1) - I_i(I_i+1)$ arises from a coupling between the $K = 2$ and $K = 0$ bands; even very small admixed amplitudes resulting from this coupling may contribute significantly to the E2 transitions between the bands, since these admixed amplitudes involve the strongly enhanced E2 matrix elements characteristic of transitions within a band. This circumstance accounts for the relatively large value of the term proportional to $I_f(I_f+1) - I_i(I_i+1)$ in Fig. 16; higher-order terms do not have a similar enhancement and are therefore expected to be much smaller.

---

expressed in terms of a power series in the rotational angular momentum. The form of this expansion, given in the upper part of the figure, follows from the symmetry of the deformation. The coefficients in the expansion, derived from the empirical energies, are given for each band. It is seen that successive coefficients decrease by a factor of order 10^3, and hence the convergence is rather rapid for the range of angular momenta in the figure. (The rate of convergence is similar to that observed in the H_2 molecule.)

topes, as illustrated in Fig. 13. The intensity of this transition provides a measure of the deformation of the pair field, also referred to as the order parameter.

PHASE TRANSITIONS INDUCED BY ROTATIONAL PERTURBATION; YRAST REGION

The very detailed studies of the nuclear rotational motion in the three-dimensional space of our daily-life experience have provided a large body of information concerning the response of the nuclear structure to the rotation of the average deformed field in which the nucleons move. In particular, these studies have revealed the major effect of the superfluidity on the collective rotational flow and the resulting moments of inertia.

The perturbations of the intrinsic structure produced by the rotation increase rapidly with angular momentum, and an exciting new frontier is opening as a result of the possibility of transferring to the nucleus such large amounts of angular momentum that the structure undergoes major modifications.

The nuclear spectrum as a function of angular momentum is illustrated schematically in Fig. 17. The states with lowest energy for given angular momentum are referred to as the yrast line; in this region of the spectrum, the nucleus though highly excited may be considered as "cold," since almost the entire excitation energy is expended in generating the angular momentum. The available beams of heavy ions provide powerful tools for producing nuclear states with 20, 50, or perhaps 100 units of angular momentum. In this angular momentum range, one can envisage a number of different phase changes in the nuclear structure including the disappearance of superfluidity (in analogy to the breakdown of superconductivity by a magnetic field), the loss of axial symmetry (an effect related to instability phenomena in rotating stars), as well as a variety of different fission processes engendered by the centrifugal forces. In other situations, the quantal effects associated with the shell structure may lead to discontinuities in the yrast line, associated with a maximal alignment of the angular momenta of the particles within each shell.

Already the first glimpse into the region $I \approx 15\,\hbar$ to $20\,\hbar$ has revealed striking new phenomena as illustrated in Fig. 18, which shows the moment of inertia as a function of the rotational frequency along the yrast line of ^{162}Er. The interpretation of the pronounced structure in

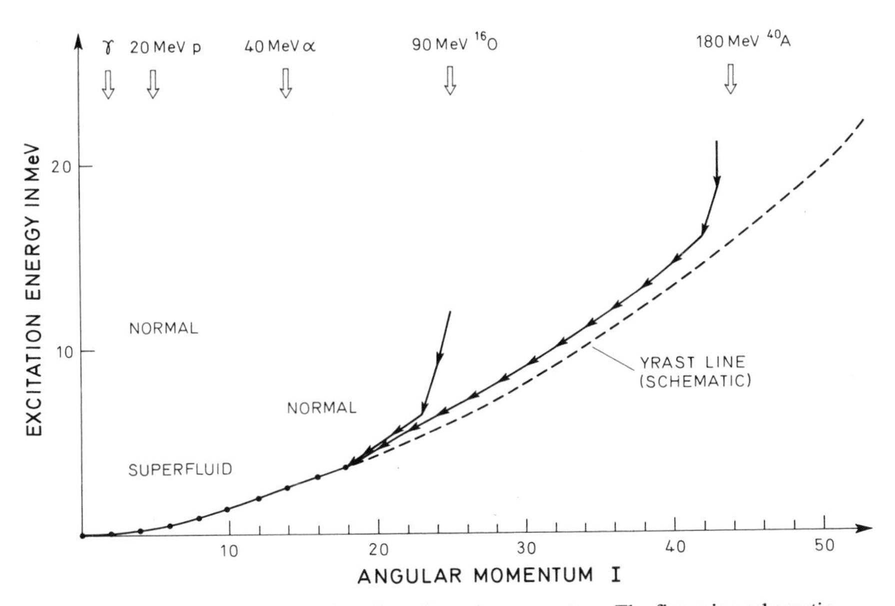

FIGURE 17 Nuclear spectrum as function of angular momentum. The figure is a schematic representation of some of the important landmarks in the nuclear spectrum as a function of angular momentum I. (The magnitudes of the energies are chosen to correspond to a nucleus with mass number $A \approx 160$). The lowest state of given angular momentum is referred to as the "yrast" level. The low-angular-momentum portion of the yrast line ($I \leqslant 18$) corresponds to the known levels of ^{162}Er (see Fig. 18); the tentative continuation of the yrast line is based on the assumption that, for high angular momentum, the nucleus will rotate with a moment of inertia corresponding to rigid rotation. In the region above the yrast line, the density of levels increases rapidly with excitation energy (see, for example, Fig. 11). The low-energy and low-angular-momentum states exhibit strong pair correlations of the type that are characteristic of a superfluid phase. With increasing angular momentum and energy it is expected that these correlations will become less effective and that the system will go over to a "normal" phase. The arrows at the top of the figure indicate the magnitude of the angular momentum that is brought into the nucleus in a grazing collision with a number of different projectiles currently available. In a typical reaction, the initially formed compound nucleus is in a highly excited state, which "cools" through a sequence of neutron emissions; the evaporated neutrons are light and of low energy (the nuclear temperature is of the order of 1 MeV) and thus carry away very little of the angular momentum. When the system arrives at an excitation energy with respect to the yrast line that is less than the neutron separation energy (≈ 5 MeV), the subsequent decay takes place by a cascade of gamma rays. The figure shows two illustrative (hypothetical) gamma cascades leading eventually to the yrast levels in the region of $I \approx 20$.

Fig. 17 is not yet definitely established, but it is possible that we are encountering the disappearance of superfluidity, which is expected to lead to a normal phase rotating with a moment of inertia equal to that for rigid rotation. We here face the challenging task of analyzing a phase transition as revealed by the properties of the individual quantum states of the system.

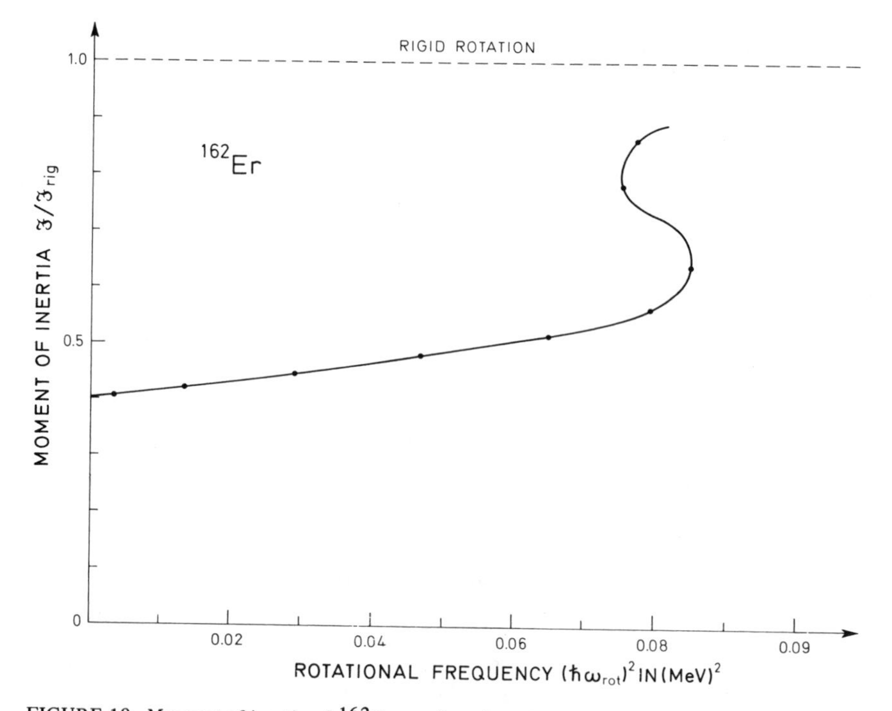

FIGURE 18 Moment of inertia of ^{162}Er as a function of rotational frequency. Within a rotational sequence, the energy is a smooth function of the angular momentum; from the observed energies, it is thus possible to derive the rotational frequency, which is the derivative of the energy with respect to the angular momentum. In the figure, the frequency ω_{rot} has been obtained by interpolating linearly in the variable $I(I+1)$ between successive members of the rotational band. The moment of inertia $\mathcal{J}$ is obtained in a similar manner from the canonical definition, as the ratio of angular momentum to rotational frequency. The figure gives the observed moment of inertia as a function of ω_{rot}^2 for the ground-state rotational band of ^{162}Er [A. Johnson, H. Ryde, and S. A. Hjorth, *Nucl. Phys. A179*, 753 (1972)]. For low rotational frequencies, $\mathcal{J}$ varies as a linear function of ω_{rot}^2, with very high acuracy. The irregular variation of $\mathcal{J}$ in the figure occurs for $I \approx 16$ and provides evidence for a major change in the nature of the rotational motion.

CONNECTIONS TO OTHER FIELDS

In the present report, it has only been possible to touch upon a few of the areas of activity in present-day nuclear-structure research, but the examples may give an impression of the depth of penetration into the nuclear world that has been achieved by the ingenuity and resourcefulness of the experimental effort and illustrate the richness of the phenomena that have been revealed to the explorers. These examples may also indicate some of the perspectives that can be seen at the various frontiers and the opportunities that lie ahead for vastly expanding the range of phenomena accessible to investigation. Already with the tools that are now becoming available or are technically feasible, we appear to be on the threshold of a major increase in the scope of the field; the heavy-ion beams will make it possible to study nuclear matter in quite new regimes, and the high-energy probes are providing microscopes for exploring the texture of this matter at new orders of resolution. On the "inner frontier," the greatly increased precision and flexibility of nuclear spectroscopy will constitute a challenge for sharpening the concepts employed in the description of the nuclear many-body system.

At many points, we have also alluded to the intimate connection between the nuclear phenomena and those encountered in other domains of quantal physics. These connections that stem from the ubiquitous role of many-body problems in the present stage of physics have become increasingly apparent in recent years and have been of inspiration not least to the nuclear physicists, who find themselves at an intermediate position on the quantum ladder. The efforts to view the developments in the various domains of quantal physics as a unified whole may in the future, to an even greater extent, become a stimulus for continued progress over the broad front. Not least, it may be important to give proper emphasis to this perspective in the presentation of our science to the coming generations of participants.

It has not been possible within the framework of this report to deal with the vast network of applications of nuclear science and techniques. The access to energy sources hitherto confined to the sun and stars has opened opportunities and provided challenges of truly revolutionary dimensions. Less dramatic, but perhaps of no less profound significance, is the increasing extent to which nuclear processes and techniques provide tools for probing and transforming both living and inanimate matter at the various levels of organization. Such applications have become indispensable in almost all the modern fields of natural science and in many areas of industry, and indeed contribute a major component to the arsenal of tools available for attacking problems in our modern society. These potentialities of nuclear science give an additional dimension of contact between the participants in the research effort and the community at large.

VICTOR F. WEISSKOPF *is an Institute Professor and head of the physics department at the Massachusetts Institute of Technology. He served as Director General of the European Organization for Nuclear Research (CERN) from 1961 to 1965 and is a member of the U.S. National Academy of Sciences, the French Academy of Science, the Austrian Academy of Science, and the Danish Academy of Science. He is a Vice President of the International Union of Pure and Applied Physics.*

10 *What Is an Elementary Particle?* / VICTOR F. WEISSKOPF

The idea that matter consists of some simple and unchanging elementary constituents is deeply ingrained in our way of thinking. We observe that matter appears in enormous varieties of different realizations, qualities, shapes, and forms, transforming and changing from one into others. In these changes, however, we observe many recurrent properties, many features that remain unchanged, or if changed, return under similar conditions. We find constancies and regularities in the flow of events; we recognize materials with well-defined properties, as water, metals, rocks, or living species; we conclude that there must be something unchanging in nature that causes these recurrent phenomena. This is the origin of the idea of elementary particles. Newton has expressed it very succinctly:

All these things being considered, it seems probable to me that God in the beginning formed Matter in solid, massy, hard, impenetrable, moveable Particles, of such Sizes and Figures, and with other Properties, and in such Proportion to Space, as most conduced to the End for which he formed them; and that these primitive Particles being Solids, are incomparably harder than any porous Bodies compounded of them; even so very hard, as never to wear or break in pieces; no ordinary Power being able to divide what God himself made in the first Creation. While the Particles continue entire, they may compose Bodies of one and the same Nature and Texture in all Ages: But should they wear away, or break in pieces, the Nature of Things depending on them would be changed. Water and Earth, composed of old worn Particles and Fragments of Particles would not be of the same Nature and Texture now, with Water and Earth composed of entire Particles in the Beginning. And therefore that Nature may be lasting, the Changes of corporeal Things are to be placed only in the various Separations and new Associations and Motions of these permanent Particles.

Newton recognized the problem of elementary particles: they must have well-defined specific, unchanging properties; in his time, this quality could only result from being "incomparably" hard. Such elementary units were in fact discovered during the eighteenth and nineteenth centuries, when the chemists found that all matter is made up of 92 different species of "atoms," a term that is the Greek equivalent of Newton's "incomparable hardness."

The problem took a quite different turn, however, when in 1911 Rutherford took apart what "no ordinary power [is] able to divide," by showing that the atom consisted of electrons and a nucleus. The ever-recurring, specific, and unchanging properties of the atom had to be explained in view of that fact that it can be changed and taken apart. A solution to this problem was given by the discovery of the role of the quantum of action in the behavior of matter in the small. It led to the development of quantum mechanics, which is based on the particle–wave duality in the behavior of electrons and other small entities. The wave properties give rise to typical recurrent structures and patterns whenever such entities are held together by attractive forces. One of the consequences is the fact that a system of bound particles can be found only in a number of discreet quantum states. In each of these states the energy and many other properties are well defined and fully determined by the nature of the system, its forces and symmetries. The recurring, specific features of the atoms are explained by the existence of these quantum states: whenever the object assumes one of these states—in particular the state of lowest energy, the ground state—it will have the same identical properties associated with this state. The energy threshold between the ground state and the next higher ones plays an essential role: if the interactions with the surroundings are energetically small compared to this threshold, the system always remains in the ground state—it acts like an elementary particle as if it were "incomparably hard"; it has well-defined properties, mass (energy), charge (if it is an ion), angular momentum, fixed electric and magnetic moments, form factors, etc. Only when the interactions become more energetic than this threshold would the system lose its "elementarity"; then the specificity of its original quantum state would be destroyed, and it would exhibit different properties.

A typical spectrum of quantum states of an atom (thallium) are shown in Fig. 1. The energies and shapes of the quantum states are de-

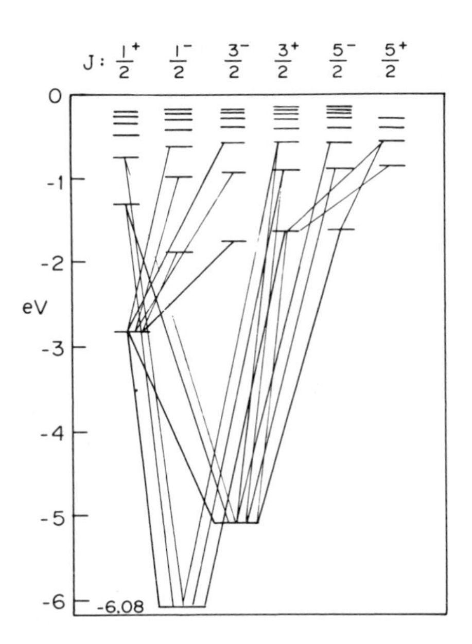

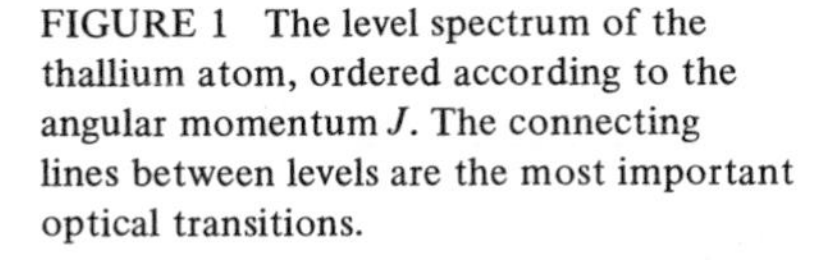

FIGURE 1 The level spectrum of the thallium atom, ordered according to the angular momentum J. The connecting lines between levels are the most important optical transitions.

termined by the quantum mechanics of electrons attracted by the Coulomb field of the nucleus; they are the energies of the order of a few electron volts and dimensions of the order of 10^{-8} cm. Thus, the electron volt is the energy scale of atomic physics, and the angstrom (10^{-8} cm) is the scale of atomic sizes.

The quantum states are grouped according to their angular momentum J. Each state is $(2J+1)$-fold degenerate, a consequence of the spherical symmetry of the electric force in the atom. The excited states have a finite lifetime; transitions to lower states take place, which, in isolated atoms, are accompanied by the emission of light quanta, carrying away the balance of energy and also one unit of angular momentum.

It also follows from the quantum mechanics of electrons in the Coulomb fields of nuclei that two or several atoms are attracted to each other. In many cases, the attraction is strong enough to form molecules and solids. The force of attraction between two atoms depends in its details on the type of atoms; the general form of the potential of this force is shown in Fig. 2 as function of the distance between the nuclei. The distance is measured in angstroms, the energy in electron volts. We notice a potential minimum of a few electron volts at about one unit of distance and a strong repulsion at smaller distances. We also notice that the force goes rapidly to zero outside a range of a few units. It is important to realize that this force is a direct quantum-mechanical consequence of the electrostatic attraction or repulsion between the electrons

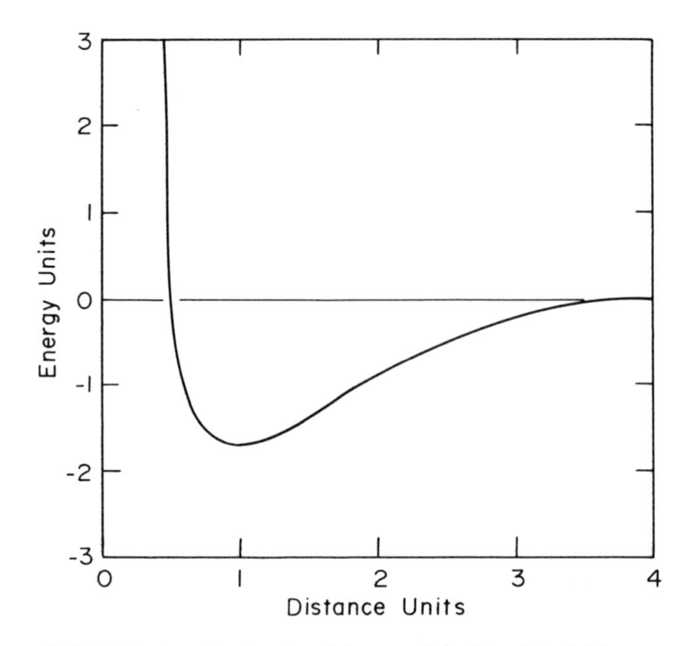

FIGURE 2 A qualitative plot of the potential of the chemical force between two atoms and of the nuclear force between two nucleons. The energy units are eV in the first case and about 20 MeV in the second case. The distance units are measured in angstroms (10^{-8} cm) between the nuclei in the first case and in fermis (10^{-13} cm) between the nucleons in the second case.

and nuclei involved. Chemistry is reduced to physics in the sense that the chemical bound is recognized as a manifestation of the fundamental Coulomb force acting between atomic constituents.

The quantum mechanics of electrons and nuclei describes and explains a large part of our natural environment. The behavior of gases, solids, liquids under terrestrial conditions, the phenomena of chemistry, and even the phenomena of living matter can be understood in principle on that basis and understood as the behavior of two types of particles—electrons and atomic nuclei—which attract or repel each other by a simple force proportional to the inverse square of their distances.

Quantum mechanics has given the answer to the problem of how composite systems can show typical recurrent properties, but it is only a partial one. It is predicated upon the assumption that the constituents—electrons and nuclei in the atomic case—are in themselves "elementary," that they are endowed with typical well-defined and unchangeable properties.

Are the atomic nuclei and the electrons incomparably hard? Certainly in the former case the answer is negative. Within six years after his atomic experiments, Rutherford showed that a nucleus can be transformed, and Chadwick's discovery of the neutron in 1932 (20 years after Rutherford's discovery of the atomic structure) revealed the composite nature of nuclei; they consist of protons and neutrons–nucleons–held together by a newly discovered agency, the nuclear force.

A new world of phenomena was discovered, the world of nuclear excitations, of nuclear reactions, of new kinds of radiation, of radioactivity, and of nuclear fission and fusion. This world is dormant under terrestrial conditions, where energy exchanges are way below the nuclear thresholds. It comes to life only in our laboratories when the requisite energies are applied to matter. It is most active, however, in the center of stars where it produces stellar energy and in star explosions where it is responsible for the production of atomic nuclei. A few of these natural nuclear processes are still going on here on earth; natural radioactive elements are the last embers of the stellar explosion that created terrestrial matter.

Closer studies of a nuclear force revealed that its details depend on the relative orientation of the spins of the nucleons between which it acts and the symmetry properties of the quantum state, but the general forms of the potential are also shown by Fig. 2; here the unit of distance is the fermi (1 F = 10^{-13} cm), and the unit of energy is about 20 MeV. In contrast to the electric attraction, this force has a finite short range of about 2 F; the potential has a minimum of about 20 MeV at about 1 F and becomes strongly repulsive at smaller distances.

The behavior of a system of neutrons and protons under the influence of such a force can be predicted by quantum mechanics, and the experimental investigation of nuclear processes bore out these predictions. In the years after 1932, nuclear physicists were engaged in a repeat performance of the application of quantum mechanics to bound systems, nucleons instead of electrons. Again we find discreet quantum states in nuclei (Fig. 3), but the larger mass of the constituents and the stronger force causes the energy scale to be in units of MeV, rather than eV, and the spatial dimension to be in fermis rather than angstroms.

Let me point out a typical difference between the effects of the chemical and nuclear force. The motion of the nuclei in a molecule under the influence of the chemical force is such that the corresponding de Broglie wavelengths are considerably shorter than the molecular dimensions. Hence the positions of nuclei within the molecules are reasonably well defined, causing the well-known "architecture" of molecular structure–steric arrangements, lattices, helices, etc. In the nuclear struc-

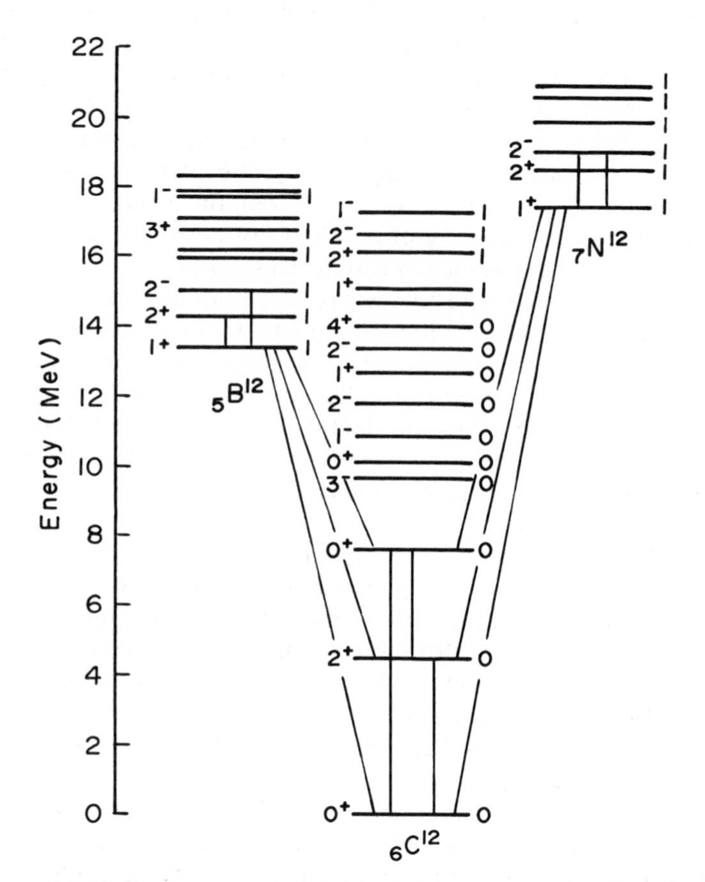

FIGURE 3 The level spectrum of the 12-nucleon system. The numbers on the left of each level give the angular momentum J and parity, the numbers on the right give the isotopic spin. The connecting lines are the most important transitions; the vertical ones are electromagnetic, the skew ones are lepton pair emissions.

ture the situation is different: the wavelengths of the nucleon motion within the nucleus under the influence of the nuclear forces (kinetic energies of about 30 MeV) are of the same order as the range of the force and not much smaller than the size of the system. Hence nuclear structure will not exhibit the steric structure observed in molecules; the positions of the nucleons are spread over distances comparable with the dimensions of the nucleus.

An important new feature appears in the transitions from one nuclear quantum state to another. Again we find the emission and absorption of light quanta, but there appears a new way of getting rid of the energy of an excited state: the emission of an electron–neutrino or positron–neutrino pair. It is the radioactive decay of nuclei (β-radioactivity) and is ascribed to an effect called "weak interaction." The fundamental process is the transition of a proton into a neutron with the emission of a positron plus neutrino or the transition of a neutron into a proton with the emission of an electron plus an antineutrino:

$$n \to p + e^- + \bar{\nu}, \quad p \to n + e^+ + \nu.$$

The neutrino and the antineutrino are massless particles that always accompany electrons or positrons in weak interactions.

Such emissions are connected with a change of charge of the nucleus. Protons are transformed into neutrons or vice versa. Hence we should consider the two nucleons no longer as two different particles but rather as two quantum states of the same entity, the neutron being higher in energy and able to perform a transition to the proton state. The opposite transition takes place within certain nuclei, when the difference in nuclear binding or electrostatic effects depresses the energy of the neutron below that of the proton.

In the spectra of nuclear quantum states the ordering quantum number again is the angular momentum J, an effect of the spherical symmetry of the forces. But there is a new important symmetry here: the nuclear force does not distinguish protons and neutrons; the nuclear structure exhibits a symmetry with respect to the exchange of a proton with a neutron. This symmetry results in the appearance of a new quantum number I, which is called the "isospin." It has some formal analogy to the ordinary spin; in both cases we deal with the consequences of an alternative between two fundamental states: proton and neutron here, spin-up and spin-down of an object with a half-unit spin there. There are $2I+1$ states of similar nuclear structure that form a multiplet of isotopic spin I. They differ from each other only by the fact that neutrons are replaced by protons or vice versa. Each substate has the same number A of nucleons and almost the same nuclear energy; they differ in charge (number of protons), and therefore they belong to different nuclei. It is reasonable, however, to consider all nuclei with the same A to be different states of the same entity, since the proton and the neutron are but two states of the same particle.

After the discovery of nuclear structure, the world seemed to exhibit a relatively simple fundamental structure; it seemed to consist of very few elementary particles: the proton, the neutron, the electron, the neutrino, and the light quantum. This situation, however, turned out to be deceptive. It seemed suspicious that the nuclear force does not appear to be as simple and fundamental as the Coulomb force between two charges; it rather resembles the chemical force between atoms, which is an effect derived from the Coulomb forces between the atomic constituents. This analogy would point toward some internal structure of the nucleon containing the fundamental agent responsible for the nuclear force. Furthermore, other particles—mesons, hyperons, muons—have been discovered in the cosmic radiation. Finally, in 1952, 20 years after Chadwick's discovery of the neutron and 40 years after Rutherford, Fermi and his collaborators found a short-lived highly excited state of the proton, and the neutron, the so-called Δ-state. The nucleon seemed to be excitable into different quantum states; it therefore could no longer be considered as elementary, it must have some internal structure.

This was the beginning of a series of discoveries that opened up another quite different world of phenomena, which becomes active if matter is subject to energies of the order of several hundred MeV or higher. It is a world full of new and unexpected phenomena, such as excited nucleons, heavy electrons, and strange mesons—a world whose elementary structure is still unknown.

Before we proceed with the description of this world, a few words must be said about the properties of particles under relativistic conditions, that is, when they enter into interactions with energies comparable to or higher than their rest masses. This is one of the features by which the new subnuclear world distinguishes itself from the previous phenomena. The energies involved in the excitation and decomposition of atoms and of nuclei are negligibly small compared to the mass energies of the constituents. Atomic excitation energies are only about 10^{-5} times the electron mass and less than 10^{-8} times the mass of the nucleus; nuclear excitations deal with energies that are roughly a thousandth of the nuclear mass. This is no longer so in the subnuclear processes. The excitation energy of the Δ-state is about one third that of the proton and higher than the mass of the lightest meson. The excitation energies of other excited states of the nucleon are near or even higher than its mass energy. Therefore, in any subnuclear process we are dealing with interaction energies comparable to the mass energies of the participants. There are remarkable consequences of this fact, which are closely related to the existence of antimatter.

In 1932, not only was the neutron found, but Anderson and Neddermeyer discovered an electron with positive charge. This discovery confirmed one of the most dramatic conclusions drawn from Dirac's equation of the relativistic electron. This equation is one of those great intellectual insights that contain much more than the author suspected at the outset. It exhibits a fundamental symmetry, which later on turned out to hold for any particle, not only for electrons—the doubling of natural phenomena: to every particle there must exist an antiparticle of equal mass and spin but of exactly opposite electromagnetic and such other properties as charge and magnetic moment. Hence, for any form of matter there exists a corresponding form of antimatter. The depth of this idea must be appreciated by considering the fact that the world as we face it practically does not contain any samples of antimatter. There are ways, however, in which such samples can be created. These are the processes of pair creation in which a pair of a particle and an antiparticle appear when sufficient energy is applied. The opposite process, the mutual annihilation of an antiparticle with a particle (transformation into some form of energy), causes the quick disappearance of any sample of antimatter when it comes in touch with ordinary matter.

The processes of creation and annihilation of particle–antiparticle pairs are rare phenomena in the atomic and nuclear world, but they play an important role in the world of subatomic phenomena. Antiparticles are copiously produced in high-energy collision processes, and particle–antiparticle pairs will be involved in the dynamic structure of the entities. The latter circumstance is a consequence of the fact that the interaction energies are of the order of the mass energies of the particles; hence, these interactions are strong enough to produce pairs in the neighborhood of a given particle. It therefore is surrounded by a cloud of "virtual" pairs, and the system can be considered as a conglomeration of an indefinite number of particles and antiparticles. In such cases, it is not possible to speak of a definite number of constituents in a given entity, as we do when we count electrons and nuclei in atoms, or protons and neutrons in nuclei.* This is a new feature, which complicates the interpretation of the subnuclear phenomena; we are not used to dealing with such situations.

We now return to the excited states of the nucleon. In the years after the discovery of the previously mentioned Δ-state, many other excited states were identified, and it was recognized that some of the new particles found previously in cosmic-ray events were in fact excited states of the nucleon. The term "baryon" is used for the nucleon and its excited states. Figure 4 represents the "spectrum" of the most important

*The rudiments of this phenomenon are already present in atoms, in the form of the so-called polarization of the vacuum. The Coulomb force produces a slight accumulation of virtual electron–positron pairs in the neighborhood of the atomic nucleus, within a distance of a Compton wavelength of the electron (10^{-11} cm). However, the effective number of these pairs is small ($\sim Z/137$), because of the weakness of the Coulomb force.

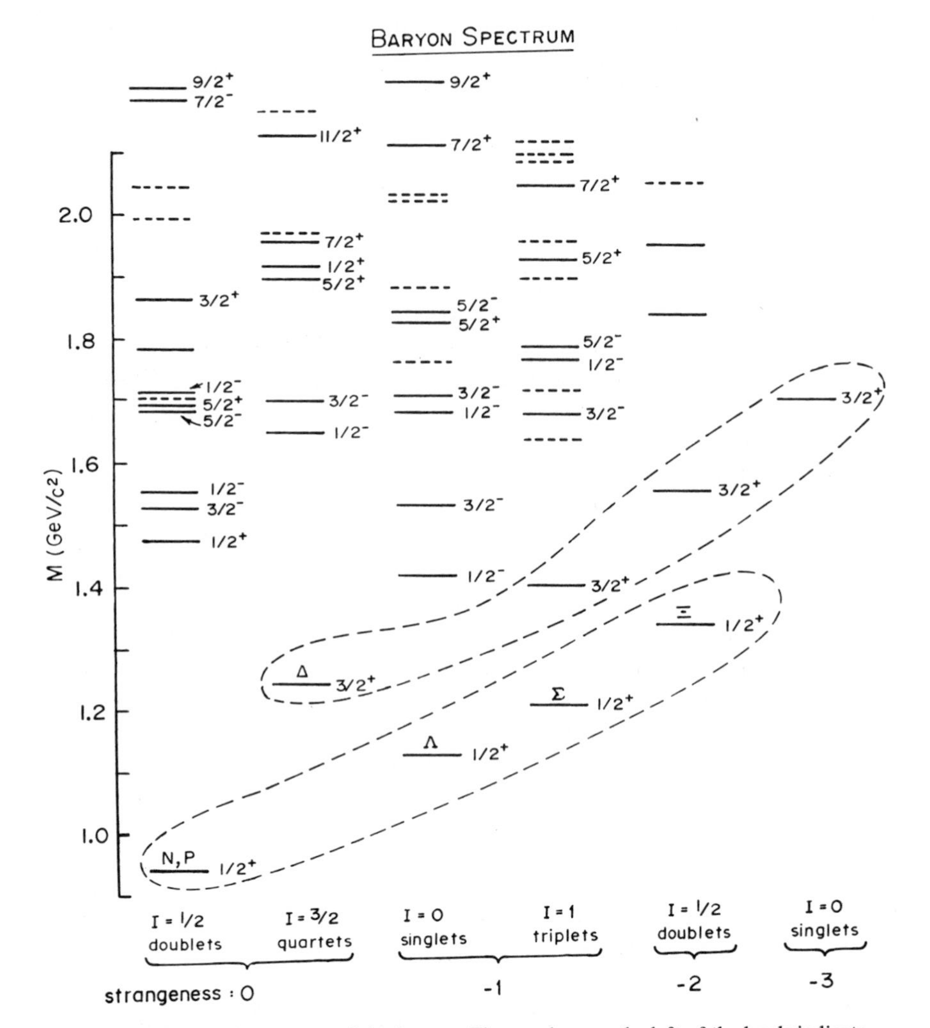

FIGURE 4 The levels spectrum of the baryon. The numbers on the left of the levels indicate the angular momentum quantum number and the parity.

ones of these quantum states. In certain ways this spectrum is similar to the spectra of atomic and nuclear systems, in some ways it is different. The energy scale here is of the order of 10^8 eV thousand times larger than the nuclear scale. The ground state is an almost degenerate doublet: the neutron and the proton. Again we can ascribe the quantum numbers J and I to the levels. The former is the angular momentum; all baryons have half-integer J-values, and each level is a multiplet $2J + 1$ sublevel.

The existence of these degenerated multiplets is based on the same principles as in atoms and nuclei; it is nothing but a reflection of the fact that there is no preferred orientation in space for the baryon system. The isospin I plays a similar role here as in the nuclear spectrum and gives rise to almost degenerated multiplets. The neutron–proton doublet is the simplest example; it is only a special case of a more general symmetry: other states also form doublets, triplets, and quartets of states with different charges but almost the same energies and other properties. Furthermore, there appears a new quantum number S, the "strangeness"; the levels are separated into groups with $S = 0, -1, -2, -3$. The significance of this grouping will become clearer later on.

Some important new features of this spectrum show up in the transitions between the quantum states (Fig. 5). They are transacted by the emission or absorption of light quanta as in the other cases and by the emission of electron–neutrino pairs with the corresponding changes of charge. Here already we observe something new: whenever such pairs are emitted, there is also an emission of another electron together with a neutrino; this new type of electron—the muon—is 207 times heavier than the normal one, but in all other respects it seems to be identical and follows the same laws on interaction. The accompanying neutrino or antineutrino is also massless but, according to recent experimentation, is not identical with the neutrino that goes with the ordinary electrons. The muon had been discovered before among the products of the cosmic radiation; it is an unstable particle and decays into a muon-neutrino with the emission of an electron–neutrino pair. The reasons for the existence of a heavy electron are completely unknown. It is one of the many unsolved riddles in this realm of subnuclear phenomena.

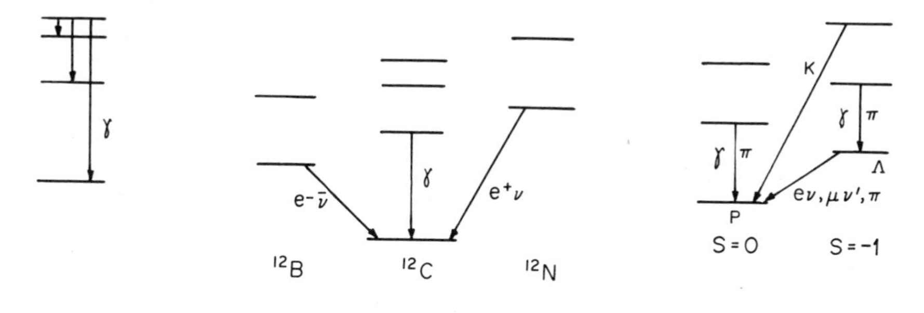

FIGURE 5 The different kinds of transition between the levels of excitation in the different spectra. The symbol γ refers to photons.

The most important new energy "currency," which is copiously exchanged in transitions between baryon states, is the *meson.* Let us first describe the situation as it seemed about 12 years ago: At that time we knew of two types of meson, the pions and the kaons. They are massive entities, the pion mass being about 1/7 of the proton mass, the kaon mass is a little more than half of the proton mass; some mesons are charged and some not; the pions form a triplet with charges +1, 0, −1, the kaons form doublets with charges +1 and 0 (there is also a doublet of "antikaons," having the charges −1, 0). Now the significance of the strangeness quantum number S will emerge: the pions are emitted or absorbed only in transitions between states of the same strangeness, the kaons or their antiparticles only between states differing by one unit of S. One therefore considers the pions as particles with $S = 0$, and the kaons as carriers of one unit strangeness. (The antikaons, of course, carry one unit of negative S.) There is a law of conservation of strangeness in the emission and absorption of mesons that results in the above regularities. At the same time, the total electrical charge must also remain conserved so that the emission or absorption of charged mesons is tied to a change of charge.

Most of the meson emissions and absorptions are "strong" compared with other types of such processes. For example, the widths of most excited levels in the baryon spectrum are of the order of a tenth or at least a hundredth of their distance; that is much more than, say, the ratio in atomic levels decaying via electromagnetic interactions. We therefore call the interactions between mesons and baryons "strong interactions."

The large mass of the mesons has an interesting consequence. The lowest baryon states with a strangeness different from zero are all at higher energy than the ground state (proton, neutron). Still they cannot perform the transitions to the proton by the emission of a kaon, because the mass of the kaon is larger than the energy differences. Hence those states are stable against strong interactions. They are not completely stable, however, since they can perform those transitions by other means. Not by the emission of light quanta, because the electromagnetic interactions were found also to obey strictly the conservation of strangeness, but by the emission of lepton pairs, that is, by weak interactions that are found not to obey that rule. (We introduce the term "lepton" for the two types of electron and their neutrinos.) Things are further complicated by the fact that not only do lepton pairs seem to be emitted in weak interactions but also mesons. In this case, however, the S-selection rule is not valid, and π-mesons are emitted weakly in strangeness changing transitions. Those low-lying states with strangeness $S \neq 0$ are relatively long-lived—their lifetimes measure 10^{-10} sec—whereas states that emit mesons via the strong interactions have lifetimes of 10^{-23} sec or less. This is why the former states are more easily observed and have received special names such as Λ, Σ, Ξ, Ω particles.

The assumption of two types of meson, π and K, turned out to be too simple. Experiments have shown that many more types with different masses and angular momenta are emitted and absorbed by the baryons; one has today an extended spectrum of mesons as shown in Fig. 6. The

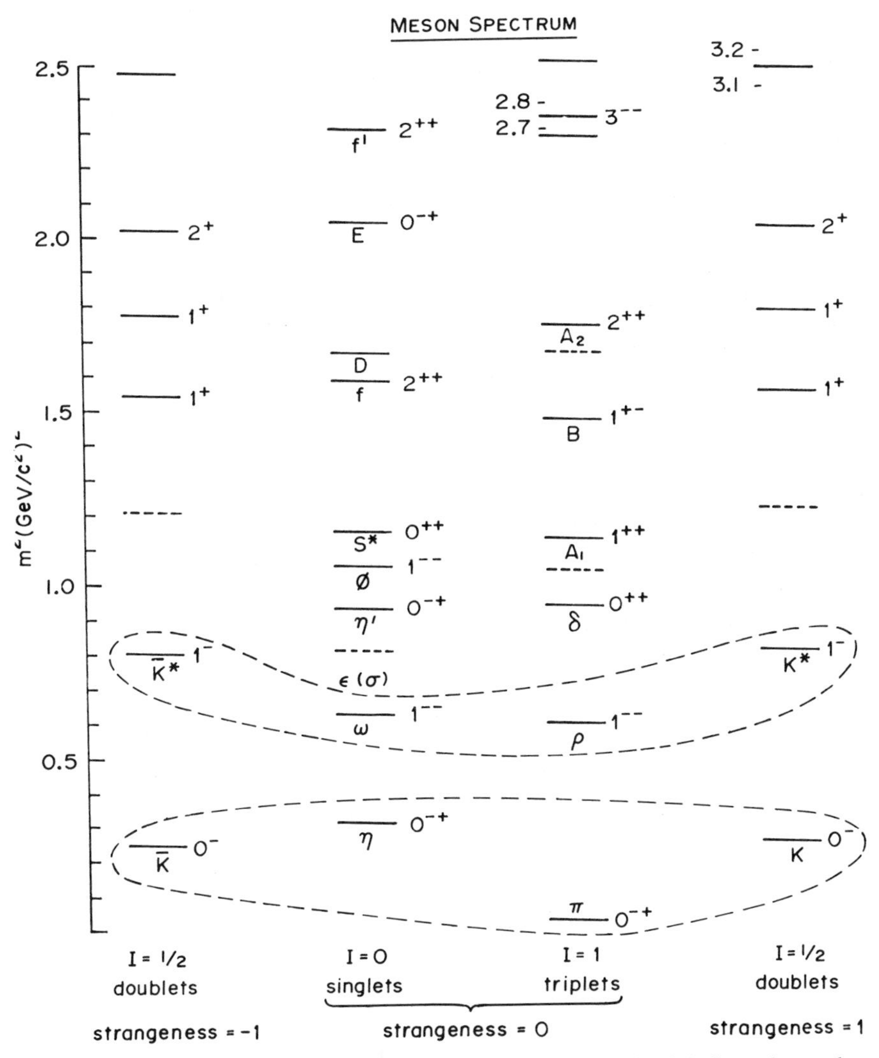

FIGURE 6 The spectrum of mesons. The numbers on the left of the levels indicate the angular momentum quantum number, the parity sign, and, in some cases, the charge-conjugation sign.

situation is the same, however, with respect to the strangeness selection rule: all known mesons can be grouped in three groups, those with zero strangeness, the brothers of the pion; those with one positive unit of strangeness, the brothers of the kaons; the antiparticles of the latter group, which have negative strangeness. These new types of mesons may be considered as excited states of the pion and kaon. They are indeed short-lived and decay mostly into the pion or kaon by the emission of other pions and kaons. The angular spin values are all integer in this spectrum.

Even the ground states of the meson spectrum, the pion and the kaon, have a finite lifetime. We have considered them to be a new energy currency; as such they can be transformed into other forms of energy. For example, the uncharged pion decays into two light quanta, the charged pion and the kaons into lepton pairs (Fig. 7).

The excited states of the nucleon of the meson are often referred to as "new" particles. This is the origin of the frequently heard assertion that the high-energy physicists have discovered up to a hundred elementary particles. Actually they should not be considered as new particles; they are new forms of realization of the baryon or the meson; their existence points toward a yet unknown internal dynamics of these entities.

All the diverse interactions, decays, and transformations of the different entities can be interpreted as obeying certain important conservation laws. The following characteristic numbers are strictly conserved and do not change in any process: the total electric charge Q, the "baryon number" B, and the "lepton number" L. The baryon number is the total number of nucleons minus the total number of antinucleons

$$\pi^{\circ} \longrightarrow 2\gamma \qquad\qquad K^{+} \begin{cases} \rightarrow (e^{+}\nu) \\ \rightarrow (\mu^{+}\nu) \end{cases}$$

$$\pi^{+} \begin{cases} \rightarrow (e^{+}\nu) \\ \rightarrow (\mu^{+}\nu) \end{cases} \qquad\qquad K^{-} \begin{cases} \rightarrow (e^{-}\bar{\nu}) \\ \rightarrow (\mu^{-}\bar{\nu}') \end{cases}$$

$$\pi^{-} \begin{cases} \rightarrow (e^{-}\bar{\nu}) \\ \rightarrow (\mu^{-}\bar{\nu}') \end{cases} \qquad\qquad \eta \longrightarrow 2\gamma$$

$$\mu^{+} \longrightarrow (e^{+}\nu) + \bar{\nu}'$$

$$\mu^{-} \longrightarrow (e\ \bar{\nu}) + \nu'$$

FIGURE 7 Some meson transformations into other forms of energy.

(excited or not). The lepton number is the total number of electrons (heavy or light) and neutrinos minus the total number of antielectrons and antineutrinos. The antiparticles are always counted negative. The conservation of B and L prevent, for example, the decay of a proton into a positron with the emission of a photon; it therefore guarantees the stability of the material world.

The mesons and photons carry no baryon or lepton number; they are entities with $B = 0$ and $L = 0$; no conservation law prevents their appearance or disappearance.

We have given a sketchy and incomplete description of the new realm of natural phenomena that opens up when matter is exposed to very high energies. We did not describe the many reactions that occur when energetic mesons or nucleons collide with nucleons; nor the violation of right- and left-symmetry nor even the violation of the world–antiworld symmetry, observed in weak interactions. It is a strange world, full of new forms of matter, new transformations and reactions, with an unexpected richness and variety. We are at the very beginning of the exploration of this unknown part of the universe; only 12 years ago, practically none of the many quantum states of the baryon and meson were known, and today we have no idea how far these spectra will continue into higher energies. Experimentation is difficult since we are dealing with matter under extreme conditions.

Can we understand what is going on here in the same way as we understand atomic and nuclear structure? So far we cannot; we are trying to apply the concepts that were useful for the understanding of atoms and nuclei, such as quantum states, quantum numbers, interactions, in short the language of quantum mechanics and field theory. This language was particularly successful in describing the interactions of electrons with the electromagnetic field (quantum electrodynamics). Charged particles with half-integer spin interact with the field quanta—the photons—with a relatively weak interaction measured by the small coupling constant $e^2/hc = (137)^{-1}$. The fundamental process of this interaction is a change of state of the electron accompanied by the emission and absorption of a photon.

Some tests of quantum electrodynamics were extremely impressive, such as the explanation of the Lamb shift and similar radiative effects, which reproduce the observed effects with an accuracy of one part in 10^{+8}. One was tempted, therefore, to describe and explain the new phenomena with the same concepts and methods; it is the only language we know today when speaking about such phenomena, but is it the language spoken in this newly discovered world?

At first glance, the language seems to fit and the phenomena fall into place, but when one tries to sharpen the focus, difficulties appear all over the scene. Let us try to illustrate this state of affairs. The analogy between electrodynamics and the mesonic phenomena has its origin in the uncanny vision of Yukawa in 1934. He noted that the introduction of a light quantum with a finite rest mass m would change electrodynamics such that the Coulomb field would no longer extend to infinity but would acquire a finite range h/mc. He connected the observed short range of the nuclear force with this consideration and concluded that the nuclear force is transmitted by a field whose quanta are particles with a mass that he estimated from the range to be of the order of 100 MeV. Thus he foresaw the existence of mesonic phenomena from the tiny evidence of a nuclear range, merely by exploiting the language of quantum electrodynamics.

Particle theory is still under the spell of this idea. The analogy between the electron–photon interaction and the nucleon–meson interaction is striking. The production of mesons by decelerated protons (Fig. 8) reminds us strongly of the production of x rays by decelerated

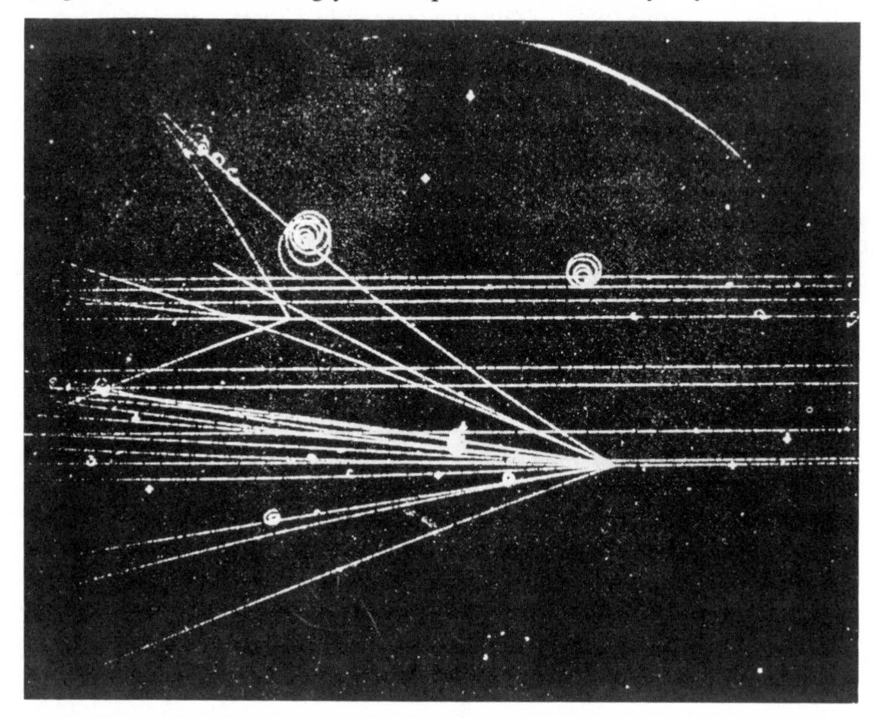

FIGURE 8 Bubble chamber photograph of meson production by fast protons colliding with hydrogen atoms.

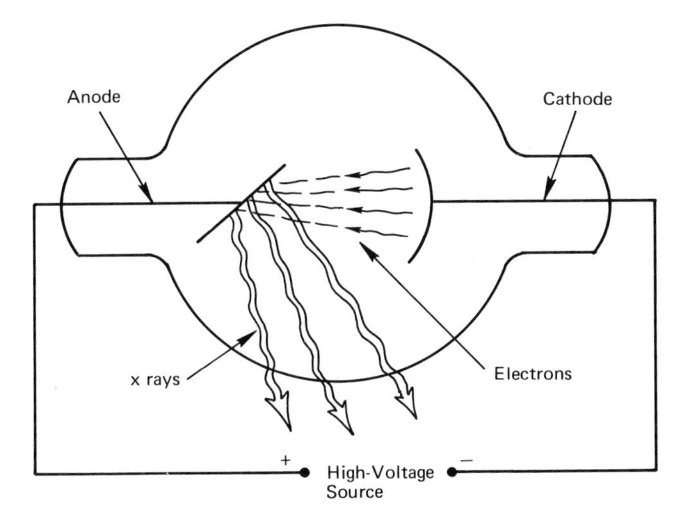

FIGURE 9 Schematic sketch of x-ray production by fast electrons colliding with anode.

electrons (Fig. 9). There are, of course, some differences besides the mass, such as the fact that there are charged and "strange" mesons and that mesons have different spins. The coupling constant of the meson–nuclear interaction seems to be much larger—after all, we have "strong interactions."

A large coupling constant presents serious difficulties to the quantitative exploitation of a field theory. All the methods used in quantum electrodynamics are based on the smallness of e^2/hc. But it is intriguing to contemplate how a large coupling constant of the order unity or higher could perhaps lead to phenomena such as the observed ones.

Let me remind you of three items in electrodynamics: the scattering of light by light, the decomposition of the field of a charge into virtual light quanta, and the nature of positronium. Let us find out what happens with these items when the coupling constant is large. The first is a very weak phenomenon, its cross section is of the order $(e^2/hc)^4$ smaller than the square of the Compton wavelength. If the coupling constant were of the order unity, there would be strong forces between light quanta, which perhaps (if they were attractive) would cause the quanta to join together and form stable or metastable bundles of some definite mass and spin. We then would have a larger variety of field quanta, a situation reminiscent of the meson spectrum.

A moving charge is surrounded by an electric and magnetic field. Quantum electrodynamics describes this situation by stating that a charge produces "virtual" light quanta in the surrounding space. The average number of these quanta is low, it is connected with the small value of the coupling constant. If the latter were of the order unity or higher, there would be a large number of field quanta enveloping the charged particle like a cloud. There would be many different ways in which they arrange themselves with respect to spins and quantum states, and, therefore, we may expect not only one type of particle but many different possible realizations of a particle, with different masses and different other characteristics, a situation reminiscent of the baryon spectrum.

A discussion of the nature of positronium is most useful for the understanding of a situation created by a strong coupling constant. Positronium is a bound system consisting of a positron and an electron. It has a spectrum similar to a hydrogen atom, but it has a finite lifetime because it decays by annihilation into two (sometimes three) photons. We are justified in considering it an excited state of the light quantum since it is capable of a transition into a light quantum with the emission of photon. Conversely (consider the inverse process), a photon can be "excited" into positronium by "absorption" of a photon. In electrodynamics, a photon is very different from positronium and its excited states. What would be the situation when the coupling constant is large? We can only guess, since there is no way known how to solve the equations of quantum electrodynamics with large coupling between particles and fields. The difference between the photon and the positronium states would become less pronounced; the splitting between the positronium states would be much larger. The fact that the photon is transformable into an electron pair would cause it to become a conglomeration of electron pairs; it may even give rise to a rest mass of the photon. All together, the distinction between photons and particle–antiparticle pairs will be wiped out to a great extent; we may face a spectrum of many possible combinations of field and pairs, a situation again reminiscent of the meson spectrum (Fig. 10).

It seems as if a strong coupling between particles and fields gives rise to phenomena similar to the ones that were found in nature. We find ourselves in an exasperating situation: the mathematical and conceptual difficulties of strong coupling theories have not permitted us to verify or disprove such ideas. Is field theory the language germane to the new realm of phenomena, or is its application nothing but an abortive effort to formulate the unknown in a wrong language?

The mystery becomes even more striking by the following observations. Some properties of excited baryons suggested strongly the existence of a substructure of three constituents commonly referred to as

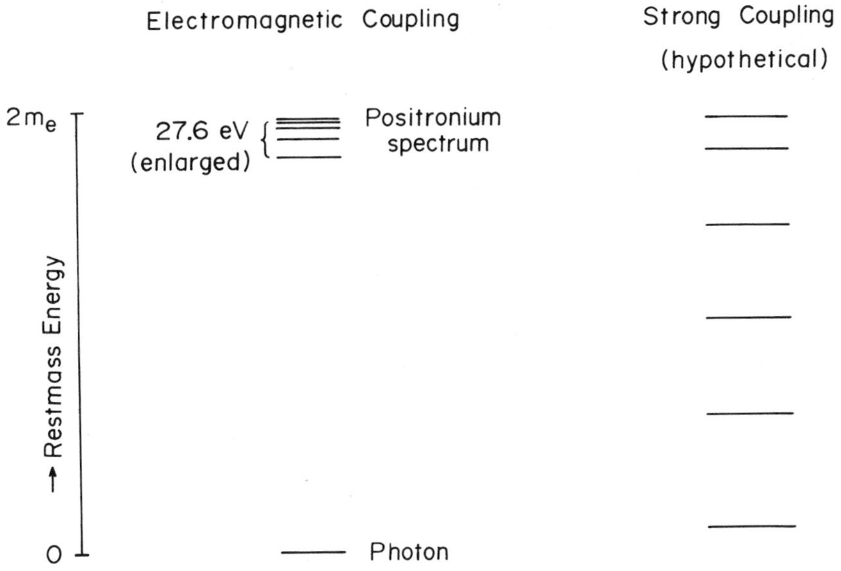

Electromagnetic Coupling

Strong Coupling (hypothetical)

$2m_e$

27.6 eV (enlarged)

Positronium spectrum

→ Restmass Energy

0

Photon

FIGURE 10 The quantum states of the photon, in terms of their rest mass. The photon is at zero; the rest mass of positronium is equal to two electron masses minus the binding energy of the specific state. The ground state is bound by two Rydberg units (27.6 eV). If the coupling were strong, the positronium spectrum would extend to lower values and there would be no clear distinction between the photon and positronium.

"quarks." The baryon spectrum exhibits some of the unmistakable traits of a spectrum belonging to a three-particle system, and many observations can be easily interpreted as coming from a bound system of three quarks. The quarks, however, would have to carry fractional electric charges, 1/3 and 2/3 of the natural unit; there would be three different types of quark, each with a spin of 1/2, one of which would be endowed with strangeness and would be contained only in those baryons that carry a nonzero strangeness quantum number.

Furthermore, the meson spectrum exhibits traits pointing to a particle–antiparticle system made up of the same types of quarks. Clearly, it is tempting to assume that baryons and mesons are indeed composite systems made up of these quarks. In such a model, the fundamental particles and field quanta would not be the nucleon and the meson but rather the quark and some new field acting between the quarks whose quanta are sometimes referred to by the descriptive but ugly name of "gluons." We should not be overly surprised that the nuclear field quanta—the mesons—appear in this model as particle–antiparticle pairs. As it was discussed before, in a strong-interaction theory there is no genuine difference between particle pairs and field quanta.

The description of the subnuclear phenomena in terms of quarks has certain fascinating aspects. The quantum states of baryons and mesons can be classified along some symmetry rules based on the equivalence of the triplet of fundamental quarks. By assuming that the interquark forces are almost the same between any pair of quarks irrespective of their type, one expects a grouping of the baryon and meson states into certain multiplets with similar energies, decay schemes, and other properties. Octets, decuplets, and other groupings are expected, which should exhibit remarkable symmetries in their properties that are based upon the threefold variety of the constituent quarks. Such multiplets have indeed been identified. The symmetry that gives rise to these regularities carries the official name of SU_3. There is more than that: some scattering experiments of highly energetic electron beams by protons have shown features that can (but need not) be interpreted as pointing toward some charged units of very small dimensions within the proton. This interpretation is deduced from the observation that electrons are scattered at rather large angles, much larger than one would expect if the charge of the proton were smoothly distributed over its volume. If this interpretation is correct, these experiments, which were performed by a team from the Massachusetts Institute of Technology and the Stanford Linear Accelerator Center would be the analogue of the Rutherford experiment of 1911, in which the large-angle scattering of alpha-particles by atoms led to the conclusion that the positive charge is concentrated in a small region in the nucleus of the atom.

So far quarks have never been found as free particles. A number of reasons point to the possibility that they do not exist at all as real particles and that the success of the quark model is an indication of some yet unknown mechanism that superficially exhibits the properties that lead to the quark hypothesis. One of these reasons is the fact that free quarks have never been observed even in cosmic rays, where there is practically no limit to the available energy. The other is the abnormal statistics of quarks; they do not seem to obey Fermi statistics as particles of one-half spin are supposed to do.

What will be the future development in this field of physics? When we start speculating, we are limited to the use of the only language we have—the language of field theory—be it the correct one or not. We are bound to think in terms of some particles and some interactions between them. We concluded previously that pair creation and annihilation must play a dominant role in subnuclear matter. An interesting consequence seems to appear. We pictured an entity such as the proton as a conglomerate of particles and antiparticles of some kind. The excitation of such an entity to higher and higher states should involve the creation of more and more virtual pairs within that system. New degrees of freedom are constantly created. In the ordinary world of atoms, molecules, solids, or nuclei, the transfer of energy is generally connected with an increase of kinetic energy of a fixed number of constituents; we speak of heating up the system. Occasionally, it happens that new degrees of freedom are created as, for example, in the melting of a solid. In this case, the supply of energy does not lead to an increase of temperature, until the substance is completely molten.

A similar situation seems to exist in a subnuclear entity. When energy is supplied, instead of raising the energy of each constituent, we create new ones. On the basis of this picture we expect that subnuclear matter cannot be heated to a higher temperature than a certain limit T_0 (kT_0 is about 300 MeV), and the "molten vacuum" is a mix of mesons and baryons and their antiparticles. In this case, however, there is an infinite supply of the substance to be molten; the temperature T_0 would be a limit that cannot be transgressed.

If this situation is indeed realized, the level density of a baryon or meson should increase exponentially with energy. In any "normal" (nonrelativistic) material system the level density increases in the limit of high energies as the exponential of a power of the energy less than unity. We will be able to test this most unusual property of subnuclear matter with the new accelerators and storage rings that are now available or will be soon. We will see whether the level density continues to increase strongly when one goes to still higher energies. Already today we know some facts that point in this direction. In high-energy collisions it is observed that the emerging secondaries have surprisingly small momenta perpendicular to the direction of the collision. These momenta are of the order of 1/3 GeV/c and remain so even if the momenta of the incident particles are of the order of many GeV/c. This may be an indication of a limiting temperature if one is allowed to consider the perpendicular momenta as an indication of the internal kinetic energies. Future studies of the momenta distributions of fragments in very-high-energy collisions may shed more light on the strange properties of highly excited subnuclear matter. So far one has found interesting relations also among the momenta of fragments in the direction of collision. One speaks of "scaling," which means that the distribution of momenta seems to be independent of the collision energy in the limit of high energies, if the momenta are expressed in units of the relative momentum of the incident particles.

The language of field theory may perhaps be more appropriate for the discussion of weak interactions for which the coupling constant appears to be small at presently attainable energies. There exist interesting speculations indicating that the energies available in the near future may lead to a richer world of weak-interaction phenomena. These speculations are based on a growing recognition that electromagnetism and weak interactions are two realizations of the same fundamental interaction. We may be about to touch an essential nerve of nature here, similar in importance to the century-old recognition of the electromagnetic nature of light. The weak interactions would be transmitted by a field in the same way as the electromagnetic interactions are transmitted by the electromagnetic field. These two types of field would in fact be two "components" of a common fundamental field. If this is so, there should exist a field quantum of weak interaction corresponding to the light quantum but endowed with a mass and a charge, since the nature of weak interactions requires these prerequisites. The existence of that particle—the intermediate boson—has been hypothesized before. The new theories initiated by S. Weinberg in 1967 point toward a mass of this particle of the order of 50 GeV; these are the energies at which weak interactions are expected to be comparable in strength with electromagnetic ones. That value puts it within the possibility of detection in the near future. The ideas may be correct or not; in any case, they indicate that we can expect new surprises in the field of weak interactions.

These and similar speculations lead one to believe that the energy regions opened up by the new facilities will lead to a number of important discoveries. Theoretical predictions should not be relied upon too heavily; they only serve as an indication that new phenomena may be found. In the past, the new discoveries always were richer and more profound than the theorists had foreseen. When Columbus left Cadiz in 1492, his theorists told him that he would reach India; he discovered a new continent.

What is an elementary particle? What is the answer to our original question, in view of the development of our knowledge of the structure of matter? Elementarity appears to be a relative concept, depending on the kind of phenomena under consideration (see Table 1). The chemist and the atomic physicist correctly consider the electrons and the atomic nuclei and the photon as elementary. They appear as entities with well-defined and unchanging qualities, as long as the energy exchanges are well below, say, 10^4 eV. For the nuclear physicist, the elementary particles are protons, neutrons, electrons, neutrinos, and photons; they are the unchanging entities as long as the energy exchanges remain well below 10^8 eV. The discoveries of the subnuclear phenomena have shown that the proton and the neutron are no longer elementary and that a number of additional entities such as excited protons, mesons, and heavy

TABLE 1 Conditional Elementarity

Elementary particles		
Atoms	(for kinetic gas theory)	(1/10 eV)
Nuclei, electrons	(for atomic physicists)	(1-100 eV)
Nucleons, electrons, neutrinos	(for nuclear physicists)	(10^5-10^7 eV)
Nucleons in excited states Quarks ? Light and heavy electrons Neutrinos Mesons	(for high-energy physicists) (10^7-10^{10} eV)	

electrons appear when the energy exchanges reach the GeV scale. Today it is not yet clear whether nature on this level will or will not reveal again a new series of elementary particles or whether the phenomena will have to be interpreted in a different way.

It is historically interesting that these three progressive steps toward a deeper understanding of the fundamental structure of matter were initiated by discoveries made almost exactly 20 years apart: the discovery of the nuclear atom by Rutherford in 1911, the discovery of the neutron by Chadwick in 1932, and the discovery of the excited Δ-state of the proton by Fermi and collaborators and its interpretation by Brueckner and Watson in 1952. Table 2 describes the main features of each of these steps, the corresponding elementary particles, the forces that act between them, the relevent sizes and energies, the quantum numbers, and the types of transitions.

The layers of elementarity that were discovered until now are consequences of the laws of quantum mechanics. These laws require the exis-

TABLE 2 Steps in the Structure of Matter

System	Elementary Particles	Interaction	Size cm	Energy eV	Field Quarks
Atom Molecule	Electrons Nuclei	Electro- magnetic	10^{-8}	10^{-1}-10^{-3}	Photons
Nucleus	Protons Neutrons	Nuclear Weak	10^{-12}	10^5-10^7	Photons Lepton pairs ($e\nu$)
Baryon Meson	? (Quarks)	Strong Weak	10^{-13}	10^8-10^9	Photons Lepton pairs ($\mu\nu$), ($e\nu$) Mesons
Quarks ? Electron?	?	?	?	?	?

tence of energy thresholds for the excitation of the internal structure of any mechanical system; when the energy transfers are less than this threshold, the system appears as an elementary entity. But so far the laws of quantum mechanics could be applied only to the atomic and nuclear systems; the behavior and the internal structure of baryons and mesons has yet defied a complete description in terms of a quantum-mechanical system.

Particle physics today is in the exploratory stage rather than in the explanatory one. In the last 20 years, the most important achievements in this field have come mainly from the ingenuity of the accelerator builders and of the experimental physicists rather than from new theoretical insights. The work of theoretical physicists was mainly directed at the systematic ordering, formulation, and classification of the experimental results. The rapidly developing art of accelerator construction, of particle detection, and of data evaluation brought to light an enormous amount of new phenomena, a rich world of unexpected particles, reactions, and transformations. Matter under conditions involving energy exchanges of several GeV was found to behave very differently from what we know and understand to happen at lower energies. Such conditions probably are dominant only in collapsing stars, in exploding galaxies, or perhaps at the very beginning of the expanding universe and—due to the inventiveness of our engineers and physicists—also at the target areas of our large accelerators.

We have every reason to believe that the next period of 20 years, from 1972 to 1992, will be as fruitful as the previous one in revealing new and unexpected phenomena. We are only at the very beginning of the exploration of a new continent of material behavior, a continent in which matter and antimatter are intimately interrelated and where we expect to find the deeper roots of material structure. It may be the place where the physics of the very small and of the very large will find their common source, where the connections between gravity and the other forces of nature will become apparent, and where the answers to the question of the origin of matter are buried. We are penetrating this world, we are transferring it into our laboratories. It seems to be far removed from the one in which we live, as it always seemed when a new realm of phenomena had been discovered in the past, but it is a world most relevant for our quest to penetrate the fundamental basis of our existence.

KIP S. THORNE *is a professor of theoretical physics at the California Institute of Technology in Pasadena. In the half dozen years since Dr. Thorne received his PhD, he has done research on a large variety of astrophysical and cosmological subjects and taught not only at the California Institute of Technology but at Princeton, New Jersey; Varenna, Italy; and Les Houches and Paris, France.*

11 *Gravitation Theory* / KIP S. THORNE

It is now 57 years since Albert Einstein published the final, definitive version of his theory of gravity—general relativity. During the first 47 years of its existence, general relativity interacted sadly little with the rest of physics or with astronomy. However, the last decade has seen a marked change—slowly at first, but with gathering momentum.

The isolation of general relativity between 1915 and 1963 was caused by the smallness of the relativistic corrections to Newtonian gravity in "normal" astrophysical systems. Consider a gravitationally bound system of mass M and characteristic size L. Analyze that system using Newton's theory of gravity first and then using Einstein's theory. The predictions of the two analyses will agree to within a fractional difference:

$$f \sim \frac{GM}{c^2 L} \sim \frac{1}{c^2} \times \left| \begin{array}{l} \text{characteristic value of Newtonian} \\ \text{potential in system} \end{array} \right|, \tag{1}$$

where G and c are Newton's gravitation constant and the speed of light, respectively. For the earth, $f \sim 10^{-9}$; for the inner regions of the solar system, $f \sim 10^{-7}$; for the sun, $f \sim 10^{-6}$; for a white-dwarf star, $f \sim 3 \times 10^{-4}$; for the galaxy, $f \sim 10^{-6}$; for that part of the universe scrutinized by Hubble in his discovery of the universal expansion, $f < 10^{-2}$. Thus, prior to 1963, experiment and observation could not distinguish Newtonian gravity from relativistic gravity, except in three tiny phenomena: perihelion shift, light deflection, and gravitational red shift.

And then came the revolution:

1963

Discovery of quasars by optical and radio astronomers, working jointly. Theorists were stymied. No adequate model to explain quasars could be found—or has been found, even today. But almost all attempts at models have been driven, by the observational data, to consider massive energy sources with small linear sizes. Such models are always constrained by relativistic considerations to have $L > 2GM/c^2 = (3\text{ km}) \times (M/M_{sun})$. Otherwise the energy source would be inside a black hole and useless. Even with $L > 2GM/c^2$, the models often entail relativistic gravitational phenomena and often are predicted to produce big black holes (the most relativistic of astronomical objects) when they "die."

1964

Discovery of cosmic microwave radiation—radiation most naturally interpreted as photons, created in free–free atomic transitions when the universe was less than a day old and was filled with hot, exploding gas—photons scattered extensively from the moment of emission until the universe was about a million years old and then perhaps free to propagate, unimpeded, except for red shift, until absorbed by antennas on earth. This microwave radiation gives information about homogeneity, isotropy, and temperature in regions of the universe so large that Newtonian gravity is inapplicable ($f \sim 1$) and so long ago that in extrapolating back from now to then one must not use Newtonian theory.

1967

Discovery of pulsars in 1967—and by late 1968, general agreement that pulsars must be rotating neutron stars with maser beams shining off their surfaces or off their magnetospheres. The theorist who would build Newtonian models for pulsars must be careful: for neutron stars $GM/c^2 L$ can be as large as 0.3, and the Newtonian potential at the stellar center can exceed $1.0\ c^2$. Thus, relativistic models can differ from Newtonian models by factors of 2. Observation and theory do not agree well enough yet to make such differences crucial, but perhaps in 5 years. . . .

1969

Announcement by Joseph Weber that he believes he is *detecting bursts of gravitational waves*. Subsequent scrutiny of his experiments by dozens of experts has revealed no concrete grounds for disbelieving Weber's claim; the experiment looks "clean." Theorists are thrilled: gravitational waves could never exist in Newtonian theory; such waves are a purely relativistic phenomenon; moreover, they should be a powerful probe of stellar collapse, of supernovae, of black holes. But theorists are disturbed: attempts to build models of sources for Weber's waves fail even more miserably than attempts to build models for quasars.

1971

Discovery of compact x-ray sources—and, by mid-1972, convincing evidence that the x rays are generated when gas flows from a normal star onto a compact companion star—a companion that in some cases is surely a neutron star but in others may be a black hole. At last, prospects are bright for observational studies of black holes—objects that owe all of their existence and structure to relativistic aspects of gravity.

The revolution is approaching its tenth year. General relativity has become a standard working tool of theoretical astronomers.[1] And we all wait for the next startling observational discovery—with the hope that the next decade will be as upsetting as the last, and the fear that it might not be.

EXPERIMENTAL TESTS OF GENERAL RELATIVITY

As we try to build relativistic models of the universe, of quasars, of pulsars, of gravitational-wave sources, and of compact x-ray sources, we suffer a sense of uneasiness. Our model building relies on Einstein's theory of gravity. But is that theory really correct in the domain where we apply it? Or might the correct relativistic theory of gravity be that of Dicke, Brans, and Jordan—or that of Whitehead—or Papapetrou—of Belinfante and Swihart—or . . . ?[2]

It would be naive to expect that the very astronomical phenomena in which relativistic gravity is crucial will provide a definitive answer. In cosmology, in quasars, in pulsars, in compact x-ray sources, and in sources of gravitational waves, gravitational effects are inextricably interwoven with the local behavior of matter and magnetic fields. There is little hope of separating them sufficiently to get *clean* tests of the nature of gravity. The astrophysical enterprise must be largely one of using the laws of gravity as input and trying to get out information about what the matter and fields are doing.

The greatest hopes for clean tests of relativistic gravity lie in solar-system experiments using today's rapidly advancing space and laboratory technology. Table 1 compares the size of relativistic effects in the solar system with the precision of current technology. This comparison

TABLE 1 Technology of the 1970's Confronted with Relativistic Phenomena in the Solar System[a]

Quantity to Be Measured	Magnitude of Relativistic Effects	Precision of a Single Measurement in the Early 1970's
Angular separation of two sources on the sky	Solar deflection of starlight (1) if light ray grazes edge of sun, 1".75 (2) if light ray comes in perpendicular to earth–sun line, 4×10^{-3}"	(A) Angular separation of two stars with optical telescope: ~ 1" (B) Angular separation of two quasars with radio telescope (differential measurement from day to day, not absolute measurement) in 1971, ~ 0".1 in mid-1970's, ~ 10^{-3}"
Distance between two bodies in solar system	(A) Perihelion shift per earth year (1) for Mercury, 118 km (2) for Mars, 15 km (B) Relativistic time delay for radio waves from earth, past limb of sun, to Venus (one way), 8×10^{-5} sec = 24 km (C) Periodic relativistic effects in earth–moon separation (1) in general relativity, 100 cm (2) in Jordan-Brans-Dicke theory, 500 cm	(A) Separation of another planet (Mercury, Venus, Mars) from earth by bouncing radar signals off it, ~ 0.3 km (B) Separation of a radio transponder (on another planet or in a spacecraft) from earth by measuring round-trip radio travel time, ~ 3×10^{-8} sec = 10 m = 0.01 km (C) Earth–moon separation by laser ranging, 15 cm
Difference in lapse of proper time between two world lines in solar system	(A) Clock on earth vs clock in synchronous earth orbit, $\Delta t/t \sim 5 \times 10^{-10}$ (B) Clock on earth vs clock in orbit about sun, $\Delta t/t \sim 10^{-8}$	Stability of a hydrogen maser clock, $\Delta t/t \sim 10^{-13}$ for t up to one year

[a]For a review of both the technology and the relativistic effects, see Davies.[19]

shows that atomic clocks, lunar laser ranging, spacecraft tracking, interplanetary radar, and very-long-baseline radio interferometry are capable of measuring relativistic effects with precisions ranging from one part in 10 to one part in 10^5. Unfortunately, to translate that capability into concrete experiments is an exceedingly difficult task.

Despite the difficulty of the task, great progress is being made. Table 2 shows how far we have come since the days of "three standard tests." Described in Table 2 are 20 different *types* of experiments, of which 14 have already been performed in a definitive manner, and 6 are in progress or on the drawing boards. The theoretical significance of these experiments is spelled out in recent review articles by Will[2,3] and in a forthcoming relativity textbook (MTW).[1] In brief, experiments devised and performed in the last two years have disproved eight theories of gravity that were previously viable–including the theory constructed by Whitehead in 1922 as an alternative to general relativity. So far as I know, the only remaining theories that are self-consistent, complete, and in agreement with experiment are general relativity; the Dicke-Brans-Jordan scalar–tensor theory and generalizations of it due to Bergmann, Wagoner, and Nordtvedt[2]; and several new theories, invented within the last 12 months by specialists in the theoretical interpretation of experimental relativity (Nordtvedt, Will, Hellings, Ni, Lee, Lightman).

BLACK HOLES

General relativity theory has been probed and manipulated by theorists for 57 years. After so long, one might expect it to be nearly as well understood–and nearly as dead a medium for basic theoretical research–as Maxwell's electrodynamics. Not so. By its nonlinearity, the theory produces astounding phenomena–and hides them with ease for decades from the prying minds of theorists.

Perhaps the most astounding phenomenon explored by theorists in the last five years is the *black hole*.[4] Black holes were first discovered in the equations of general relativity in 1938, by Oppenheimer and Snyder. From then until about 1967 little new was learned about them. But these last five years have seen progress analogous to the progress on the airplane between the Wright brothers (1903) and the first jet (194x).

This progress has been produced, in large measure, by an infusion of new mathematical techniques that bear the name "global methods."[5] Global methods enable one to prove powerful theorems about black holes without making any idealizations whatsoever. (One need not assume spherical symmetry, or axial symmetry; one need not assume that space is empty of matter; one need not assume local thermodynamic

TABLE 2 Experimental Tests of Relativistic Gravity

Type of Test	Quantity to Be Measured (References where Reviewed)	Current Experimental Results and Future Prospects	Theoretical Significance of Experiment
I. Tests of Equivalence Principles	1. Universality of free-fall (difference in gravitational acceleration on different test bodies at same point). "Eötvös-Dicke Experiment" (Dicke, 1970[20]; MTW[1] § 38.3)	Platinum and aluminum fall toward sun with same acceleration to accuracy $\Delta a/a < .1 \times 10^{-12}$ (V. B. Braginsky and V. I. Panov, 1972)	Tests "weak equivalence principle." Shows that neutrons and protons experience same acceleration to 1 part in 10^{11}; electrons and nucleons same to $3/10^9$; virtual positrons and nucleons same to $1/10^5$; nuclear binding energy and nucleons same to $1/10^8$; electrostatic energy and nucleons same to $3/10^{10}$. Weak equivalence principle requires same acceleration for all forms of energy, except gravitational. Almost all theories of gravity incorporate weak equivalence principle
	2. Constancy of nongravitational physical constants [MTW[1] § 38.6]	Sample results: a. Electromagnetic fine-structure constant α is constant on earth to accuracy $\left\lvert \frac{1}{\alpha}\frac{d\alpha}{dt} \right\rvert \leqslant \frac{5}{10^{15}\ \text{years}}$ (F. Dyson, 1972) b. α is same on earth today as it was 2×10^9 years ago in galaxies 2×10^9 light years away, to accuracy $\lvert \Delta\alpha/\alpha \rvert \leqslant 0.002$ (J. Bahcall and M. Schmidt, 1967)	Tests "Einstein equivalence principle," i.e., tests whether nongravitational laws of physics are the same in every freely falling (local inertial) frame, everywhere and everytime in universe. All "metric theories of gravity" incorporate Einstein equivalence principle

	3. Gravitational red shift (MTW[1] §7.4 and §38.5; Davies, ed., 1971[18])	a. Light climbing a height h against earth's gravitational acceleration g is red shifted by $(\Delta\nu/\nu)/(gh/c^2) = 1.00 \pm 0.01$ (Pound, Rebka, Snider, 1965) b. NASA plans, in 1975, to fly a hydrogen-maser clock in a sub-orbital flight to test gravitational red shift. Projected accuracy is 2×10^{-5} (a factor 500 better than above experiment!) (Vessot)	Tests "Einstein equivalence principle" (see above). Can be interpreted as a test of "geodesic motion"; can also be interpreted as a verification that space-time is curved
II. Tests for Existence of "Directly Coupling" Long-Range Fields	1. "Special relativity experiments" that probe local Lorentz metric (MTW[1] §38.4)	Sample results: a. High-precision verification of Lorentz invariance in the kinematics of nuclear and particle reactions b. 2 percent test of time dilation for muon decay in a storage ring	Show that space-time is endowed with a second-rank, symmetric metric tensor that (i) "governs" particle kinematics and (ii) "governs" the measurements of atomic clocks—both in the familiar manner of special relativity
	2. Hughes-Drever experiment (Dicke, 1964[21]; MTW[1] §38.7)	To high precision there is no splitting of the "m_J substates" of atomic nuclei in vacuum (Hughes, Robinson, Beltran-Lopez, 1960; Drever, 1961)	Tests for the existence of a long-range tensor field (in addition to the metric) that couples directly to matter. Shows that such coupling is $\sim 10^{-22}$ as strong as the coupling to the metric
	3. "Ether-drift experiments" (Dicke, 1964[21]; MTW[1] §38.7)	To high precision there is no sign that the yearly variations in earth velocity affect laboratory, nongravitational experiments (Turner, Hill, and others)	Test for the existence of a direct-coupling, long-range vector field (in addition to electromagnetism); show with moderately high precision that no such field exists

TABLE 2 (Continued)

Type of Test	Quantity to Be Measured (References where Reviewed)	Current Experimental Results and Future Prospects	Theoretical Significance of Experiment
III. Post-Newtonian Experiments	1. Deflection of light and radio waves by sun (MTW[1] §40.3, Box 40.1)	a. Current results on deflection of quasar radio waves (1.″75 arc at limb of sun): $\frac{\text{(Observed deflection)}}{\text{(Einstein prediction)}} = 1.0 \pm 0.1$ (Several different groups) b. Expected precision, in mid or late 1970's, using transworld interferometry: ± 0.001	Measures the curvature of space produced by the sun. Is potentially a powerful tool for distinguishing between general relativity and scalar–tensor theories of the Dicke-Brans-Jordan type
	2. Relativistic delay in roundtrip travel time for radar beams passing near sun (Shapiro, 1972[22]; Davies, ed., 1971[18]; MTW[1] §40.4, Box 40.2)	a. Current results (same result obtained by bouncing radar off planets and by transponding off Mariner VI and VII spacecraft; typical relativistic delay is ~ 200 μsec): $\frac{\text{(Observed relativistic delay)}}{\text{(Einstein prediction)}} = 1.01 \pm 0.04$	Measures the curvature of space produced by the sun. That curvature forces radar beam to travel farther than it would in flat space, thereby causing "relativistic delay." Like deflection, this is potentially a powerful tool for distinguishing between general relativity and scalar–tensor theories
	3. Perihelion shifts of planetary orbits (Shapiro, 1972[21]; MTW[1] §40.5 and Box 40.3; Dicke, 1973[23])	a. Current results, for Mercury (shift ~ 43″ arc per century): $\frac{\text{(Observed shift)}}{\text{(Einstein prediction } \textit{if} \text{ sun is spherical)}} = 0.98 \pm 0.01$ (Shapiro and colleagues) b. Additional radar data on Mercury and other planets should enable one, in 1970's, to distinguish between perihelion shifts produced by solar quadrupole	Measures the nonlinearity in the superposition law for gravitational fields of different parts of sun. The interpretation of this experiment is "muddied" by uncertainties about the solar quadrupole moment—which could be so large as to produce a perihelion shift for Mercury of 4″ arc per century (Dicke, 1973[23])

	and by relativity, with precision ~ 0.″4/century (1 percent of relativistic effect)	
4. Sidereal–periodic earth tides (Will, 1973[2]; MTW[1] §40.8)	a. Experiments have not yet separated tides with period 12 sidereal hours from those with period 12 solar hours. However, the unseparated tides agree with the sidereal-free, Newtonian predictions to precision $\Delta g/g \simeq 10^{-9}$ b. By 1974 a clean separation will be made with precision $\Delta g/g < 10^{-9}$	Many gravitation theories (but *not* general relativity or scalar–tensor theories) predict earth tides with a period of 12 sidereal hours and with magnitudes ranging from $\Delta g/g \approx 10^{-7}$ (same magnitude as tides due to moon and sun) on downward. Such tides are caused by a "polarization" of the gravitational constant–which in turn is caused by (i) motion of earth relative to mean rest frame of universe or (ii) "Machian" influence of galaxy. The absence of such tides rules out a wide class of theories (Nordtvedt and Will)
5. Annual variation in earth rotation rate (Will, 1973[2]; MTW[1] §40.8)	Observed annual variation of $\Delta\Omega/\Omega = 4 \times 10^{-9}$ is explained, to precision ~15 percent, by shifting atmosphere	Many gravitation theories (but *not* general relativity or scalar–tensor theories) predict an annual variation in gravitational constant G due to annual variation in velocity of earth through universe. This variation in G produces a variation in the gravitational pull of earth, which in turn causes its rotation rate to vary by annual amounts $\Delta\Omega/\Omega$ as large as 3×10^{-8} (Nordtvedt and Will)
6. Chemical dependence of gravitational attraction (MTW[1] §40.8)	The same masses of bromine and flourine produce the same gravitational attraction to a precision $\Delta a/a \leqslant 5 \times 10^{-5}$ (Kreuzer)	Can be interpreted as a test whether "active gravitational mass" divided by "passive gravitational mass" is dependent on chemical composition. Some theories (but *not* general relativity or scalar–tensor theories) would predict such a dependence

TABLE 2 (Continued)

Type of Test	Quantity to Be Measured (References where Reviewed)	Current Experimental Results and Future Prospects	Theoretical Significance of Experiment
	7. "Nordtvedt effect" for earth–moon orbit (Will, 1973[2]; Davies, ed., 1971[19]; Dicke, 1970[20]; MTW[1] §40.9)	Now being measured using laser-ranging to moon; a precision of ~30 cm is expected	Most gravitation theories (including scalar–tensor theories but excluding general relativity) predict that gravitational energy responds to an external gravitational field in a different manner than do other forms of mass-energy. Hence, the gravitational acceleration that acts on a massive body in an external field is a function of its gravitational binding energy. Since the earth is more bound than the moon, it falls toward the sun with a different acceleration than the moon. This leads to a "polarization" of the earth–moon orbit by the sun and thence to a periodic anomaly in the earth–moon separation (Nordtvedt effect). This periodicity has magnitude ranging from ~100 m for some plausible theories down to ~1 m for the current version of Dicke-Brans-Jordan theory ($\omega \sim 5$) to zero for general relativity
	8. Three-body effects in earth–moon orbit (Davies, ed., 1971[19]; MTW[1] §40.6)	Now being sought using laser-ranging to moon; results in early 1970's will be of marginal value, at best	All gravitation theories predict periodic variations in the earth–moon separation due to three-body (earth–moon–sun) interactions. The Nordtvedt effect is one such periodicity. The others have amplitudes ranging from 100 cm (for the least interesting one) on downward

	9. Precession of gyroscope in orbit about earth (Davies, ed., 1971[19]; MTW[1] §40.7)	Experimental apparatus now in design stage at Stanford University; experiment expected in mid to late 1970's. Accuracy aimed for: 0."001 arc per year (Everitt, Fairbank, and colleagues)	Most gravitation theories (including general relativity) predict two sources of precession for an orbiting gyroscope: (i) motion of gyroscope through curved space around earth (precession ≈ 7" arc per year); (ii) dragging of inertial frames by angular momentum of earth (precession ≈ 0."07 arc per year). The two effects can be separated cleanly by suitable choices of orbit and of gyroscope orientations. The quantitative predictions differ from theory to theory
IV. Gravitational-Wave Experiments	1. Verification that gravitational waves exist (Weber, 1973[23]; Press and Thorne, 1972[9]; MTW[1])	Joseph Weber presents persuasive evidence that he is detecting bursts of cosmic gravitational waves; but nobody else has yet (Sept. 1972) completed a confirming experiment	All relativistic theories of gravity predict the existence of gravitational waves and agree in order of magnitude on the power radiated by typical astrophysical sources. However, the predicted powers are such that, with conventional views of the nature of the galaxy and universe, one cannot explain Weber's events. See text
	2. Measurement of polarization of waves (Weber, 1973[24]; Eardley *et al.*, 1973[25])	Weber sees "transverse, spin-2" polarization; he has also searched for, but failed to find, "spin 0" polarization	General relativity permits only "transverse, spin-2 polarization." Scalar–tensor theories permit "transverse, spin-2" and also "spin 0" (scalar waves). The most general "metric theories" (theories satisfying Einstein equivalence principle) permit six independent polarization states. The experimental determination of which polarization states actually exist will be a powerful test of gravitation theories
	3. Measurements of speed of propagation of waves (Press	By 1980 one may be able to detect bursts of gravitational waves	General relativity and scalar–tensor theories predict that light and gravitational waves

TABLE 2 (Continued)

Type of Test	Quantity to Be Measured (References where Reviewed)	Current Experimental Results and Future Prospects	Theoretical Significance of Experiment
	and Thorne, 1972[9]; Eardley *et al.*, 1973[25])	from supernovae in the Virgo cluster. By comparing the arrival time for a burst with the time of the optical flareup (measurable, e.g., to within ~10 days), one could compare the propagation speeds for light and gravitational waves with an accuracy of $\Delta c/c \sim 10^{-9}$	propagate with the same universal speed ("null geodesics in curved space-time"). However, many other theories (e.g., "Lorentz-symmetric theories") predict propagation speeds that would differ, in the region between earth and the Virgo cluster, by $\Delta c/c \sim 10^{7}$. Thus, a measurement of $\Delta c/c$ to high precision would be a powerful test of gravitation theories
V. Cosmological Observations	1. Large-scale features of universe, e.g., homogeneity, isotropy, expansion, evolution (Peebles, 1972[10]; Zel'dovich and Novikov, 1973[1]; MTW[1])	a. Measurement of expansion rate using red shifts and magnitudes of galaxies b. Measurement of high-precision isotropy of cosmic microwave radiation c. Strong evidence for evolutionary effects in quasar population (Schmidt)	The "correct" theory of gravity should explain, in a natural manner, the large-scale homogeneity, isotropy, and expansion of the universe. General relativity and scalar–tensor theories succeed, by and large. Continuous creation theories have difficulty. How successful other theories are is not clear; few cosmological models have been attempted using them
	2. Time variation of gravitational constant in solar system (Dicke and Peebles, 1965[21]; Shapiro, 1972[22])	a. Radar measurements of planetary orbits currently yield $\lvert\dot{G}/G\rvert < 4 \times 10^{-10}\ \mathrm{yr}^{-1}$ (Shapiro and colleagues) b. Additional radar data by 1976 should bring the accuracy down to $\sim 3 \times 10^{-11}\ \mathrm{yr}^{-1}$ c. Geophysical data suggest (but do not convincingly show!) $\dot{G}/G \sim -3 \times 10^{-11}\ \mathrm{yr}^{-1}$ (Dicke and colleagues)	Most theories, but *not* general relativity, predict time variation of G, caused by the expansion of the universe. The typical magnitude of the change is $\lvert\dot{G}/G\rvert \lesssim 3 \times 10^{-10}\ \mathrm{yr}^{-1}$ Typical cosmological models in the Dicke-Brans-Jordan theory predict $\dot{G}/G \simeq 3 \times 10^{-11}\ \mathrm{yr}^{-1}$

equilibrium or quasi-equilibrium; one need not assume that magnetic fields are "frozen" into the plasma; . . . .)

Let me give you a taste of global methods and their results, beginning with the definition of a black hole.

In defining a black hole one begins with the concept of "future null infinity"—denoted $\mathscr{I}^+$(script I plus, pronounced "skreye plus"). Future null infinity has a precise mathematical definition; but for us a heuristic physical definition will suffice: $\mathscr{I}^+$ is that region of space-time toward which outgoing light rays, radio waves, and other "null geodesics" propagate.* The universe is divided into two regions: the "external universe" and black holes. The external universe is that region which can send signals (timelike or null curves) to $\mathscr{I}^+$.The black holes are that region which cannot send signals to $\mathscr{I}^+$. (See Fig. 1 for an example.)

Not only are black holes unable to communicate with $\mathscr{I}^+$, they also cannot communicate with any event in the external universe—for if they could, then the communicating signal could be retransmitted from that event to $\mathscr{I}^+$.

The boundary that separates the external universe from the black holes is called the *horizon*. The horizon can be regarded as the "union of the spacetime trajectories of the surfaces of all black holes." Figure 2 is a space-time diagram showing the external universe [denoted $J^-(\mathscr{I}^+)$], several black holes [shaded regions], and the horizon [denoted $\dot{J}^-(\mathscr{I}^+)$].

Any spacelike three-dimensional surface $\mathcal{S}$ slicing through space-time describes the state of the universe at one particular moment of "time." The intersection of such a slice with the black holes consists of several independent components (two in the cases of $\mathcal{S}_2$ and $\mathcal{S}_3$ in Fig. 2). Each such intersection is an individual black hole, at the "moment of time" $\mathcal{S}$. In Fig. 2 there are no black holes at the "time" $\mathcal{S}_1$; between $\mathcal{S}_1$ and $\mathcal{S}_2$ two black holes form by the gravitational collapse of stars; between $\mathcal{S}_2$ and $\mathcal{S}_3$ those two holes collide and coalesce, and a third hole is formed by the collapse of a star.

Global methods have been used to prove a number of powerful theorems about the structure and evolution of black holes. A key building block in most such theorems is the following information, due largely to Penrose, about the structure of the horizon. (See Fig. 1.) All curves that

*Actually $\mathscr{I}^+$ is well defined only in asymptotically flat space-times. In this sense, my claim that one needs no idealizations was a lie. In applying global methods to black holes one typically replaces the correct cosmological boundary conditions—as far away as the most distant galaxy or farther—by an asymptotically flat, "island-universe" boundary condition. Obviously, such an idealization is harmless so far as the local, black-hole physics is concerned—until and unless one starts asking about the creation of black holes in the initial "big-bang" explosion or their destruction in the final stage of recontraction of the universe. There and only there might the results described below break down.

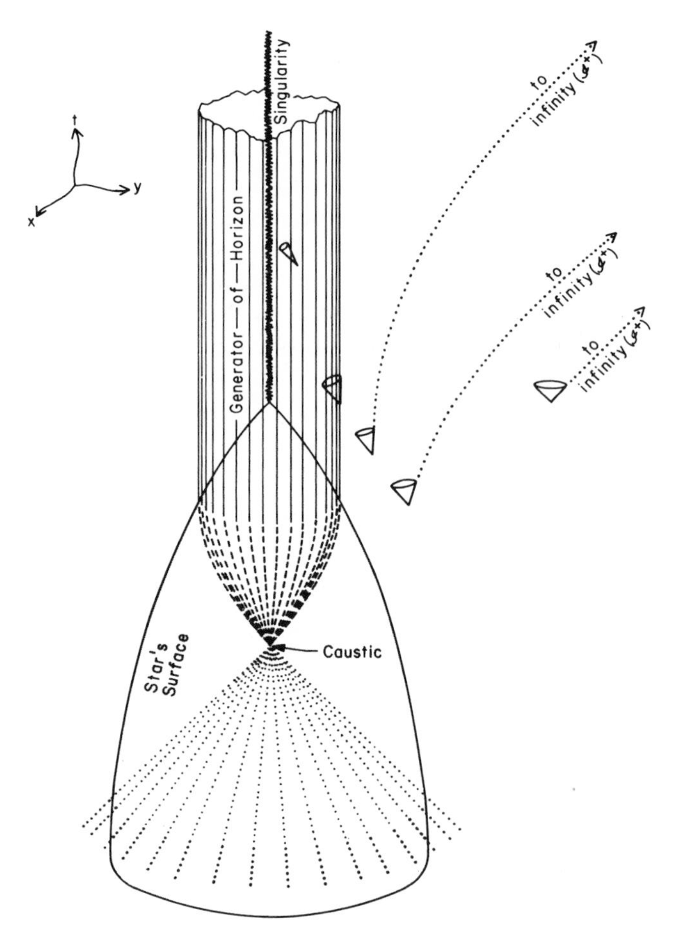

FIGURE 1 The gravitational collapse of a spherical star to form a black hole (space-time diagram with one spatial dimension suppressed). The gravitational pull of the star and hole, i.e., the curvature of space-time, tilts the light cones inward as shown. Except for that tilting, this diagram is perfectly analogous to those used in special relativity. The text discusses various features of the diagram. In the diagram all dotted curves are light rays (null geodesics). The dashed curves, which become solid after exiting through the star's surface, are light rays that generate the horizon ("null geodesic generators"; see text). The horizon is the surface separating the external universe (outside) from the black hole (inside). Deep down inside the black hole is a space-time singularity.

lie in the horizon are either spacelike or null. All timelike curves that pass through the horizon move from the external universe (in the past) into the black hole (in the future); no timelike curve can ever exit through the horizon. At each event on the horizon the light cone is tangent to the horizon. The line of tangency is a "null-geodesic genera-

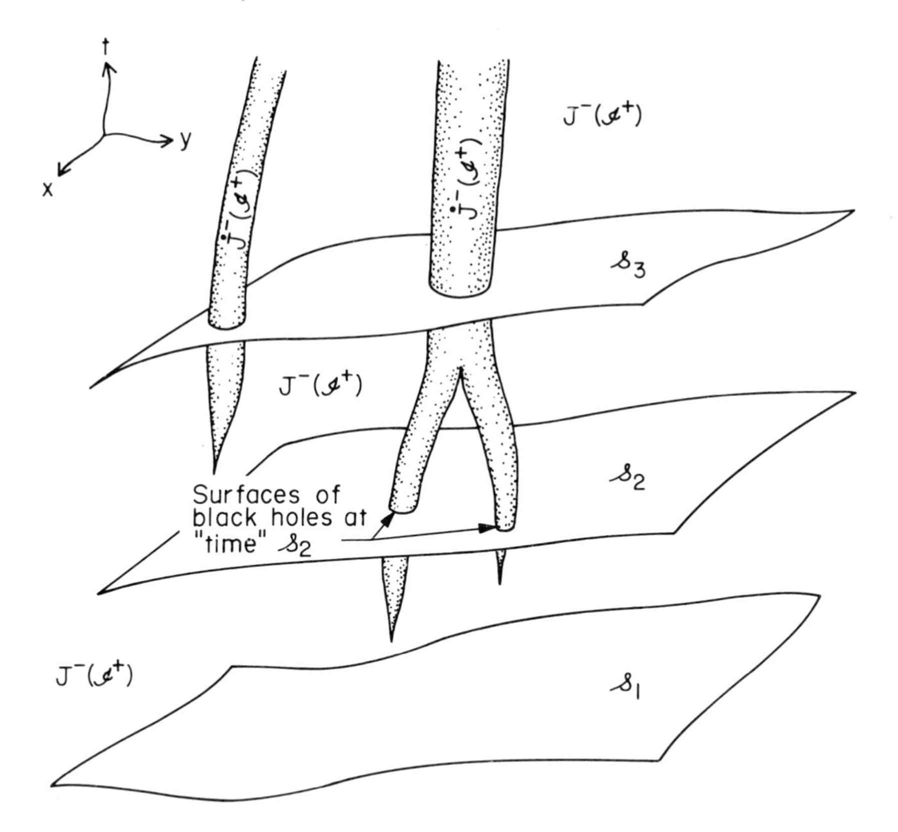

FIGURE 2 Space-time diagram depicting three black holes that form by the collapse of three stars and the collision and coalescence of two of those holes to form a single large hole. In this diagram the external universe is denoted $J^{-}(\mathcal{J}^{+})$ (J^{-} means "region that can send timelike or null curves to," and $\mathcal{J}^{+}$ means "future null infinity"). The horizon is denoted $\dot{J}^{-}(\mathcal{J}^{+})$ (the dot over the J means "boundary of"). The black holes are the region inside the horizon. The "sheets" $\mathcal{S}_1$, $\mathcal{S}_2$, and $\mathcal{S}_3$ are three-dimensional surfaces, each to the future of the preceeding one, that slice through space-time. They can be regarded as the states of the universe at three different "moments of time." The text discusses various features of this diagram.

tor" of the horizon in this sense: (1) Through each event on the horizon passes at least one such generator, and it is a null geodesic (possible world line for a photon). (2) Generators can enter the horizon from the external universe, but once on the horizon they can never leave it—either by exiting into the external universe or by going down the hole. (3) Generators can cross each other at the events where they climb onto the horizon [such events are *caustics* ("pucker points") of the horizon]. But no generator can cross any other generator after climbing onto the horizon. Thus, through every noncaustic event there passes precisely one generator.

Among the theorems proved using these properties of the horizon, and using other "global geometric" results, are these:

Theorem 1 (Penrose and Hawking)—Once created, a black hole can never be destroyed.

Theorem 2 (Hawking)—Although two black holes can collide and coalesce to form a single black hole (cf. Fig. 2), a single black hole can never bifurcate to form two black holes. Bifurcation is impossible no matter what one does to the hole—zap it with a laser gun, drop a neutron star down it, ...

Theorem 3, the second law of black-hole dynamics (Hawking)—In any isolated region of space (region from which no black holes exit) the sum of the surface areas of all black holes can never decrease. Put more precisely, if one slices the horizon by a family of spacelike three surfaces, $\mathcal{S}_1, \mathcal{S}_2, \mathcal{S}_3, \ldots$, each to the future of the preceding one (Fig. 2), then the sum of the surface areas of the intersections will be no smaller on $\mathcal{S}_2$ than on $\mathcal{S}_1$, and no smaller on $\mathcal{S}_3$ than on $\mathcal{S}_2$, and so forth. Typically, when a lump of matter falls down a black hole, it creates a caustic at which new generators climb onto the hole's horizon and increase its surface area. Typically, when two black holes collide, new generators climb onto the horizon, and again the surface area goes up.

In describing these powerful theorems, I failed to tell you about a crucial, unproved hypothesis that underlies them all—the "hypothesis of cosmic censorship." This hypothesis says that the universe possesses no "naked singularities," i.e., any space-time singularities that might exist are hidden inside the horizon, rather than residing in the external universe, exposed to the scrutiny of man. If naked singularities were to exist, they might perform all sorts of wonderful or horrible feats (create matter and eject it into the universe, destroy black holes, destroy spaceships, create Weber's bursts of gravitational waves, . . .). Precisely what they might do is unknown because, by definition, a space-time singularity is a region where the laws of physics as we know them break down, i.e., a region where space-time ceases to possess a local Lorentz structure, perhaps as a result of infinite curvature (infinite gravitational forces).*

All attempts to prove the hypothesis of cosmic censorship have failed. But all attempts to construct counterexamples have also failed. (A counterexample would be any solution to Einstein's equations, in asymptoti-

*This description of a space-time singularity is admittedly heuristic. A more precise description will be given below.

cally flat space-time, in which the collapse of a star or some other object produces a singularity with no surrounding horizon.)

Not all progress in black-hole theory has come from global methods. Conventional techniques of differential geometry and differential-equation theory have also produced major results. Perhaps the most important of these is the following.

Consider a newly formed black hole of mass M (this mass being measured by application of Kepler's laws to the orbits of distant planets). Follow its evolution for a time much longer than

$$\tau = GM/c^2 \sim (10^{-5} \text{ sec})(M/M_{\text{sun}}).$$

In this time the black hole might vibrate violently, emitting strong beams or bursts of gravitational waves. But after this time, it will presumably settle down into a quiescent, stationary state. Much later, when it collides with a star, the hole may vibrate violently again; but afterward, in a time somewhat longer than τ, it should settle down once again into a new stationary state. Can one construct a catalogue of all possible stationary states? The answer is yes—and the catalogue is exceedingly short. Theorems due to Israel, Carter, and Hawking[6] point strongly to the following conclusion (I expect the minor missing links in the proof to be filled in during the next year or two): Stationary black holes form a unique three-parameter family. The parameters are mass M (as measured by application of Kepler's laws), angular momentum J (as measured by precession of gyroscopes in the neighborhood of the hole), and electric charge Q (as measured by difference in attractive pull of the hole on electrons and positrons). *All* other features of a stationary black hole are determined uniquely when one knows its M, J, and Q. This absence of other degrees of freedom is described graphically by Wheeler's phrase "a black hole has no hair."

Not only is the structure of a stationary black hole determined uniquely by M, J, and Q (we think); but we even know that structure today, in all of its elegant mathematical detail (see Chapter 33 of MTW). Thus, analyses of stationary black holes are free from the types of uncertainty that plague neutron-star physics (strength of crust, equation of state, superconducting or not?) and that plague the physics of the earth, sun, planets, and stars.

Any black hole with significant charge in the real universe would presumably neutralize itself by pulling in opposite charges from interstellar space. Thus, the stationary holes studied by astrophysicists have $Q = 0$ and are characterized uniquely by M and J alone.

Many astrophysical processes involve small perturbations of stationary black holes. [Examples: (1) the late stages of the dynamical evolution by which a young hole settles down into its stationary state (Fig. 3); (2) the accretion of gas onto a stationary hole in a binary star system; (3) the swallowing of a star by a supermassive hole in the nucleus of a galaxy—with an accompanying emission of gravitational radiation; (4) the gradual extraction of a hole's rotational energy by one process or another.] During the summer of 1972, Teukolsky successfully separated the variables in the perturbation equations that govern such processes. (The variables had previously been separated only for the special case of a nonrotating, spherical hole.) With the general, separated equation now available, I anticipate major new insights into the astrophysics of black holes.[7]

At the time of this lecture, vigorous astronomical searches for black holes are under way. These searches rely on observable phenomena that should occur when black holes interact with surrounding matter.[8]

The most promising phenomena are those in a close binary system, where one component is a hole and the other is a normal star. One should see a telltale periodic Doppler shift in the spectrum of the normal star as it and the hole circle their common center of mass. In addition, gas pulled off the normal star or ejected by it should spiral into orbit about the hole, forming a hot disk that radiates intense x rays.[8]

Since the winter of 1971–1972 x-ray astronomers and optical astronomers have been studying several compact binary systems that emit x rays. It now seems likely—but it is not yet (September 1972) proved—that some of these systems contain holes and resemble the above model. If they do, then in the next few years "black-hole physics" may get transformed into "black-hole astronomy."

OTHER GRAVITATIONAL PHENOMENA

There are many other recent developments in gravitation physics that I should describe: the firming up of the theory of the emission and propagation of gravitational waves; current estimates of the roles of gravitational radiation and radiation reaction in the universe; designs and prospects of various gravitational-wave detectors; the exciting hope that by 1980 we can detect gravitational waves from several supernova explosions per year in the Virgo cluster of galaxies—and that we can thereby scrutinize the deep interiors of supernovae[9]; current ideas and results about the evolution of the universe beginning with the "big-bang" explosion (general relativity *insists* on such an origin—it forbids a steady state); and including the issue of whether and why there might be more baryons than antibaryons in the universe; the issue of why the "big-bang"

FIGURE 3 The dynamical evolution by which a nonrotating ($J = 0$) uncharged ($Q = 0$) hole with a lump on its surface gets rid of the lump and settles down into a stationary, spherical state. This diagram and the following scenario are based on detailed perturbation-theory calculations by Price.[26] [Adapted from Box 34.2 of MTW.[1]]

SCENARIO:

• When star begins to collapse, it possesses a small nonspherical "lump" in its density distribution.

• As collapse proceeds, lump grows larger and larger (instability of collapse against small perturbations—a phenomenon well known in Newtonian theory).

• The growing lump radiates gravitational waves.

• Waves of short wavelength ($\lambda << GM/c^2$), emitted from near horizon, partly propagate to infinity and partly get backscattered by the "background" "Schwarzschild" curvature of space-time. Backscattered waves propagate inward through horizon formed by collapsing star.

• Waves of long wavelength ($\lambda >> GM/c^2$) emitted from near horizon get fully backscattered by space-time curvature; they never reach out beyond radius $r \sim 3GM/c^2$; they end up propagating "down the hole."

• Is lump on star still there as star plunges through horizon, and does star thereby create a deformed (lumpy) horizon? Yes, according to calculations.

• *But* external observers can only learn about existence of "final lump" by examining deformation (quadrupole moment) in final gravitational field. That final deformation in field does not and cannot propagate outward with infinite speed. (No instantaneous "action at a distance.") It propagates with speed of light, in form of gravitational waves with near-infinite wavelength (infinite red shift from edge of horizon to any external radius). Deformation in final field, like any other wave of long wavelength, gets fully backscattered by curvature of space-time at $r \lesssim 3GM/c^2$; it cannot reach external observers. External observers can never learn of existence of final lump. *Final external field is perfectly spherical, lump-free, Schwarzschild geometry*!

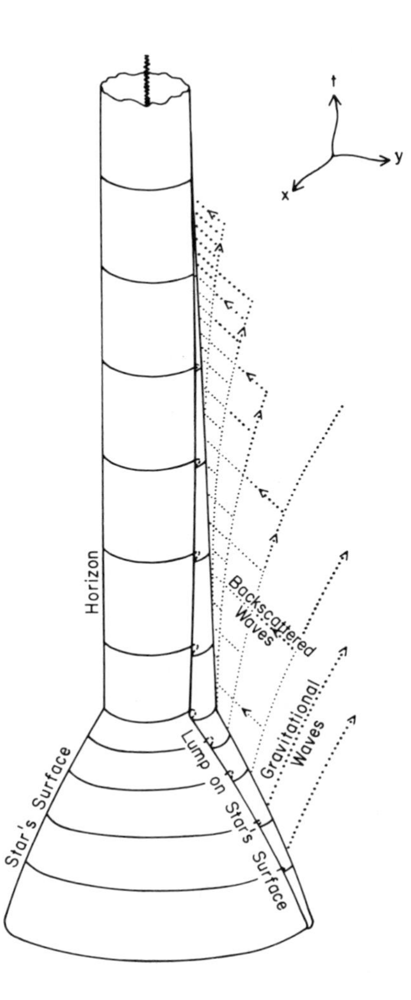

• Even in region of backscatter ($2GM/c^2 < r \lesssim 3GM/c^2$), final external field is lump-free. Backscattered waves, carrying information about existence of final lump, interfere destructively with outgoing waves carrying same information. Result is destruction of all deformation in external field and in horizon!

• Final black hole is a spherical, "Schwarzschild" black hole!

produced 10^9 times more photons than baryons; the problem of why the universe is so extremely homogeneous and isotropic; theoretical models to explain how galaxies and clusters of galaxies condensed out of the primordial gas, and why they have the observed masses and angular momenta[10]; progress in creating a marriage between quantum theory and gravitation theory, including the concepts and fruits of superspace; the keys to superspace provided by studies of minisuperspace; twistors; canonical approaches to the quantization of gravity; the creation of particle–antiparticle pairs by strong gravitational fields; and attempts to derive the gravitational constant and the cosmological constant from quantum gravity considerations.[11]

But of all the topics I should describe, one prods me so strongly that I must devote a few words to it before closing. This is the issue of singularities in the structure of space-time.

SPACE-TIME SINGULARITIES

Space-time singularities are of particular interest because our physical intuition cries out against their existence. They are also interesting because they appear to play important roles in cosmology ("initial" and "final" states of universe), in gravitational collapse (Might naked singularities be created? Do singularities reside inside black holes?), and in the quantization of gravity (Will the correct quantum theory of gravity predict singularities analogous to those of classical general relativity, or will quantization convert the classical singularities into well-behaved—albeit quantized—regions of space-time?).

What is a space-time singularity? From a mathematical standpoint perhaps the most satisfactory definition is one devised in 1970 by B. G. Schmidt.[12] Put in heuristic terms, Schmidt's highly technical definition goes something like this: In a space-time manifold consider all spacelike geodesics (paths of "tachyons"), all null geodesics (paths of photons), all timelike geodesics (paths of freely falling observers), and all timelike curves with bounded acceleration (paths along which observers are able, in principle, to move). Suppose that one of these curves terminates after the lapse of finite proper length (or finite "affine parameter" in the null-geodesic case). Suppose, further, that it is impossible to extend the space-time manifold beyond that termination point—e.g., because of infinite curvature there. Then that termination point, together with all adjacent termination points, is called a "singularity." (What could be more singular than the cessation of existence for the poor tachyon, photon, or observer who moves along the terminated curve?)

Between 1964 and 1970, Penrose, Hawking, and Geroch used global methods to prove a variety of powerful theorems about the existence of singularities.[13] These theorems all have the following form: (1) if gravity is correctly described by general relativity or by the Dicke-Brans-Jordan theory or by some other member of a certain class of theories; (2) if some particular causality condition is satisfied (the condition varies from theorem to theorem; an example is "the impossibility of time machines," i.e., "no closed timelike curves"); (3) if a certain energy criterion is satisfied (typically, that all observers everywhere in the universe see nonnegative energy density); and (4) if appropriate initial conditions are satisfied (see below), then space-time must evolve a singularity.

All the above hypotheses–except perhaps the "initial conditions"–seem eminently reasonable. Indeed, I would be rather disturbed by a failure of any of them, with one exception: as space-time tries to evolve a singularity in accord with the theorems, the extreme conditions produced may cause a breakdown in classical gravitation theory, i.e., space-time may become quantized (see below).

What initial conditions do the theorems require? One theorem[11] takes as initial conditions the existence of a "trapped surface"–a phenomenon that we expect to occur inside the horizon of any stationary black hole. Thus, this theorem strongly suggests that singularities exist inside black holes. Another version of the same theorem takes as initial conditions several observed large-scale features of the universe, including its expansion.[10] From this version one extrapolates backward in time and concludes that the universe must have possessed a singularity in the distant past (presumably at the time of the big-bang creation).

What are the physical properties of the singularities predicted by these theorems? The definition of a singularity provides few clues, so one must search elsewhere. Thus far the search has produced no definitive answer. However, when the answer does come, it will probably be one of the following:

Possibility 1 The typical singularity is a region of infinite tidal gravitational forces (infinite curvature), which crush to infinite density, and which tidally tear apart, all matter that comes near. A number of examples of such singularities are known. Moreover, there are hints in the work of Belinsky, Khalatnikov, and Lifshitz[14] that the generic singularity is, indeed, of this type; more precisely, that it is a "mixmaster singularity," which ergodically distorts matter in one direction and then another, infinitely many times, while simultaneously crushing the matter to infinite density.[15]

Possibility 2 The singularity is a region of space-time in which timelike or null geodesics terminate, not because of infinite tidal gravitational forces or infinite crushing, but because of other, more subtle pathologies; see, e.g., Ref. 16.

Possibility 3 The singularity may be sufficiently limited in "size" and influence that all or most of the collapsing matter successfully avoids it. In the case of cosmology this might produce an oscillating universe, with a "harmless" singularity at each "moment" of maximum contraction. In the case of a black hole, the matter cannot explode back outward through the horizon that it went down; the horizon is a one-way membrane and will not let anything back out. Instead, the matter may reach a stage of extremal compaction and then re-explode into some other region of space-time (multiply connected space-time topology—"wormhole"). One might not take such a possibility at all seriously, except that there exist analytical solutions for collapsing, charged spheres that do re-explode in precisely this manner.[17]

Possibility 4 Various combinations of the above.

If, as one suspects today, the classical singularities are of a very physical, infinite-curvature type, then one must face up to John Wheeler's "issue of the final state" in its most raw and disturbing form. Wheeler, when faced with the issue argues that infinite-curvature singularities necessarily signal a breakdown in classical general relativity—a breakdown forced by the onset of quantum gravitational phenomena. Whether quantization of gravity will actually save space-time from such singularities one cannot know until the "fiery marriage of general relativity with quantum physics has been consummated."[18]

The work reported in this paper was supported in part by the National Science Foundation under grants GP-27304 and GP-28027 and by the National Aeronautics and Space Administration under grant NGR 05-002-256.

REFERENCES

(*Note*: In such a short paper I cannot possibly give proper credit and citations to the original literature. Therefore, I have confined most of my references to the review literature, where citations of the primary literature abound.)

1. For textbook introductions to general relativity as an astrophysical tool see Ya. B. Zel'dovich and I. D. Novikov, *Relativistic Astrophysics Vol. 1: Stars and Relativity* (University of Chicago Press, Chicago, 1971); *Relativistic Astrophysics Vol. 2: The Universe and Relativity* (University of Chicago Press, Chicago, in preparation). Also C. W. Misner, K. S. Thorne, and J. A. Wheeler, *Gravitation* (W. H. Freeman, San Francisco, 1973), cited henceforth as MTW.

2. For catalogues of relativistic gravitation theories and comparative analyses of them, see W.-T. Ni, "Theoretical Frameworks for Testing Relativistic Gravity. IV. A Compendium of Metric Theories of Gravity and Their Post-Newtonian Limits," *Astrophys. J. 176*, 769 (1972); C. M. Will, "The Theoretical Tools of Experimental Gravitation," in B. Bertotti, ed. (Academic Press, New York, 1973), *Proceedings of Course 56 of the International School of Physics "Enrico Fermi*," also available in preprint form as Caltech Orange Aid Preprint 289.
3. C. M. Will, "Einstein on the Firing Line," *Phys. Today 25*, 23 (Oct. 1972).
4. For brief introductions to black holes see R. Ruffini and J. A. Wheeler, "Introducing the Black Hole," *Phys. Today 24*, 30 (January 1971); R. Penrose, "Black Holes," *Sci. Am., 226*, 38 (May 1972). For a textbook introduction see Chapters 31–34 of MTW.[1] For a definitive monograph see B. DeWitt and C. DeWitt, eds., *Black Holes: Proceedings of Summer 1972 Session of Ecole d'Ete de Physique Theorique* (Gordon and Breach, New York, in press).
5. The definitive monograph on global methods is S. W. Hawking and G. F. R. Ellis, *Singularities, Causality, and Cosmology* (Cambridge U.P., Cambridge, in press). More brief overviews will be found in R. Penrose, "Structure of Space-time," in *Battelle Rencontres: 1967 Lectures in Mathematics and Physics*, C. DeWitt and J. A. Wheeler, eds. (W. A. Benjamin, New York, 1968); R. P. Geroch, "Spacetime from a Global Viewpoint," in *General Relativity and Cosmology: Proceedings of Course 47 of the International School of Physics "Enrico Fermi,"* R. K. Sachs, ed. (Academic Press, New York, 1971), S. W. Hawking, lectures in DeWitt and DeWitt, eds.[4] and Chap. 34 of MTW.[1]
6. See B. Carter, lectures in DeWitt and DeWitt, eds.[4] for a review of the theorems and proofs.
7. There is no review article on past studies of perturbations of nonrotating holes or on the phenomena that should accompany perturbations of rotating holes. However, most of the primary literature is cited by J. M. Bardeen, lectures in DeWitt and DeWitt, eds.[4]; W. H. Press and S. A. Teukolsky, "Floating Orbits, Supperradiant Scattering, and the Black-Hole Bomb," *Nature 238*, 211 (1972); R. A. Breuer, J. Tiomno, and C. F. Vishveshwara, "Polarization of Gravitational Synchrotron Radiation," in press. (Available from University of Maryland.) M. Davis, R. Ruffini, J. Tiomno, and F. Zerilli, "Can Synchrotron Gravitational Radiation Exist?" in press. (Available from Princeton University.) D. M. Chitre and R. H. Price, "Polarization and spectra of gravitational synchrotron radiation," in press. (Available from University of Utah.)
8. For reviews of these phenomena see P. J. E. Peebles, "Gravitational Collapse and Related Phenomena from an Empirical Point of View, or, Black Holes Are where You Find Them," *General Relativity and Gravitation 3*, 63 (1972); I. D. Novikov and K. S. Thorne, "Astrophysics of Black Holes," in DeWitt and DeWitt, eds.[4]
9. For reviews of these and other aspects of gravitational-wave physics see W. H. Press and K. S. Thorne, "Gravitational-Wave Astronomy," *Ann. Rev. Astron. Astrophys. 10*, 335 (1972); also Chap. 35–37 of MTW.[1]
10. These and other aspects of general relativistic cosmology are reviewed by P. J. E. Peebles, *Physical Cosmology* (Princeton U.P., Princeton, N.J., 1972); by Zel'dovich and Novikov,[1] and in Chaps. 27–30 of MTW.[1]
11. For a beautiful and comprehensive review of all aspects of the quantization of gravity see C. J. Isham, "Quantum Gravity," in press. (Available from Department of Theoretical Physics, Imperial College, London, S.W. 7.)

12. See Hawking and Ellis[5] for details and review.
13. For a detailed, pedagogical introduction to the singularity theorems and to the techniques by which they are proved, see Hawking and Ellis.[5] For the simplest, most understandable proof of any singularity theorem, see R. P. Geroch, "Singularities," in *Relativity*, M. Carmeli, S. I. Fickler, and L. Witten, eds. (Plenum Press, New York, 1970). For a review of the observational data that show our universe to obey the "initial conditions" of one of the theorems and the resulting conclusion that the initial big bang did possess a singularity see S. W. Hawking and D. W. Sciama, "Singularities in Collapsing Stars and Expanding Universes," *Comments Astrophys. Space Phys. 1*, 1 (1969).
14. V. A. Belinsky, I. M. Khalatnikov, and E. M. Lifshitz, "Oscillatory Approach to a Singluar Point in the Relativistic Cosmology," *Usp. Fiz. Nauk 102*, 463 (1970). English translation in *Adv. Phys. 19*, 525 (1970).
15. For a review of mixmaster singularities, see Box 30.1 of MTW.[1]
16. C. W. Misner, "Taub-NUT Space as a Counterexample to Almost Anything," in *Relativity Theory and Astrophysics. I. Relativity and Cosmology*, J. Ehlers, ed. (American Mathematical Society, Providence, R.I., 1967).
17. For reference see § 34.6 of MTW.[1]
18. J. A. Wheeler, "Geometrodynamics and the Issue of the Final State," in *Relativity, Groups, and Topology*, C. DeWitt and B. DeWitt, eds. (Gordon and Breach, New York, 1964).
19. R. W. Davies, ed., *Proceedings of the Conference on Experimental Tests of Gravitational Theories, November 11–13, 1970, California Institute of Technology*, JPL Technical Memorandum 33-499 (JPL, Pasadena, Calif., 1971).
20. R. H. Dicke, *Gravitation and the Universe* (American Philosophical Soc., Philadelphia, Pa., 1970).
21. R. H. Dicke, *The Theoretical Significance of Experimental Relativity* (Gordon and Breach, New York, 1964); also published in *Relativity, Groups, and Topology*, C. DeWitt and B. DeWitt, eds. (Gordon and Breach, New York, 1964).
22. I. I. Shapiro, "Testing General Relativity: Progress, Problems, and Prospects," *General Relativity and Gravitation 3*, 135 (1972); also published in Ref. 19.
23. R. H. Dicke, in *Proceedings of Course 56 of the International School of Physics "Enrico Fermi,"* B. Bertotti, ed. (Academic Press, New York, in press).
24. J. Weber, in *Proceedings of Course 56 of the International School of Physics "Enrico Fermi,"* B. Bertotti, ed. (Academic Press, New York, in press).
25. D. Eardley, D. Lee, and A. Lightman, 1973, in preparation. (Will be available from Caltech.)
26. R. H. Dicke and P. J. E. Peebles, "Gravitation and Space Science," *Space Sci. Rev. 4*, 419 (1965).
27. R. H. Price, "Nonspherical Perturbations of Relativistic Gravitational Collapse. I. Scalar and Gravitational Perturbations," *Phys. Rev. D 5*, 2419 (1972).

H. GUYFORD STEVER *is the director of the U.S. National Science Foundation. He came to this post after a distinguished career in government and in university teaching and as president of the Carnegie-Mellon University. He has served as president of the American Institute of Aeronautics and Astronautics and holds membership in the Royal Aeronautics Association, the American Academy of Arts and Sciences, and the U.S. National Academy of Engineering.*

12 *Some Perspectives on "Physics in Perspective"*/

H. GUYFORD STEVER

I feel privileged to be addressing IUPAP on this occasion, your first General Assembly to be held in the United States and also the year of your fiftieth anniversary. Let me begin by welcoming all of you to Washington, a city where both the strong and weak forces of physics and politics often meet and interact. In fact, with a little imagination, one might view the Washington area as a huge high-energy physics laboratory with our circumferential highway as a type of synchrotron, with particle sources at National and Dulles Airports, designed to eject its physicist travelers into the chambers of the government, where they collide with others of different mass and charge and make strange tracks. These tracks are then puzzled over by the press and the public for some clue as to why these people behave the way they do. I admit that this analogy is not very precise—particularly if you observe our highways at rush hour. I hope your accelerators operate with less energy dissipation and greater precision.

While I recognize that IUPAP is our most prestigious *international* physics organization, I want to spend a part of my remarks discussing a recent report* on the state of physics in the United States, which, nevertheless, should be of interest to all physicists. This is the report called *Physics in Perspective*, recently released by the National Academy of Sciences after three years of preparation by the Physics Survey Com-

*A summary of the findings of the Physics Survey Committee is presented as the following chapter in this volume.

mittee of the National Research Council. D. Allan Bromley, the Chairman of the Committee, and the many panel members who worked on this report deserve a great deal of praise for this extensive and perceptive document. It is one of the most thorough and thoughtful surveys of the state of any science that has ever been done in this country. And my own comments tonight are only some preliminary, personal reflections on its general observations.

Physics in Perspective—the Bromley report—is a document that should be read and seriously considered by many people: by physicists who want a good overview of the state of physics beyond their own discipline; by scientists and engineers in industry, government, and the universities who want a better appreciation of what this science is contributing and can further contribute to their work; and by leaders and decision-makers in and out of government who must recognize the role of physics in the world today and the support it needs to sustain it.

There has been some early criticism of this report on the grounds that, having been prepared by physicists, it lacks some objectivity and tends to be self-serving. But what would have been the alternative to its authorship? As the Preface of the report indicates: "More than 200 active members of the U.S. physics community have participated in the Survey." What group is better qualified to describe the state of physics, its accomplishments, its apirations, and its needs? The physicists have presented their viewpoints and argued their case effectively. It is now up to the reader of their report to supply some of the objectivity—the outsider's point of view—that might be missing. Let me attempt to do some of this, although I am no stranger to the world of physics and cannot claim total objectivity.

The Bromley report recognizes, and rightfully so, the enormous contribution of physics to basic and applied science, to technology, and to the advancement of society. To me, the great importance of physics—in the past, today, and for the future—is summed up in these introductory words in the report's Recommendations:

> Science is knowing, and the most lasting and universal things that man knows about nature make up physics. As man gains more knowledge, what would have appeared complicated or capricious can be seen as essentially simple and, in a deep sense, orderly. To understand how things work is to see how, within environmental constraints and limitations of wisdom, better to accommodate nature to man and man to nature.

Most people fail to realize this basic importance and function of physics, just as they probably do not recognize how many other scientific disciplines are an outgrowth of the knowledge of physics and depend increasingly on it for their growth. Also, few people realize how directly dependent we have become on the contributions of physics. It would be most revealing—and certainly have an impact on the public—if we were to have people imagine that our society recalled from use, even momentarily, all physics discoveries of recent decades. If this were done, I think most people would be astounded at the extent to which their modern world would be paralyzed or disappear. Few recognize how dependent our communications, transportation, industry, and even our arts and entertainment have become on technologies born in the physics laboratory.

But even more significant is how important these discoveries and research now in progress are going to be in our future. There are few human endeavors now, whether they are involved with energy, materials, environmental problems, urban or rural development, human health, and even education, that will not benefit directly or indirectly from advances in the various disciplines of physics. The *Physics in Perspective* survey lists some 15 "high-leverage" areas of physics, disciplines where the proper support might soon lead to breakthroughs that could contribute significantly to the solution of some of our major problems today. Plasma physics, nuclear physics, solid-state physics, and cryogenics are only some of the fields essential to developing new methods of generating and transmitting power. Solid-state physics, spectroscopy, acoustics, activation analysis, and a variety of detectors—optical, infrared, x-ray, gamma-ray, particle, and so forth—will give us new concepts of how to deal with environmental conditions, as well as the methods and new instrumentation essential to analyzing and monitoring them. They are also playing a growing role in medical diagnosis and treatment today, and we have only begun to see their use in this role.

At the frontiers of biology—in fields such as cellular and molecular biology—physics and the products of the physics laboratory have become increasingly important. The same is true in our investigations of large-scale natural phenomena in geophysics, oceanography, and atmospheric science. As far as most sciences are concerned, physics is a cornucopia of ever-increasing ideas relevant to their field in one way or another. And it is the fundamental science necessary to provide man with knowledge he must have to satisfy his future needs as well as his intellectual curiosity. *Physics in Perspective* makes this case magnificently.

The question, then, of the support of physics is not a question of its worth to science or its worth directly or indirectly to society. The question, I think, is one of the necessary level of support within a certain time scale and of the distribution of that support. And that is a question that looms large in the world of science because scientists are so close to its ramifications. In relating physics to national priorities, the Bromley report attempts to put this matter in perspective also, but other viewpoints are desirable as well.

A realistic level of support, considering the competitition of the other sciences and social priorities, cannot be based on the historical growth of physics in recent decades. You may recall—and the Bromley report does so—that financial support of physics in the United States from the 1940's to the mid-1960's increased at an average annual growth rate of 15 to 20 percent. The growth in number of physicists trained during this period reached a doubling time of ten years. Also, the cost of doing physics research—the facilities and equipment—rose phenomenally. All this created a great momentum in the world of physics, and when support was cut back in the late 1960's, due to changes in emphasis in national programs, the shock to the physics community was profound. And the aftershocks still linger.

The hardship that these cuts inflicted notwithstanding, it is doubtful that the support of physics would have continued to grow in that manner much longer. And if it had, would the resulting scientific productivity have approached anything near that rate of investment? But more worthwhile than analyzing what might have been is to look ahead realistically. And here, I believe, the picture is far from depressing.

What are some of the indications that lead me to feel this way? Today there is a new recognition of the need for more and better applied science and technology, which we know is most dependent on more and better basic research. This is reflected in the fact that federal funds for research are rising once more, albeit slowly in comparison with their earlier exponential growth. I think this will continue, but support will be more selective, more subject to the pressures of priorities. One result will be that programs of some apparent value may be sacrificed to give preference to activities in more promising subfields. There is a calculated risk in sacrificing any program in basic research, of course, because that program may have held the key to some essential knowledge not available when needed later by another field. The Bromley report covers well the questions and risks involved in selecting priorities. In fact, its discussion of this subject—setting up a method of establishing a set of merit criteria by which one may evaluate scientific subfields—is unique and could be very valuable to those responsible for funding science.

In the years ahead, the allocation of funds to science, and within its disciplines, is going to raise difficult questions. Nevertheless, we will have to make hard choices, and fields that show the most "ripeness" and promise will no doubt win out in these decisions.

Though physicists will probably not receive the increasing rate of support that they enjoyed in previous years, they still have great new tools and will be receiving funds to carry out research with these, even if at a somewhat lower level than hoped for. Success will prompt increased funds, but at present it is difficult to convince many scientists—let alone Congress and the public—that further heavy investments in physics should be made at the expense of other important and exciting fields such as biology, geology, and the social sciences.

Even though there is an underlying belief in the importance of basic research, the short-term relevance and contributions of physics—or any science for that matter—to national and human needs will have to be emphasized by its practitioners and sold to the public, Congress, and federal agencies. In this regard, the importance of public awareness of physics should not be overlooked. The recommendations in the Bromley report for increased participation of members of the physics community in informing the public about physics and the work of physicists should be emphasized. It is particularly important to make the public aware of how physics relates to providing the knowledge and tools necessary to solve today's and tomorrow's human problems. The Bromley report reveals much of this, and well. I am pleased to see that the report is being read and commented on by persons other than physicists and their fellow scientists. Also, it is gratifying to note that its arguments are being received favorably by the press, as evidenced, for example, by an editorial in the *Wall Street Journal* that seemed impressed by the case the Physics Survey Committee had made for increased support of physics.

An interesting and worthwhile recommendation made in the report is that "all federal agencies with missions that rely to some extent on basic physics should accept the support of physics research as a direct responsibility." In some respects, this parallels the Administration's new directive encouraging the mission agencies to increase their support of basic research in fields related to their mission. This is a concept I feel deserves strong support. It also bears some similarity to recommendations in the recent British report on "The Organization and Management of Government R&D"—the "Rothschild Report"—which urges a customer–contractor relationship between a government department and the research establishment that might supply it with valuable scientific knowledge.

Concerning the manpower issue in physics, there is no doubt that the system for keeping a proper supply of trained physicists must be watched, particularly to assure the best training and career opportunities for the most talented students. At the moment, the growth of physics manpower is dropping. This is due to several factors—trained physicists leaving the field because of lack of employment opportunities, physics students shifting to other fields, and a continued drop in enrollment of new students in physics. Considering the employment situation, the Physics Survey Committee of the National Academy of Sciences is probably wise in urging the academic community to limit temporarily the number of young physicists entering the field. But they are also wise in cautioning against a general overreaction to the employment situation now which could lead to a serious shortage of trained physicists in the late 1970's.

While federal funds and federal programs may play a significant role in the future health of the physics community, equally important will be what the physics community does itself to adjust its ranks and, particularly, to attract and inspire talented students. Unless it wants to lose them to other fields, it will have to tap the bright young minds that could look elsewhere for challenging, meaningful work and show them the vitality and relevance of physics.

Up to this point, I have focused on the Bromley report because I think it is a most important document concerning the current state and future of physics. Though it relates primarily to physics in the United States, I think there is much in it of meaning to physicists in other countries, many of whom have, or will be having, problems similar to those we face. The report also devotes an important chapter to the international aspects of physics, which is the subject I want to address in the balance of my remarks.

While there have been eras of single-nation dominance in physics, its research has always been international and is obviously so today. There are several ways of indicating this. One is to note the percentages of entries in *Physical Abstracts* according to geographic distribution. If we do this, we see that roughly 35 percent fall in North America; 35 percent in Western Europe; 25 percent in Eastern Europe; and 5 percent in Asia and Africa.

As members of the International Union of Pure and Applied Physics, which this year celebrates its fiftieth anniversary in the business of internationalizing physics, I am sure you do not have to be reminded of the enormous effort that goes into bringing together physicists from all over the world. Your organization alone is now responsible for arranging some 20 international conferences a year. These conferences and the growing work at international centers of physics, such as the International Centre for Theoretical Physics and the laboratories of CERN and Dubna, represent a great testimony to the vitality of physics as an international enterprise.

I am sure we will see this type of cooperation continue to grow to the benefit of all involved. This will happen for a number of reasons, not the least of which will be to meet a combination of scientific and economic necessities. For while we have long recited the more idealistic reasons for cooperation in sharing scientific knowledge, the current state of science—the growth of "big science" whether in nuclear physics, astrophysics, or geophysics—tells us that much of our future physics research will have to employ global laboratories in one sense or another. In high-energy physics, for example, the cost of future accelerator laboratories will make multinational arrangements similar to those of the CERN laboratory essential. And with foreign research teams already scheduled to work at our National Accelerator Laboratory, this new facility will see much international activity.

Space research will also necessitate the sharing of large and costly facilities. This is already taking place in the cooperation between the European Space Research Organization and NASA in the United States. We also look forward to some close cooperation in space research with the Soviet Union, for which arrangements were concluded based on discussions between Philip Handler, President of our National Academy of Sciences, and President Keldysh of the Soviet Academy.

Our expanded geophysical research also makes international cooperation and the sharing of facilities more rational. We at the National Science Foundation are most pleased that geophysicists from several nations have conducted work on our deep-sea drilling ship, the *Glomar Challenger.* Certainly the need for more environmental research involving the oceans and the atmosphere will require more international cooperation. The International Council of Scientific Unions, of which IUPAP is a member, will be a major force in organizing the necessary international research programs, as it has in the past with the International Geophysical Year and the International Years of the Quiet Sun. I understand that plans are already under way for an International Magnetosphere Survey and an International Geodynamics Program.

Astrophysics is another area in which we will see growing international cooperation. At the National Science Foundation, we look forward with great excitement to the construction of the Very Large Array (VLA) radio telescope, a facility we will be sharing with foreign scientists through the allocation of observation time. This huge radioastronomy observatory, to be built on a 3000-acre site in New Mexico, is a good example of the high cost of "big science." The total cost of the facility is projected at $76 million, which over the years of its construction will consume a healthy portion of our NSF astronomy budget. We expect the facility to be in partial operation in 1976 and completed for full operation by 1982. It should make a major contribution to man's understanding of the laws of gravity, physical processes in interstellar gases, and the origin and evolution of the universe.

In addition to our support of large facilities such as the VLA that can be shared by scientists of all nations, we at the National Science Foundation are particularly proud of our role in managing bilateral science agreements, exchange programs, and the participation of scientists in international conferences. We believe these programs will grow, and that we will all benefit by the new knowledge they create and spread throughout the world.

I look forward to the opportunity to share with you some of the challenges and rewards that lie ahead for science in general and physics in particular. I hope I have conveyed to you my feeling that there is much ahead that will challenge and reward us.

Physics in Perspective impresses on us that "science is knowing, and the most lasting and universal things that man knows about nature make up physics." The lessons of our time tell us there is much more to know and much more that needs knowing. I conclude, then, that the future of physics will be bright.

The report, Physics in Perspective, *which was the subject of the preceding remarks by H. Guyford Stever, was widely discussed at the General Assembly meeting. The Chairman of the Physics Survey Committee of the U.S. National Academy of Sciences, which wrote the report, was* D. ALLAN BROMLEY, *chairman of the physics department of Yale University and a member of the U.S. Delegation to the General Assembly. So much interest was evoked by this report that, with the permission of the American Institute of Physics, Dr. Bromley's summary of the findings of the Physics Survey Committee, which was published in* Physics Today, 25, *23 (July 1972) is reproduced here.*

13 *Physics in Perspective*

D. ALLAN BROMLEY

Early in 1969, and as a consequence of extended discussions within and among the Committee on Science and Public Policy of the National Academy of Sciences, the Division of Physical Sciences of the National Research Council, the Office of Science and Technology, the President's Science Advisory Committee, and several of the major federal agencies that support physics, a decision was reached to initiate a survey of the U.S. physics enterprise. (The completed report of the Survey Committee is available from the Printing and Publishing Office, National Academy of Sciences, 2101 Constitution Avenue, Washington, D.C. 20418.)

This report deals principally with the opportunities that physics faces during the 1970's. These opportunities are of several kinds: for fundamental new insight into the nature and causes of natural phenomena; for the development of new devices and technologies; and for greater service, both direct and indirect, to U.S. society, of which physics forms an integral part. The problems addressed are largely those that we foresee in the realization of these opportunities.

In large measure these problems reflect significant changes that have occurred in recent years. Among these have been striking changes in the different subfields of physics, a marked change in the growth rate of federal support for physics, changes in many aspects of U.S. society, changes in an ill-defined but none the less real ordering of priorities in the public mind, and changes in both the academic and federal communities that have particularly affected the motivation and basic philosophies of much of a student generation.

WHY A NEW REPORT?

These changes, too, provide much of the answer to the very real question: Why yet another major survey report on a scientific field? In 1966, under the aegis of the National Academy of Sciences Committee on Science and Public Policy (COSPUP), a national committee chaired by George E. Pake published an extensive report on the status and opportunities of U.S. physics titled *Physics: Survey and Outlook*.

That survey was completed at a time when the growth rate of U.S. science in general, and of physics in particular, was at an all time high and the field was in a state of robust health. Under these circumstances, and within the framework of a burgeoning national economy, it was not surprising that physics was considered in a relatively narrow context that justified extrapolation of the needs and objectives of the field on the basis of internal considerations and the anticipated exploitation of most of the new opportunities that the field then presented.

As is clear from Fig. 1 these extrapolations have not been realized. Although it is extremely important for physicists to continue to emphasize these potentials, it has become necessary to address the difficult question of priorities among and within subfields of physics. Many of the frontier areas of physics, because of the scope and scale of the activity and instrumentation required, are inescapably dependent on federal support. And the competition for available federal support is increasing steadily as new attacks are mounted on major problems of national and social concern. Therefore, the Committee has attempted to define and develop contingency alternatives in an effort to ensure the most effective use of the available support throughout physics.

But there should be no misunderstanding: The opportunities and challenges for both internal growth and external service to society are still present; however, if current trends in the growth of support for U.S. physics continue, only a few of these opportunities will be realized, and the relative position of the United States in the international physics community will inevitably decline.

Physics has contributed and continues to contribute to society in a great many ways through its concepts, its devices and instruments, and, more especially, the capabilities and activities of its people. Basic to the training of a physicist is the effort to reduce complex situations to their most fundamental aspects, so that they can be subjected to mathematical tools and the philosophic rigor of natural laws. The research style and approach that characterize the physicist frequently constitute his major contribution to attacks on problems outside his own specialized field. In the past, the training of a physicist was characterized also by a breadth

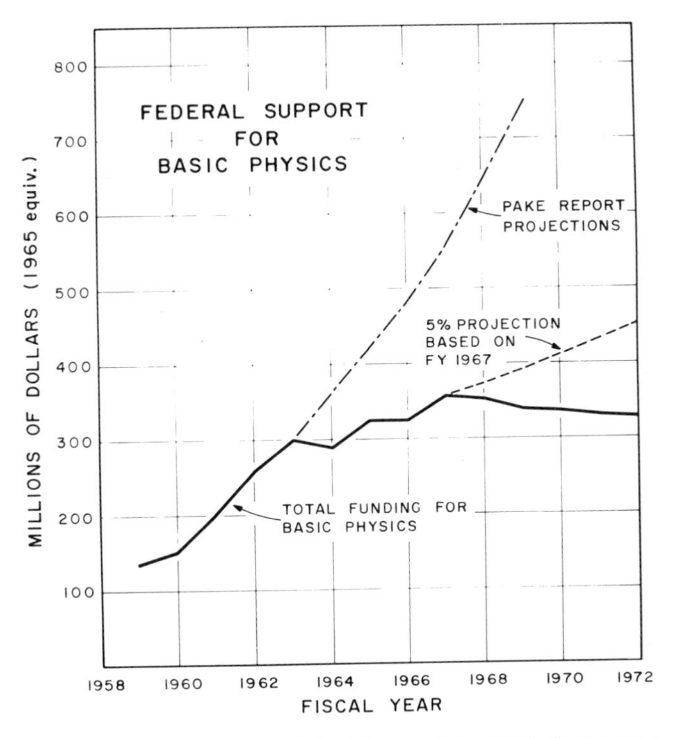

FIGURE 1 The predictions of the Pake report about U.S. Government support of basic physics were made when the growth rate of physics was at an all-time high. Actual federal support has in fact not even kept pace with inflation.

and flexibility that permitted him to range widely in his search for challenging problems not only in physics but also in other disciplines.

As the problems of society have multiplied—problems of population, poverty, and pollution, to name only three—largely coincidentally and unhappily, an erosion of these characteristics of breadth and flexibility has occurred, with a corresponding reduction in the effectiveness of physicists as partners in the solution of societal problems. This situation led the Committee to much more intensive study of educational and manpower questions in physics than has been undertaken previously and to explicit recommendations to the scientific and academic communities for steps toward the resolution of problems in these areas.

Again in contrast to earlier surveys and in recognition of the rapidly growing importance of work at the interfaces between physics and other sciences and disciplines, we have devoted extensive effort to the achievement of a better understanding of what contribution physics might make in these interface areas and what implications such activity might, in turn, have for the structure of and activity and training in the physics enterprise. Work at the interfaces ranges from the most abstruse to the most practical. An example of the former is the astrophysicist's search for understanding of the mechanism for the total collapse of a star into inconceivably small dimensions, together with the study of its light, its magnetic fields, and all other evidence of its former presence. The search for better isotopes for diagnosis and treatment of cancer, or the search for understanding of the mechanisms of nerve action, in biophysics, to alleviate ever more of the crippling afflictions that plague mankind are examples of the latter. In geophysics and planetary sciences, as a single additional example, the long-awaited understanding and prediction of major earthquakes may be at hand, with most profound practical and social consequences.

In short, in this report the Committee has attempted not only to consider the internal logic of physics but also to put physics into a much broader context and perspective both in science and technology and in the whole of U.S. society.

COMMITTEE AND PANELS

The members of the Survey Committee were appointed by Frederick Seitz, president of NAS in 1969, and by his successor Philip Handler. At about the same time, a parallel committee, chaired by Jesse Greenstein, was charged with the responsibility of surveying the opportunities and needs of astronomy. Both committees have functioned under the auspices of COSPUP, which is headed by Harvey Brooks. In developing its report, the Survey Committee early decided that it needed detailed information from panels of experts in each of a number of subfields. Several of these subfields have relatively well-defined and traditional boundaries in physics. In addition, there are several important interfaces between physics and other sciences. In the case of the interface with astronomy, which is a particularly active and overlapping one, the Physics and the Astronomy Survey Committees agreed to use a joint panel to report on astrophysics and relativity, an area of special interest to both. The broad overlap of physics with geology, oceanography, terrestrial and planetary atmospheric studies, and other environmental sciences was defined as earth and planetary physics, and a panel was established to survey it. In covering the physics–chemistry and physics–biology interfaces, the broader designations physics in chemistry and physics in biology were chosen to avoid restricting the study panels to the already traditional boundaries of these established interdisciplinary fields.

Although each panel, particularly those in the core subfields, was asked to consider the interaction of its subfields with technology, the Committee anticipated that the emphasis would be on recent developments that advanced the state of the art and on what is generally known as high technology. Therefore, to include also the active instrumentation interface between physics and the more traditional manufacturing sectors of the economy—steel, drugs, chemicals, and consumer goods, to name only a few in which many old parameters are being measured and controlled in new and ingenious ways—a separate panel was established with the specific mission of examining the entire range of U.S. instrumentation activities.

Two additional panels were appointed to centralize the statistical-data collection activities of the Survey and to address the questions of physics in education and education in physics. In addition, the Survey benefited enormously from assistance on a wide range of topics from a number of smaller working groups and individuals.

ORGANIZATION OF THE REPORT

The Committee report comprises 14 chapters in all. In this brief preview I can at best provide a kind of "road map"—some indication of the scope and direction of the Committee activities and a brief indication of how we arrived at some of our most important recommendations.

Chapter 2 is, in a sense, a summary of the Committee findings and is directed to three quite distinct audiences: the federal government, the scientific (and particularly the physics) community as a whole, and the educational community spanning the entire range from elementary through graduate and postdoctoral training. In Chapter 3 we discuss in some depth the nature of physics as a science and as a part of Western culture. We address such basic questions as: What is the science of physics? Why are physicists interested in it? And why should anyone else be interested in it?

A summary for each subfield of physics and for the interface areas considered in the Survey is included in Chapter 4. Each summary describes the present status, recent developments and achievements, and outstanding opportunities now identified, together with examples that demonstrate the vital unity and coherence of physics. This unity is the subject of the concluding section of the chapter, in which we illustrate first the remarkable similarity of concepts and theoretical techniques that are employed in condensed-matter, atomic and molecular, nuclear, and elementary-particle physics and then the even more remarkable extent to which almost every branch of classical and modern physics contributes to the understanding of the recent beautiful measurements on pulsars. In considering achievements and opportunities, we have focused on those relating to the solution of major national problems and to other fields of science and technology as well as those internal to physics. This chapter and the preceding one have been written with the hope of providing a convenient and concise overview of all of physics at a level of treatment that makes it accessible—and, we hope, interesting—to any interested layman and most particularly to students.

We have divided each subfield into program elements: These are scientific subgroupings having reasonably identifiable and unambiguous boundaries with which it is possible to associate certain reasonably accurate fractions of the total manpower and federal funding in each subfield. These program elements form the basis for our subsequent development of priorities and program emphases.

Chapter 5 is in many ways the heart of the report in that it attempts to address these very difficult questions of priority, program emphases, and levels of support. Because new developments and discoveries can change situations and priorities in physics (as indeed in any human enterprise) in rapid and unexpected fashion, the Committee thought it desirable to emphasize and develop the criteria that could be used as a basis for priority decisions rather than the specific decisions. Exceptions, of course, occur, as in the case of major facilities.

To this end we have examined in detail many of the approaches to priority determination in science that have been discussed in recent years, and we include in Chapter 5 our evaluation of the positive and negative aspects of each. From this broader examination we have evolved our own set of criteria—much modified in the course of trial application to the sets of program elements developed in Chapter 4. We have found it convenient and effective to divide these criteria into three classes: intrinsic, relating to the internal logic of a science and its fundamental bases; extrinsic, relating to its potential for application in other sciences; and structural, relating to available manpower, instrumentation, and institutions and to questions of opportunity and continuity.

We have refined these criteria by applying them, in a jury-rating sense, to the program elements of the core subfields of physics. In Chapter 5 we also illustrate the application of the first two classes of criteria. Because the structural criteria depend to a much greater extent on detailed knowledge of the individual research project and investigator or group, we have not attempted any equivalent general illustration of their use. Instead, we have selected as illustrations certain situations in which structural considerations play a predominant role; for example, situations in which major facilities, approved and placed under construction in the mid-1960's, are now becoming operational and require a step-function increase in funds for operation. If, under declining or even level support conditions, we find it necessary to obtain these incremental funds through selective termination of other ongoing programs, the much greater size and costs of the new facilities would require the elimination of much of the present high-quality activity in each of the subfields concerned.

Finally, we address the difficult questions involved in any attempt to establish a national funding level for physics—or indeed for any science or other activity that depends heavily on federal funds—and evaluate these difficulties in relation to several recently proposed mechanisms.

We do not recommend an overall detailed national physics program level. This omission reflects the Committee's conclusion that it is impossible for any such group to develop, within the relatively limited time and level of activity that are possible, either adequate information or insight to make such a detailed attempt meaningful. Nor are we convinced that it is inherently desirable for any such small group to attempt to determine national priorities at this level of detail.

What we have suggested are criteria and a mechanism for their use that may enable intercomparison among subfields, or fields, of science on a more objective basis. We have elicited from the various panels detailed budgetary projections adequate to permit interpolation to match a wide range of possible funding levels and thus have attempted to provide contingency alternatives. I shall return to these.

In Chapter 6 we attempt to describe the short- and long-range consequences for physics, science, the U.S. research enterprise, and the nation as a whole of continued deterioration of federal support. Beginning in 1968 with the abrupt change in the growth rate of support for physics, a wide variety of what were regarded as temporary or short-term mechanisms were developed to sustain individual research programs or groups until better days. Some of these measures had beneficial effects in forcing maximum efficiency in the use of available resources. But at the same time many of these have other effects—some of very subtle character—that in the long run can seriously weaken or destroy whole segments of the research community. Measures that are tolerable for a few years as stop-gap measures become frozen or institutionalized if too long continued—and often with very unfortunate consequences.

We do not claim any special position or consideration for physics, although for a variety of reasons that we discuss throughout the report its problems are among the more serious in U.S. science; we recognize that other segments of the U.S. scientific, industrial, and technological sectors have suffered major disruption also. Our discussion focuses on physics because we know it intimately. However, we believe that the observed phenomena may have much greater generality and that it is extremely important that there be greatly increased public awareness of the consequences of a continuing deterioration in the support of U.S. science.

PHYSICS AND U.S. SOCIETY

The role of physicists and physics in U.S. society is considered in Chapter 7. It illustrates, through the discussion of selected examples, the contributions that both have made and continue to make to society. It also highlights some problems that have developed recently in the interaction between physics and society. Here, too, we discuss the connections between physics and the health of the national economy through the channel of high-technology industries, in particular, and we also compare the research and development activities in several foreign countries with that in the United States.

We argue that to remain competitive in the international marketplace U.S. industry must not only increase the effectiveness and speed with which new scientific discoveries and developments are translated into technology and applications but also increase significantly the productivity of the individual worker. As sensitivity to problems of the environment and the unforeseen side effects of technology grows, it will be necessary to carry research farther, in parallel with technological developments, than has been traditional in the past; it will also be essential to bring imaginative new science and technology to bear on the problem areas arising from misuse of *existing* technologies.

Physicists have already demonstrated their ability to respond to such challenges of attacking old problems in new ways, of identifying entirely new problems, and of applying science to practical problems in entirely new and imaginative ways. We believe that pressures for such approaches will reverse the trend of the past decade wherein physicists in industry tended to be replaced by engineers. The stakes are high, but significant changes, particularly in attitude, will be required on the part of both academia and industry if the potential benefits are to be realized.

The physics community has from its earliest days displayed a truly international character. Chapter 8 considers the implications of international interaction and cooperation for U.S. physics and the contributions that the physics community can make to better international understanding and communication.

In Chapter 9 we discuss the nature of physics and its institutions and show how the evolution of science has led naturally to the development of three major foci—the universities, the industrial laboratories, and the national laboratories; we present a brief discussion of their historical evolution and of the career patterns of typical physicists in each.

In view of a continuing trend toward user-group activity in theoretical as well as in many kinds of experimental physics, we discuss in some detail the structural and organizational problems that can arise in such user-group situations and present some recommendations for their alleviation. We discuss, too, possible mechanisms for increasing the interaction among the various institutions and the changes in each that such increased interaction might require or imply.

We examined the 1968 distribution of PhD and non-PhD physicists by subfield and by employing institution. These data, like so many others in our report, have been obtained from detailed mining of the master tapes of the NSF National Register of Scientific and Technical Personnel. There are about 17,000 PhD physicists, with about 25% in condensed-matter physics, and about 20,000 non-PhD's, with many working outside the traditional areas of physics. [See S. Barisch, *Physics Today, 24,* 40 (Oct. 1971).] The industrial component, which is about 23% for PhD's and 27% for non-PhD's, is particularly important for the figure of physics, because academic employment opportunities for physicists in the 1970's will inevitably fall far short of those of the rapid expansionary period of the 1960's.

Closely linked to the growth of U.S. physics has been the development of a complex federal support structure; indeed the multiplicity of support channels has been one of the major sources of strength of U.S. physics. Because information concerning the way that this complex support structure functions has not been readily available in the past, and because we believe that much misunderstanding and misinformation are current throughout the physics community, we have provided a brief sketch of the historical origins of the present support mechanisms and of their current operation as viewed by a prospective recipient of federal support. Finally, we briefly discuss the structure and functioning of the many federal agencies that provide funds for research in physics.

How large is the U.S. physics enterprise? Figure 2 shows the distribution of federal funds for basic physics research; the characteristic flattening and slow decline since 1967 is clearly evident. Consideration of federal funding alone, however, leads to a distorted view of the total support of U.S. physics, and we have been successful in carrying out a sampling study of industrial support of basic physics research as well. Table 1 includes both federal and industrial support; it is interesting to

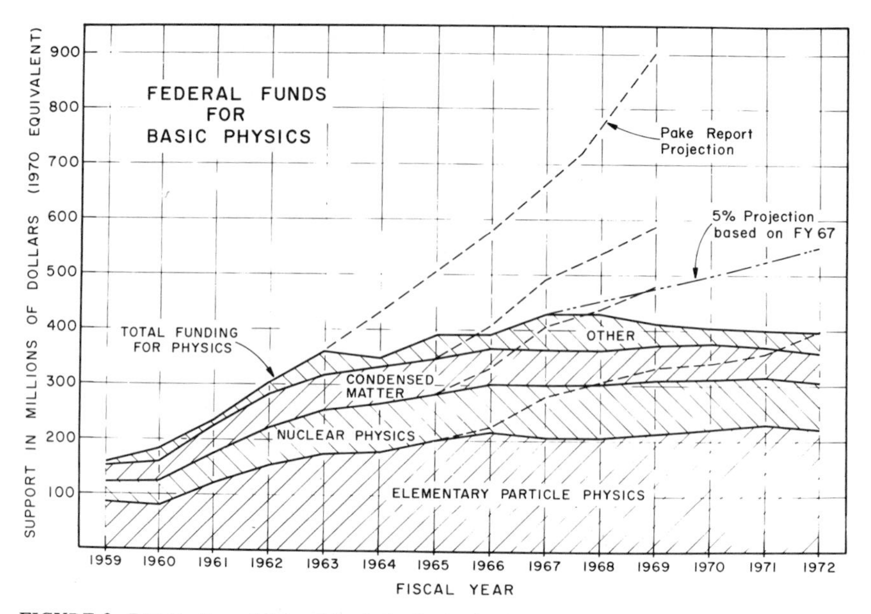

FIGURE 2 Distribution of federal funds for basic physics. Although elementary-particle physics dominates here, condensed-matter physics dominates in industrial support (see Table 1), and the federal–industrial totals for the two subfields are about equal.

note, for example, that when both sources of support are considered, elementary-particle and condensed-matter physics enjoy roughly the same total support. Nor is this the entire story. Although there is clearly a degree of double counting, inasmuch as a significant fraction of federal support funding is devoted to payment of physicist salaries, it is of in-

TABLE 1 Operating Costs for U.S. Basic Physics Subfields in Fiscal Year 1970 ($ Millions)

Subfield	Federal Funds	Percentage of Total Federal Funds	Estimated Industrial Funds	Total Federal and Industrial Funds	Percentage of Total Federal and Industrial Funds
Acoustics	14	3	1	15	3
Astrophysics and relativity	60	13	–	60	11
Atomic, molecular, and electron	13	3	7	20	4
Condensed matter	56	12	80	136	24
Nuclear	73	16	2	75	13
Elementary particles	150	33	–	150	27
Plasmas and fluids	77	17	10	87	16
Optics	12	3	7	19	3

terest to obtain a measure of the nonfederal, nonindustrial support of basic physics from the total annual U.S. expenditure on physics teaching. Making appropriate estimates for the fraction of their time devoted to physics, we find an annual expenditure from elementary through high-school teacher salaries of $310 million; college and university faculty teaching salaries total some $125 million per year.

PHYSICS AND EDUCATION

Because education is so vital to the entire physics community, and because, in turn, physics has much to contribute to education in its broadest interpretation, we have examined extensively the entire U.S. educational enterprise as it relates to both education in physics and physics in education. This examination covers the range from elementary school to graduate and midcareer education. At all levels we found significant problems that must be considered if physics is to realize its potential contributions to U.S. society, and we address recommendations to the institutions and agencies that we believe are best qualified to resolve these problems. We discovered in our study a wide variety of statistical information that was new and often surprising, at least to members of this Survey Committee. All this information, with discussion and recommendations, appears in Chapter 11.

The fraction of the total U.S. doctoral degrees in the natural sciences and engineering that have been awarded in physics has remained remarkably constant at 11% throughout the period from 1930 to 1970; even more remarkably, much the same fraction holds in both Canada and the United Kingdom under quite different societal conditions.

Where have these doctoral degrees been awarded? Figure 3 shows the pertinent data and highlights a disturbing trend. The indicated ranking here is purely on the basis of the number of PhD degrees awarded in physics and astronomy in the 1961–1965 period. Since 1970, the 20 or so established universities that have consistently educated some 50% of the total U.S. doctoral graduates, in response both to perceived employment difficulties for their graduates and to local financial problems (particularly in the case of private institutions), typically reduced the sizes of their entering graduate student classes by some 40%. Yet the total entering graduate student population changed only little. In effect, those students denied access to universities with the most qualified faculties and best facilities simply went elsewhere. Such a shift in the centroid of U.S. graduate education toward less qualified institutions can have very serious long-term consequences, and the Committee has addressed a number of specific recommendations to this problem.

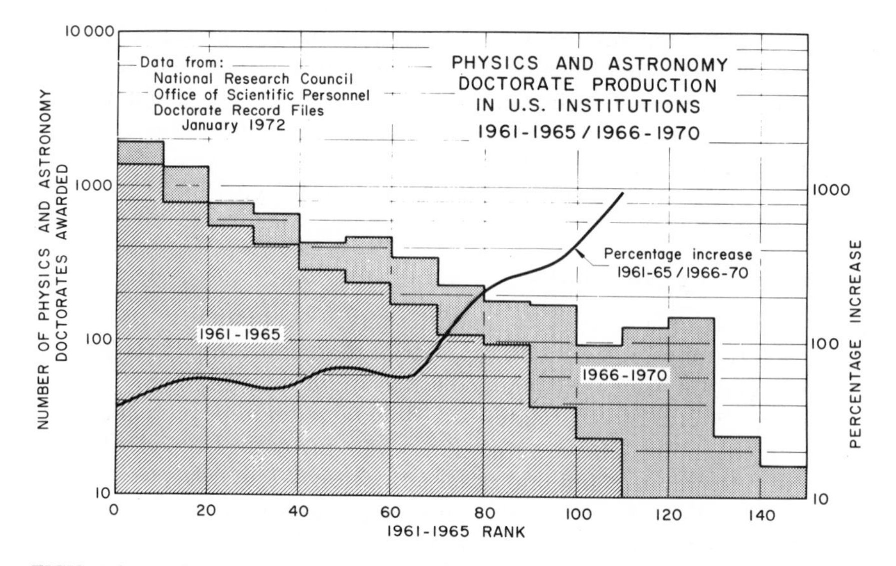

FIGURE 3 Changing patterns in PhD production become evident when academic institutions are ranked by the number of PhD's they awarded between 1961 and 1965. The leading institutions, responding to perceived employment difficulties for their graduates and to local financial problems, have cut the size of their entering graduate classes. The result, as these data from the National Research Council show, has not been a reduction in the entering population; the students have simply gone to the lower-ranked institutions.

Figure 4 illustrates, in rather striking fashion, one of the basic problems now facing physics education. Physics is an eminently nonlinear enterprise, with several equal time constants; when such a system receives an input impulse it rings. The figure clearly illustrates the first half-cycle of such a ringing, induced by the introduction in 1962 (following the Gilliland report findings) of federal programs to accelerate the training of scientists and engineers needed for the envisaged national military and space programs. Responding enthusiastically to the opportunities and challenges involved in such a national program, universities expanded their facilities and faculties and by 1968 had achieved the goals initially established for 1970. Not surprisingly the federal programs were then phased down, but the universities found it extremely difficult to follow. One of the Committee's overriding functions has been to develop recommendations aimed at damping this oscillation.

There are many inputs to such recommendations. Figure 5 shows the changing patterns of high-school program choices since 1948. Even Sputnik did not reverse the downward trend in physics. Indeed it may well be that the further drop in the 1960's reflects the removal of some of the rare highly qualified and highly effective high-school physics teachers as more attractive employment opportunities developed elsewhere. We hope that the reversal of this trend in the past few years may reverse the decrease pattern; there is already fragmentary evidence for such an effect.

Some help in predicting manpower availability can be gained by following, statistically, the history of a typical cohort of U.S. physicists. We find, for example, 460,000 twelfth graders (124,000 of them female) taking physics in 1963–1964, only 7300 undergraduate physics majors in 1966–1967 and 5500 BS physicists in 1968. Because of the enormous attrition between the twelfth-grade physics enrollments and the junior undergraduate year, it appears clear that unless further drastic reduction in the high school population occurs, the pool from which undergradu-

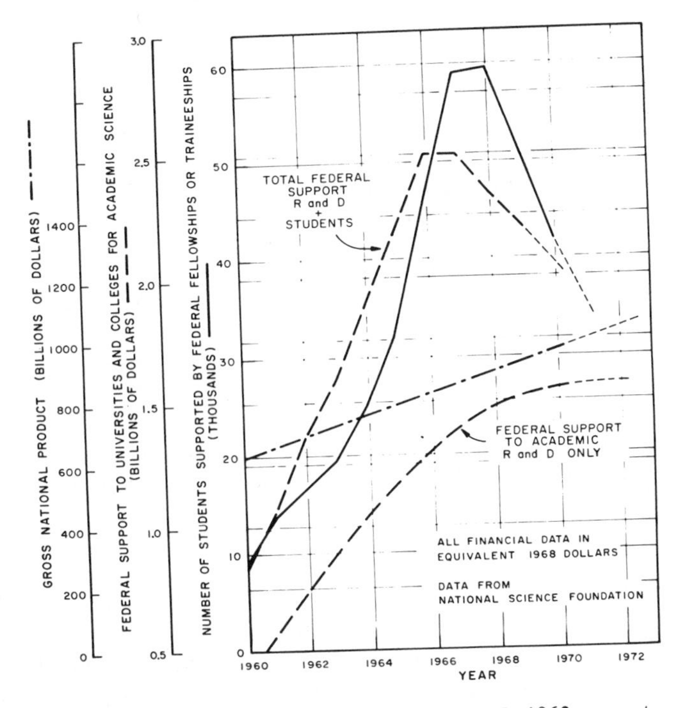

FIGURE 4 Nonlinearity of the physics enterprise. In 1962, a report urged the government to find ways to accelerate the training of scientists and engineers for the envisaged national military and space programs. The universities responded enthusiastically to the resulting increase in government fellowships and traineeships and in government aid to universities and colleges for academic science, and they have had difficulty adjusting to the current phasedown in support. A principal aim of the Survey Committee report is to avoid such oscillations in the future. (Data here are from the National Science Foundation.)

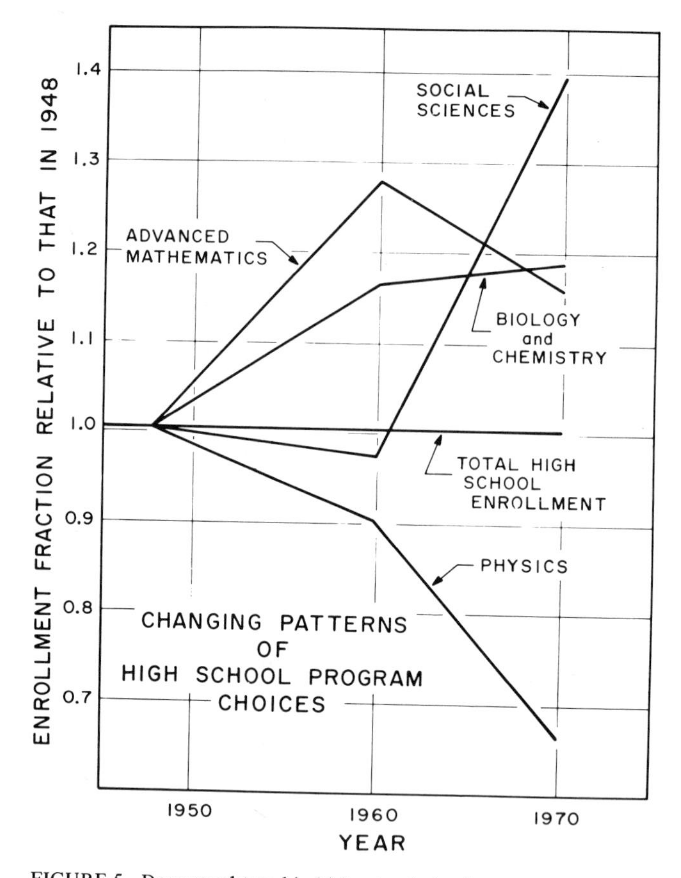

FIGURE 5 Downward trend in high school physics enrollments withstood even the effect of Sputnik, perhaps because some of the most effective physics teachers then left the high schools for more attractive jobs.

ate physics majors are drawn will remain adequately large. Of much greater immediate concern is the dramatic decrease, beginning in 1967, in the percentage of the physics baccalaureates going on to graduate work: For about two decades, roughly 53% went on to graduate work; this had dropped rather precipitously to about 37% by 1970.

On the basis of fairly stable patterns in the past we expect a production of 1500 PhD's in physics for 1974, with very similar figures holding for 1975. If we take a more pessimistic view at each stage, we obtain 600 PhD's in 1975 and extrapolate to a marked undersupply of physicists in the late 1970's, although in all our models we face a temporary oversupply situation during the next few years. The Committee has a number of specific major recommendations to alleviate the short-range manpower problems and to maintain an adequate supply of trained physicists for the nation's needs. In particular it will be important to monitor that the annual baccalaureate production not fall below roughly 5000.

There is striking evidence in these data of the rather fantastic filter that operates against women physicists in the United States in the transition from high school to college. In the question of women and of minority-group physicists, the educational enterprise is woefully inadequate.

A final important aspect of education, broadly interpreted, is public awareness of science and of its possible contributions to a better life. In the United States, efforts to inform and educate the public about science are few and usually of poor quality. The scientific community must accept a large share of the responsibility for this situation. If the present antiscience trend is to be reversed, individual scientists and scientific organizations must be prepared to spend more of their time and money than has been the case previously in fulfilling their obligation to inform a much larger segment of the U.S. public. Several of the Committee recommendations are directed to this end.

Using the extensive information that we have extracted from the National Register data tapes as well as input from our own questionnaires and those of the Manpower Committees of the American Physical Society and the American Institute of Physics, we have assembled what we believe to be the most complete information concerning the training and use of U.S. physicists yet available. We include data on mobility, sociological aspects of physics, and a brief historical study, and we project physics manpower needs with several different models. Although serious problems clearly exist in regard to employment opportunities in the traditional types of jobs that physicists have held, there is also evidence, as we have noted, to suggest that the scientific and academic communities are in some cases overreacting, and that unless some action is taken the oscillatory phenomena associated with the supply of U.S. physicists in the recent past will not be adequately damped. What emerges clearly from these studies is that the growth in tenured academic employment opportunities during the 1960's was indeed anomalous and that it is essential to take steps to broaden the motivations and interests of young U.S. physicists beyond the academic sector—as was typical before the late 1950's, for example—or serious dissatisfaction and alienation will remain. We discuss these considerations in Chapter 12.

Figure 6 is typical of the migration data that we have extracted from the National Register tapes for each physics subfield considered in the

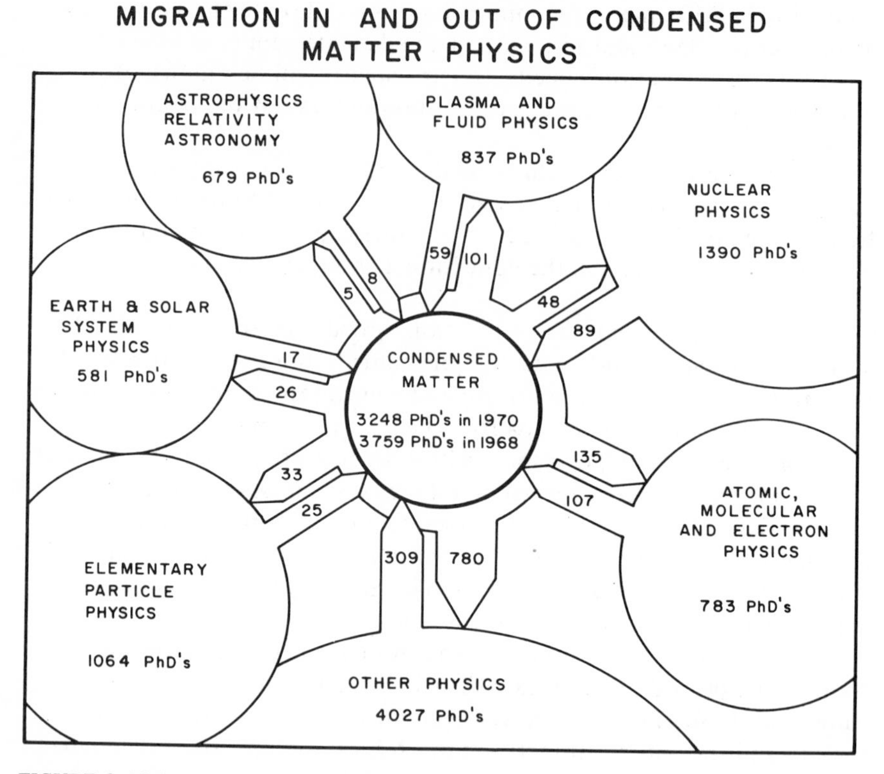

FIGURE 6 Mobility in physics is greater than is often believed. Data from the 1970 National Register of Scientific and Technical Personnel show that subfield barriers are rather permeable.

Survey. Contrary to much traditional wisdom, we find that physicists are mobile and that subfield barriers are quite permeable. Indeed we find that well over one third of all U.S. physicists have changed subfields during the past decade.

INFORMATION OF PHYSICS

Unfortunately, the available mechanisms for disseminating and consolidating the information of physics, so that it is readily and conveniently available to all potential users, have simply failed to keep pace in many areas with the rapidly increasing rate at which individual facts can be wrested from nature in modern research laboratories. In Chapter 13 we describe in detail the functioning of the various media for the communication of the information of physics. Again, we have developed an extensive body of new statistical information. We discuss these data in relation to current problems and develop specific recommendations directed toward different sectors of the physics enterprise.

We have found that traditional wisdom here too is frequently without foundation in fact. To do so we have carried out several informal operations-research studies. These have included extensive anecdotal studies with working physicists to trace the sources of their ideas and new developments and sampling studies to discover the age distribution of journals most used in large industrial, national laboratory, and university research libraries. Other sampling studies were designed to evolve an objective measure of the quality of randomly selected publications. In these, papers were submitted to panels of experts for grading; the lowest grade was achieved by papers that demonstrably set back their entire field of physics and the highest by papers that made major and lasting contributions. Several rather surprising conclusions emerged from these studies:

Primary journals are the most important single source of physics communication, with informal oral communication a close second.

The obsolescence half-life of primary-journal articles is about eight years.

Most of the primary literature of physics is apparently useful and used. A sample study in condensed-matter physics shows that 33% of the papers made additions of conceptual interest and lasting value; only 8% of the papers were judged as worthless by any member of the evaluation panel.

Any mechanism that would result in a saving of λ% of the total time the average physicist devotes to communication in physics would be worth roughly (λ/100) times $300 million to $600 million per year in the United States.

PRIORITIES AND PROGRAM EMPHASES

During the 1950's and early 1960's, U.S. science, reflecting generous public support, enjoyed a period of unprecedented growth in both quality and scope. In this period almost every competent scientist and almost every good idea could find support without undue delay. Annual growth rates of between 15 and 25% were not uncommon. The results were new knowledge, new technologies, and a large body of trained manpower commensurate with this national investment.

Such a growth rate could not continue indefinitely, and indeed, in the period since 1967 the support of many areas of U.S. science has seen marked leveling or effective decline—this is particularly true of physics. When science funding is increasing, as in the past two decades, questions of priority receive little overt attention, because worthy new projects and new investigators can be supported with little detriment to work already under way. With ample funding for new initiatives and the capacity to exploit new opportunities, it is relatively easy to maintain the vitality of the scientific enterprise; it is much more difficult to do so under conditions in which not all good ideas can find timely support and where many competent scientists cannot find professional opportunities that exploit even a part of their training. The nation is then in a position of being less able to gamble, and the cost of wrong choices becomes much higher.

Questions of priority, although often not made explicit, are an integral part of all human endeavor. Science is no exception. Scientists, science advisors, managers of science, and the scientific community must decide in one area or another what to do next and where to devote energies and resources. What areas of science are most worth pursuing? Which are most deserving of encouragement and support? These are difficult questions to which there are neither obvious answers nor, indeed, obvious methodologies for obtaining answers.

As in many other human affairs, a multiplicity of criteria must be brought to bear. Thus, perhaps, the best way to approach the question of priorities in science is to try to identify and develop the criteria by which they are made. There is now an extensive literature in this area; as a Committee we have studied this earlier work and have devoted much effort to the evolution of a set of criteria that we have found particularly useful in appraising the needs and potentials of the different subfields of physics (see Fig. 7). A number of approaches that have been proposed for the establishment of priorities are considered in the Survey report, together with what we as a Committee consider to be their positive and negative aspects.

After much discussion of the various possibilities, the Committee decided upon an approach that combines many of their features: a "jury rating" of Committee members as to the appropriate emphasis that should be applied to a given activity within the next five years, taking account of both the internal, intellectual needs of physics and their assessment of the impact of these scientific developments on other sciences, on technology, and on societal problems generally. It must be emphasized that any such rating system has a value that is relatively short-lived, because science changes so rapidly. Moreover, any group of people as small as the Physics Survey Committee is bound to represent certain prejudices or special interests that would be different for a differently selected but equally competent group of comparable size. The numbers are too small for nonobjective biases to be mutually canceling.

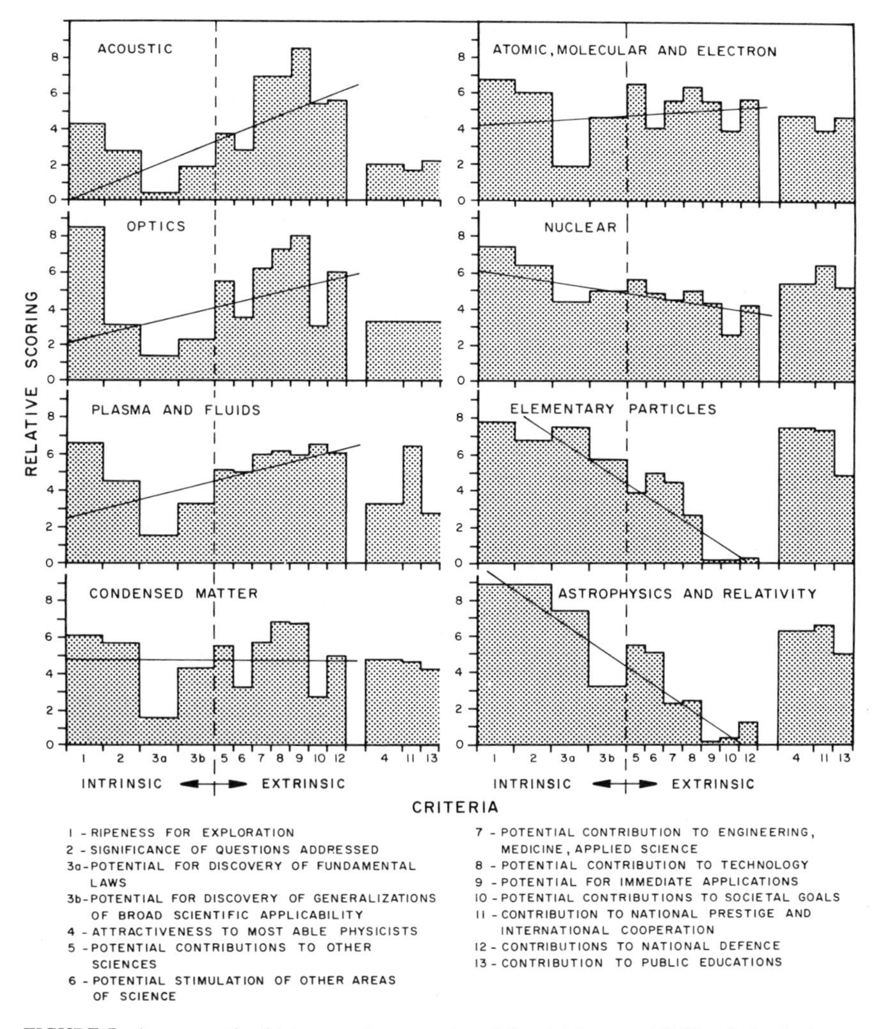

FIGURE 7 Average rating histogram shows scoring of the eight core subfields of physics in terms of the intrinsic and extrinsic criteria. (See Table 3.)

The goal has been to identify those program elements in physics that, on the basis of our criteria, should experience large *relative* growth rates. The questions of *overall* growth rates and support levels for physics and the development of contingency alternatives designed to respond most effectively to different levels of such total support are taken up below.

To illuminate the criteria through trial application to actual cases (and it should be noted that this process resulted in extensive modification of the Committee's initial criteria) and to discover the degree of consensus that existed within the Committee, matrices were prepared bearing the program elements of each subfield (see Chapter 4) as rows and the criteria as columns. (This procedure was not applied to the interface fields such as chemical physics, biophysics, and earth and space physics, which present special problems and were not considered amenable to this approach. In large measure this reflects the fact that the physics components of these interface areas, although very important, are not dominant. Lacking a more comprehensive survey of these fields to place their physics elements in better perspective, we have not considered ourselves competent to carry out a similar jury rating.)

The structural criteria were not included in this exercise because, to a much greater extent than those of an intrinsic and extrinsic character, the structural questions are of a detailed nature, applicable to each specific project, and they change rapidly with time. In arriving at ultimate program emphasis decisions, these structural criteria must be given due weight, but to apply them effectively a detailed study of the individual program elements, and of the individual research projects in them, is required. Certain exceptional cases for which the structure criteria have a direct bearing on our recommendations are discussed below.

With the reports of subfield panels in hand, and following a brief presentation of the program elements by an advocate drawn from the Committee membership, the matrix elements were rated on a 0–10 scale by each Committee member. These ratings were subsequently combined to obtain the averaged matrices for each subfield.

Figure 7 is the averaged rating histogram plotted in this fashion for the above-mentioned eight internal subfields of physics. Inasmuch as questions 4, 11, and 13 are of a somewhat different character than the remaining ones, they are presented in a separate histogram to the right of each figure. (Histograms of the Survey Committee's average jury rating of the individual program elements in each subfield appear in our report.)

Several interesting checks that give a character signature for each subfield emerge immediately from inspection of this figure. Not unexpectedly, acoustics and optics have a signature that strongly emphasizes the extrinsic criteria, whereas astrophysics and relativity and elementary-particle physics emphasize the intrinsic criteria. The remaining subfields fall between these extremes in ordering, a result entirely consistent with the intuition of the Committee members.

Most important was that, despite widely different backgrounds and interests, the spread in the ratings of the Committee members on individual matrix elements was small. In part this is a reflection of the fact that the relatively large number of criteria that the Committee chose to use makes for a more objective evaluation of each individual criterion.

Table 2 shows the first 25 of the 69 program elements considered by the Committee listed according to overall scoring. We emphasize that *within any restricted area of the listing the relative ordering should not be considered significant.*

The approach adopted herein and the criteria evolved may have rather general utility in providing a somewhat more coherent and objective evaluation of program emphases than the more subjective intuition and folklore that has tended to characterize previous attempts of the scientific community.

Thus far, consideration has been given to the distribution of a total level of funding, allocated to the national physics enterprise, among the subfields of physics without explicit consideration of what the total level might be. It is abundantly clear, however, that the distribution is inevitably a strong function of this total level.

Among the most discussed mechanisms for the establishment of long-term support levels for science have been the following:

Tie the support of science to the Gross National Product (GNP).

Because the scientific community was in a state of robust health in 1967, tie projected support levels from that point to the GNP; this leaves an obvious present deficit that would be rectified by step funding increments in the short-range future.

TABLE 2 Overall Scoring of Program Elements[a]

1. *Lasers and Masers*	14. Neutron physics
2. *National Accelerator Laboratory*	15. Nuclear theory
3. *Quantum optics*	16. *Very large radio arrays*
4. University groups–EPP	17. *X- and gamma-ray observatory*
5. *Stanford Linear Accelerator*	18. *Turbulence in fluid dynamics*
6. Nuclear dynamics	19. Superfluidity
7. Major facilities–EPP, AGS improvement	20. Infrared astronomy
8. Brookhaven AGS	21. General-relativity tests
9. Nuclear excitations	22. Oceanography
10. *Heavy-ion interactions*	23. *Atomic and molecular beams*
11. *Higher-energy nuclear physics*	24. Laser-related light sources
12. Nuclear astrophysics	25. *Controlled fusion*
13. Theoretical relativistic astrophysics	

[a]The order here within any group of, say, five program elements is not signifcant. Elements in italics are those considered to be high-leverage programs. Other high-leverage program elements are (in arbitrary order of decreasing numbers of PhD physicists in the corresponding subfield) macroscopic quantum phenomena, scattering studies in solids and liquids, biophysical acoustics, and nonlinear optics.

Because a healthy U.S. scientific enterprise is of particular importance to the well-being of our high-technology industries and these in turn to the national economic health, tie projected supported levels to the productivity of the high-technology sector of the national economy.

We examine each of these at some length and conclude that none will suffice. The problems are compounded by the fact that, in practice, funding finally made available to the subfields of physics is not interchangeable. Lack of recognition of this situation has already led to tensions within the physics community and even within subfields of physics.

CONTINGENCY ALTERNATIVES

It was recognized from the outset of the Survey that, in view of competing claims on the discretionary component of federal resources in any given year, it may not be possible to allocate to any given subfield enough support to permit it to make optimum progress. Therefore a range of contingency alternatives in each subfield was developed to provide an assessment from the physics community of ways to utilize most effectively whatever funding support becomes available—most effectively from the viewpoint of the overall health of physics and of the contribution that physics can make to U.S. society.

Accordingly, the initial charge to subfield panels requested that they develop, in as much detail as possible, programs for their subfields under various assumed funding projections: a so-called exploitation budget that attempts to exploit all the opportunities, both intrinsic and extrinsic, now perceived; a level budget—level in dollars of constant buying power; and a declining budget—declining at an arbitrarily established rate of between 6 and 7.5% per year. To obtain the hoped-for interpolation possibilities, it was necessary to evolve an intermediate growth-rate budget between the exploitation and the level budgets. In each case, the panels were asked to emphasize the costs to science and the nation of cutting back from the exploitation budget. This required very detailed examination of the internal structure of each subfield and of its opportunities and needs, as well as sharp scrutiny of the internal priorities of the subfield.

Development of such budgetary projections was more easily accomplished in some subfields than in others. In areas such as elementary-particle physics and astrophysics—and to an increasing extent in nuclear physics—the activity is largely quantized around major facilities. It is characteristic of such facilities that a large fraction of their total operating costs is invariant to the extent that the facility is maintained in operational status; support and development staffs, power for magnets, radio-frequency sources, and other systems must all be provided unless the facility is closed down. This is reflected in a very large leverage factor; that is, very small percentage changes in the overall operating budgets can be reflected as major fractions of the discretionary component of these budgets—that fraction that goes directly to the pursuit of research and not simply to keeping the doors open.

In such heavily quantized subfields, reductions below the exploitation budget typically have involved the closing down of entire facilities or, at least, major change in the style and scope of operation permitted. This results in a corresponding reduction of the manpower that the subfield can accommodate, quite apart from possible opportunities for new personnel now being trained. The dislocation and career disruptions involved here for excellent scientists and support personnel is a waste of resources, which in our opinion the nation can ill afford.

In less quantized subfields such as condensed-matter and atomic, molecular, and electron physics the effects of budgetary reductions are less obvious and the manpower problems less extreme. Because the research is much less facility-intensive, reduced funding means that the objectives of each scientist or scientific group are lowered—less work is done, fewer challenges are met, and the field slows down. Although this can proceed for a time without overt symptoms of serious trouble, trouble is there; morale drops, enthusiasm dwindles, and the subfield is less able to respond to challenges or opportunities. Quite apart from these differences, however, all subfields have concluded that a budgetary level declining at 7.5% per year would, within five years, bring the subfield below that critical point where productivity, however measured, falls dramatically.

Figure 8 shows the Committee synthesis of all the panel inputs. As discussed in detail in the report, a number of major facilities, including the Los Alamos Meson Physics Facility and the National Accelerator Laboratory, are scheduled for initial research utilization in fiscal year 1973, following initial approval in the mid-1960's and a lengthy construction period. To enable this turn-on without gross interference with the remaining programs in these fields, the Committee has recommended incremental step funding in 1973 (in fact, much of the required incremental funding did not become available in 1973 and must be deferred until 1974). Beyond this there existed a rather remarkable consensus that an 11% growth rate would permit exploitation of the opportunities in each subfield. As indicated in this figure, even such a growth rate would not bring the field, by 1977, to the level that it would have reached had a steady five percent annual growth rate in funding been possible since 1967.

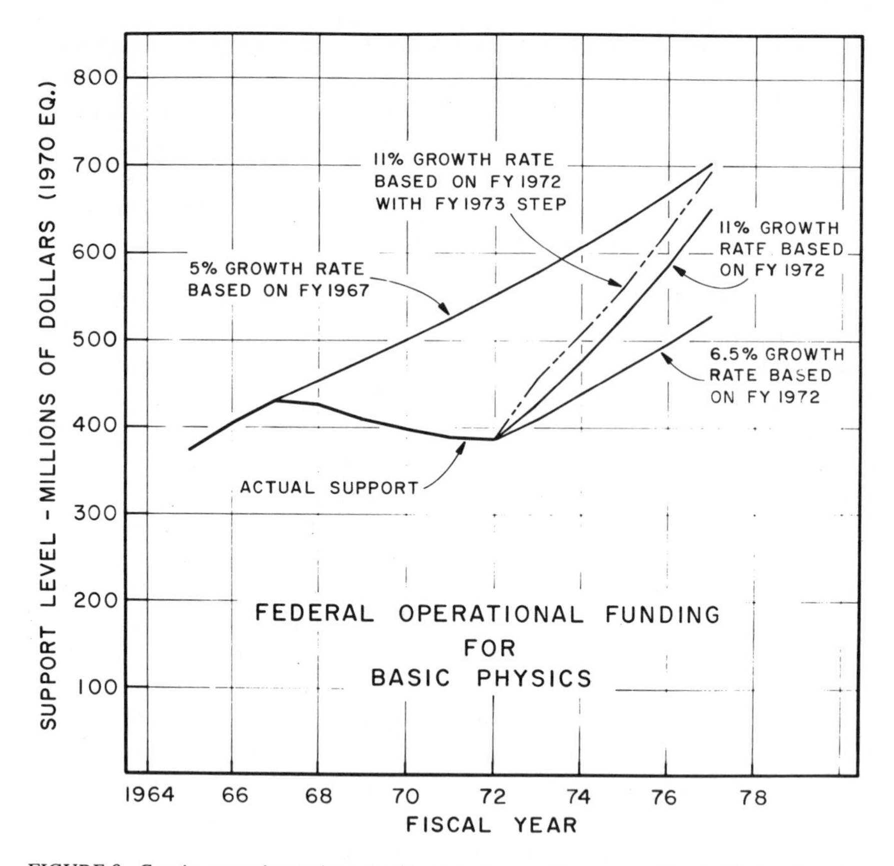

FIGURE 8 Contingency alternatives. An 11 percent annual increase, with a 1973 step-function increment, would allow exploitation of the opportunities in each subfield. Even this growth rate, however, would not bring the field to the level it would have reached had it followed a steady 5 percent growth rate since 1967.

It must be emphasized, however, that the Committee has not attempted to detail any specific national program for physics in the next five-year period in the range between the exploitation (11% growth) and flat (zero percent growth) levels of investments—both in dollars of constant purchasing power. Within each subfield, to the extent possible, this detailing has been carried out.

The fact that we do not recommend a detailed national physics program appropriate to different possible levels of funding does not reflect any unwillingness to face the difficulties inherent in any such attempt. It is unrealistic to look upon the total funding of U.S. physics as an effective reservoir from which funding for individual program elements is parceled out without having cognizance of all the internal and external pressures and constraints within both the different funding agencies and the physics community itself. An *a priori* allocation system, which parcels out a fixed amount of total funding among predefined fields of science, is likely to be stultifying of initiative and novelty.

TABLE 3 Criteria for Program Emphases

Three sets of criteria, which emerged from discussion, were refined through application to program elements in the various subfields:

Intrinsic merit

1. To what extent is the field ripe for exploration?

2. To what extent does the field address itself to truly significant scientific questions that, if answered, offer substantial promise of opening new areas of science and new scientific questions for investigation?

3a. To what extent does the field have the potential of discovering new fundamental laws of nature or of major extension of the range of validity of known laws?

3b. To what extent does the field have the potential of discovering or developing broad generalizations of a fundamental nature that can provide a solid foundation for attack on broad areas of science?

4. To what extent does the field attract the most able members of the physics community at both professional and student levels?

Extrinsic merit

5. To what extent does the field contribute to progress in other scientific disciplines through transfer of its concepts or instrumentation?

6. To what extent does the field, by drawing upon adjacent areas of science for concepts, technologies, and approaches, provide a stimulus for their enrichment?

7. To what extent does the field contribute to the development of technology?

8. To what extent does the field contribute to engineering, medicine, or applied science and to the training of professionals in these fields?

9. To what extent does the field contribute directly to the solution of major societal problems and to the realization of societal goals?

10. To what extent does the field have immediate applications?

11. To what extent does the field contribute to national defense?

12. To what extent does activity in the field contribute to national prestige and to international cooperation?

13. To what extent does activity in the field have a direct impact upon broad public education objectives?

Structure

14a. To what extent is major new instrumentation required for progress in the field?

14b. To what extent is support of the field, beyond the current level, urgently required to maintain viability or to obtain a proper scientific return on major capital investments?

15. To what extent have the resources in the field been utilized effectively?

16. To what extent is the skilled and dedicated manpower necessary for the proposed programs available in the field?

17. To what extent is there a balance between the present and envisaged demand for persons trained in the field and the current rate of production of such manpower?

18. To what extent is maintenance of the field essential to the continued health of the scientific discipline in which it is embedded?

HIGH-LEVERAGE SITUATIONS

Small changes in funding—either up or down—can sometimes be reflected in large changes in scientific productivity. In the case of major facilities, such a large fraction of the total funding is required to keep them in operation that even small fractional changes in funding are reflected as very large changes in the research component, to which scientific productivity is much more directly coupled. In fields where new breakthroughs, either in concepts or in instrumentation, have occurred, new frontiers are opened, and investment in research at those frontiers can be expected to yield high scientific return. In other fields, again because of breakthroughs in instrumentation or ideas or because the field itself in its internal development has reached a state where further investment can be expected to yield returns of high societal importance, the leverage is high.

The relative weighting or importance assigned to each type of leverage will vary from field to field and from one support agency to another. This is healthy and proper. In examining program elements as candidates for high-leverage consideration, structural criteria play an important role. It is here, for example, that continuity considerations enter explicitly.

As illustrations of various types of high-leverage situations and the utilization of our Committee-ordered listings, in combination with the subfield panel reports, the Committee selected from the subfields 15 program elements with growth potentials that warrant high priority for their support in the next five years; these are presented in Table 3.

We emphasize that the increased support recommended for these program elements should not be at the expense of other activities in the subfields, although clearly some readjustment is not only necessary but also healthy as the various program elements attain different levels of scientific maturity. At the same time, it should clearly be recognized that, should only the selected program elements be supported, the overall physics research program would be totally unbalanced.

WOLFGANG GENTNER *is the director of the Max Planck Institute for Nuclear Physics at the University of Heidelberg. As a past director of the Synchro-Cyclotron Division of the European Organization for Nuclear Research (*CERN*) and past president of the* CERN *Council, as well as being active on the Committee for Research and Technik of* EURATOM, *Professor Gentner has had great experience in the international affairs of physics. His scientific honors include membership in the Heidelberg Academy of Science, the Bavarian Academy of Science, and the Akademie Leopoldina, Halle. He is an Honorary Fellow of the Weizmann Institute of Science, Israel.*

14 *International Cooperation in Physics* / WOLFGANG GENTNER

The history of science in the Occident begins with the natural philosophy of the Greeks. Their commercial routes, which resulted from the endeavor to establish settlements along the coast lines of the Mediterranean and Black Seas, became at the same time conduits for their philosophies. Let us consider Pythagoras of Samos, who, after many voyages to the orient as well as to his native Greek country, finally settled in Kroton in lower Italy, where he founded his world-famous community. Voyages through the then known world were an integral part of the career of a great teacher.

Well into the Middle Ages, Aristotle, the teacher of Alexander the Great, was recognized as the founder of all of science in the entire Occident. The focal point of international collaboration moved in Hellenistic times from the Academy of Plato in Athens to the Occidental Academy in Alexandria, where it remained for several centuries. It was here that the astronomical, mathematical, and physical sciences, as well as poetry, reached great heights, certainly with some ups and downs, well into the seventh century under a variety of political systems. However, the first home of common European culture perished in the maelstrom of the migration of peoples and of Arabic conquests. Only part of the scientific works that were stored in Alexandria were preserved by the Arabs. When scientific research in the fourteenth and fifteenth centuries became secularized, the academies were founded anew, first in Italy, later in the remaining European countries. These academies adopted the Greek tradition of international science. Latin had meanwhile become the international language and was everywhere accepted as the common tongue of students and scholars. Today, publications in Latin, such as the abstracts of the scientific publications of the Pontifical Academy of Sciences, have become a rarity. This worldwide community of scientific endeavor was broken up by the rise of the national states within Europe. Whereas Galilei, Newton, and Kepler still published their scientific works in Latin, French became for a while the international language in the field of science, owing to the importance of the Academy of Paris. However, the growing nationalism in Europe, strongly fostered by the great French Revolution, resulted in an ever-increasing use of national languages in the sciences. It became the rule that scholars could express themselves only in their native language. The traditional traveling of students to Bologna, Pavia, Paris, or Prague, which was almost commonplace in earlier times, came to a slow extinction in the nineteenth century. Nationalistic pride produced the strongest fruits. Every state, no matter how tiny, promoted scientific publications in homegrown journals and in its native language. The ivory tower of scholars had become a veritable scientific tower of Babel.

This scene began to change gradually after World War I but more rapidly after World War II. English emerged as the modern scientific language over the whole world. I believe that the use of this new international language of scholars has been decisive for the inception of international collaboration in our century. In times past, it was necessary to have an intimate knowledge of Latin to become a student; today one must be able to read English fluently as a minimal requirement of scientific studies. Basic English has become the language of communication at scientific meetings in East and West. I have made these introductory remarks on the significance of a common language because the fiftieth anniversary of the IUPAP is being commemorated in an English-speaking country, where this problem is not nearly so apparent as it is, for example, in Europe.

When I began to collect the notes for this paper, Diana Crane of the Organization for Economic Cooperation and Development (OECD) in Paris was visiting us in Heidelberg. She gave me many valuable tips regarding the most important publications dealing with the topic of international collaboration in the sciences. Her own recent paper, "Transnational Network in Basic Science" (*Internat. Organ. XXV*, No. 3, 1971), is an excellent and fundamental contribution to the study of the contemporary state of this question. Also, members of the OECD have reported competently on problems of international cooperation in several meetings and publications. In this connection, I would like to recall the articles of Jean-Jacques Salomon, which deal in part with international scientific policy. A Scientific Policy Committee was established within OECD many years ago. This committee concerns itself with the preparation of the meetings of the ministers of science of OECD countries and with the reporting and summarizing of their results. Twenty-one nations are members of this organization, and since there are several important non-European states represented, such as the United States, Canada, and Japan, the problems are dealt with in a truly worldwide context. Already in its first report, "Fundamental Research and the Policies of Governments" (Paris, 1966), one can find an excellent treatise on several experiences in the field of international cooperation and an elucidation of many important points. The difficult questions connected with applied research and industrial research, which have far-reaching implications for the future, are also treated in this report.

Of great interest to us are the occasional failures of international collaboration in some areas of applied research, because they warn us that such collaboration requires the most careful of preparation. In this respect, I think of the protracted crises in the Euratom laboratories, which have continued for years, and which, even today, are not yet solved. Whenever economic interests collide with national interests, the utmost in care and consideration is called for. It is regrettable that considerations of national prestige and commercial egotism have not yet been exorcized, even from Western Europe; and it is these factors that interfere with the efforts to establish cooperation, especially in the field of applied physics.

Since much has already been written and said on these relevant questions within the framework of OECD and within the common market countries, I should like to restrict myself in the remainder of this paper to my own experiences, which were gathered principally during my work at the European Organization for Nuclear Research (CERN). I will also draw from many discussions, in which I was a participant, concerning other European research institutes and the International Institute for Theoretical Physics in Trieste. As a matter of convenience, I should like to restrict my remarks even further to the management of international laboratories, so that I will not mention the numerous international associations and organizations now in existence. I am deeply interested in the many problems that arise from the scene of a common working place, where people of different languages and heritage must work together and depend on one another every day. I think that the many problems that we encounter here are fascinating and worthy of our attention.

If one considers the European successes and failures as a point of departure for future planning in the East and the West, one cannot fail to draw salient conclusions for the possibilities of a worldwide common scientific policy. I will therefore single out one European laboratory that is widely considered a fine example of successful international collaboration. This laboratory is CERN in Geneva. The story of its founding has been told often, and the speeches delivered on the occasion of the tenth anniversary of the entry into force of the convention of October 1964 provide an interesting picture of the conditions in those days and of the considerations of future tasks. The latter were especially emphasized in the speech of Edoardo Amaldi, which was titled "The Future of European Cooperation."

I will attempt to analyze the causes of the success of CERN, now that this organization can look back on a history of approximately 20 years. The story of CERN's inception began, of course, several years before its formal founding. It began soon after the armistice at the termination of World War II, and it is intimately entwined with the worldwide discussion among nuclear physicists regarding their responsibilities in connection with the atomic weapons that had been employed in Hiroshima and Nagasaki.

When, on June 10, 1955, the cornerstone was laid for CERN on the grounds in Meyrin at the border of Switzerland and France, in the presence of representatives of the scientific communities of many nations, the following message was put into a hole of this cornerstone: "On this day . . . was laid the foundation stone of the buildings of the headquarters and the laboratories of the European organization for nuclear research, the first European institution devoted to co-operative research for the advancement of pure science."

The objectives and purposes of the organization are more precisely defined in its convention: "The Organization shall provide for collaboration among European states in nuclear research of a pure scientific and fundamental character and in research essentially related there to. The Organization shall have no concern with work for military requirements and the results of its experimental and theoretical work shall be published or otherwise made generally available."

The success of CERN as an international meeting place for high-energy physicists and instrumentmakers is already apparent in this document. Of great help was, of course, the considerable experience of Switzerland in matters of international organizations of all kinds. Another factor was that we physicists were just about to penetrate into the wonderful land of elementary particles, and that the world as a whole was fascinated by the discoveries in the field of cosmic rays. An entirely new world of pure research was unfolding itself before our eyes, and we might enter it provided that one could raise ample financial support needed to produce the cosmic rays artificially in earth-based laboratories, with enhanced intensities.

The war had revitalized the concept of a European community, and at CERN it seemed possible to make a genuine and meaningful contribution to science. This in essence was the theme and the message that we carried to the 12 European governments in the hope of convincing them of the usefulness and significance of such laboratories.

Now, what can we learn from this cooperation for our future goals? What aspects of the interaction have proven themselves?

All the essential decisions are made in CERN by the Council, which meets at least twice a year in Geneva. Each country has two delegates to the Council, usually one civil servant or diplomat and one physicist. This composition ensures that reporting within the Council is principally of a scientific nature. Even within the Committee of the Council, which has the task to lay the groundwork for the meetings, the scientific points of view have always been in the foreground. The Council approves the budget of the organization and appoints one, or, if needed, a number of Directors General, as well as the officers in the higher ranks. All of these are appointed for four to six years. The elections of officers are not conducted according to a system of proportionality; that is, the member states cannot demand that a number of positions, according to their financial contribution, be made available for their citizens.

There are two committees that advise the Council: the Finance Committee and the Scientific Policy Committee. The Finance Committee prepares the budget and supervises a rational and orderly distribution of large contracts from industry. Each member state has one vote in this committee, and, as expected, there are from time to time conflicts of national interests on this matter of contracts. But again, there is no system of proportionality. We always try to be true to the basic principle that the contract should be awarded to the group whose proposal combines the lowest cost with greatest expectation of technical perfection. Obviously, it is difficult to adhere strictly to this principle in every case, particularly when nationally owned or subsidized industries are involved. It is very important in these cases that the management of CERN provide the Finance Committee with clear and extensive information on its intentions and that the advertisements to the industry are prepared by top specialists. The task of the Chairman is to conduct the sessions of the Finance Committee fairly, but it is not always possible to exclude national interests, and bitter words have occasionally been spoken. Even so, such incidents have occurred rarely.

In contrast to this, in the SPC national interests scarcely collide, because of the structure of this committee. It consists of two groups of members, the first group being represented by physicists of international reputation who complement one another. If one member withdraws, a new one is elected by this group. The membership extends over a period of three years, with re-election being possible under certain circumstances. A continuity seemed to be very important to us, first, because the projects of CERN require several years of preparation, and, on the other hand, because contacts between the responsible Council members and their governments are sometimes necessary. In addition to this group of ordinary members, their number not being exactly fixed but not exceeding ten members, there is the group of *ex officio* members. These are the chairmen of the various experimental committees that are scheduling the machine time for the experimental teams.

Some explanation concerning these teams is called for. It was originally planned to provide for visitor teams from member states a certain fraction of machine time. Since, however, the preparation of experiments became technically more complicated, it was decided to work in "mixed teams." Today almost all experimental teams consist of CERN staff members as well as visitors. Such a team is not any longer built up by members of one university or one member state but, as is the rule now, is composed of representatives from different universities and different countries. Assignments of machine time then is the task of one large committee to which all proposals have to be submitted in written form.

There exist, for example, a Track Chamber Committee; an Intersecting Storage Ring Committee, which is responsible for experiments on the colliding beams; and many others. The chairmen of those committees always are experts in high-energy physics from the member states. They are replaced after fixed terms.

The committee meetings are public, and all applications for machine time are submitted to them and discussed. I should mention another committee that has played an essential part in establishing CERN programs, even though this committee had not taken on any official character. In 1963, on the initiative of several European high-energy physicists, among whom Edoardo Amaldi played an eminent role, the European Committee for Future Accelerators (ECFA) was founded.

Approximately 60 members from CERN countries, together with various subcommissions, developed a future program, not only for CERN but also for the member states in general. And these ECFA reports, carried by the physicists of the research laboratories of the member states, very essentially contributed to the realization of a new CERN program, known as CERN II. Within the framework of this program, it was especially the younger physicists who designed plans for the ISR machine, which meanwhile has been realized, as well as for the 300-GeV accelerator, which has now been approved by the CERN Council. Right now, we are having a 10-day ECFA meeting in Tirrenia, Pisa, Italy, to discuss the future research program among European users.

The chairman of ECFA is always invited to the Council meetings and is also an *ex officio* member of SPC. All these ECFA measures have been taken to give the younger physicists in the member states the necessary information that they rightly claim, and this has kept their interest in the destiny of CERN alive.

Let me now return to the role of SPC regarding the functioning of the whole organization. Political and national interests are to be excluded from SPC decisions. Members exercise their own independent judgments, and the decision of the Council always depends on the judgment of the SPC. For example, proposals made by the SPC for the election of new Directors General thus far have always been accepted by the Council.

This independence on the one hand and the authority on the other hand are by no means a matter of course. I remember an SPC meeting of another European organization, not in Geneva but in Brussels, where during the session the door opened and we were told: "Don't worry anymore about this problem. The ministers have already decided." There, you must know, the ministers had their own scientific advisors, and the decision was taken among themselves. Hence, their laboratories worked relatively inefficiently, since the members without being asked had to submit to the decisions of financial supporters, who mainly were guided by political interests. Perhaps a general outline of the aspects under which experiments are selected by the experiment committees seems to be opportune in this place.

There is no longer a place for trial and error in the field of big science. Each experiment is well considered and should decide whether a given theoretical idea is right or wrong. If there is not any valid theory, experimental data should be collected to promote the development of a theory, if at all possible. This explains why theoretical physicists in big science have gained a much stronger influence than was formerly the case in experimental physics. Both the complexity of a large experiment for which one needs large groups of specialists of different kinds, as well as degree of complexity of the attending theory, have increased. Therefore, large research plants today are managed often by a theoretician. This development can be observed in the United States as well as in the Soviet Union or in European organizations.

But this development, no doubt, entails a great danger for experimental research. As long as a universally valid theory does not exist—in the field of elementary particles, for example, we have not gotten this far—any theoretician unwittingly will try to select experiments under the aspects of further development of his own theory. He might, for example, say one day that new experiments will not be necessary anymore, for his theory is complete and explains everything. Therefore competitive institutes in the United States, the Soviet Union, and Western Europe are strongly needed, for one single worldwide institute could result in the danger of stagnation.

Let us return to research problems of the international community and especially to the situation given in Europe. At the beginning, I discussed the foundation of CERN and mentioned that in 1954 twelve countries of Western Europe ratified the foundation document for this laboratory. Although from the very beginning UNESCO, especially the director of the Science Department, Pierre Auger, advocated a participation of the East European countries, but negotiations with the Soviet Union proved difficult, and waiting for the end of those negotiations would have caused very long delay. Also, the Soviet Union herself had build up big research centers of nuclear physics shortly after the termination of World War II.

But even after its foundation, the Council of CERN remained interested in a cooperation with Soviet laboratories. In 1956, first talks with the directors of the institute in Dubna took place concerning the possibility of the exchange of scientists and the cooperation in planning and constructing big accelerators. Shortly afterward, a first Russian delegation was sent to Geneva. We were informed of plans concerning a foundation of "East-CERN." It was intended to establish an international laboratory in Dubna, the equipment there being provided by the Soviet Union as a gift to the new organization. The foundation was soon realized, and many of the experiences of CERN were incorporated in the statutes. Though it was also 12 nations who formed a research community in Dubna, the character of this institution in decisive points was different from CERN. In Dubna, the Soviet Union played the role of a powerful partner. In contrast to Dubna, CERN members were represented by big and small countries, with the big countries like England, France, Italy, and Western Germany contributing comparable shares, and the scientific level at the universities being in good balance. In addition, the joint laboratory is located in a small country with a long tradition of neutrality.

Promoted by common scientific interest in nonpolitical high-energy physics, mutual visits of CERN and Dubna co-workers soon became a matter of course; and an agreement on regular exchange of scientists and engineers came into force. But this exchange gradually became somewhat one-sided, since the development of Dubna unfortunately lagged behind CERN. The high-energy accelerator operated in Dubna was built when the modern principles for such techniques were still unknown. In CERN, however, the energy as well as the intensity of the beam was essentially higher. In addition, CERN made larger financial investments in further development and expanded continuously the experimental equipment. Hence the interest of Dubna scientists, including those from the member states, in working in Geneva was larger than vice versa. This situation changed after it became known that the Soviet Union was planning a new national laboratory with a modern accelerator twice as big as the CERN accelerator to be built up on the other side of Moscow. This new Russian 70-GeV accelerator, to be built on a national basis in Serpukhov was met with large interest throughout the scientific world. In 1965, then, first talks on scientific and technical cooperation between CERN and Serpukhov were initiated. In December 1966, the Council of CERN authorized the Director General to negotiate an agreement with the Soviet Government. This agreement between the State Committee of the USSR for the Utilization of Atomic Energy and CERN was formally signed in Moscow on July 4, 1967.

The agreement of Serpukhov provides for the following:

1. The equipment for the proton beam extraction will be provided by CERN. This includes magnets and vacuum chamber.
2. CERN will construct particle separators that will stay in Serpukhov for at least ten years.
3. In compensation for this, CERN is entitled to propose experiments that become part of the research program. (One experiment is intended to be carried out in the first year of operation by a mixed team of CERN and its member states and scientists from the Soviet Union. CERN members are allowed to participate in all meetings dealing with those experiments.)
4. Finally, close cooperation should take place in the field of bubble chambers. A big bubble chamber is provided by France, and the photographs taken with it will be mutually available.

This project has been realized without any difficulty. The Serpukhov accelerator went into operation, and a first scientific paper was published jointly by CERN and Serpukhov.

As formerly was the case between the CERN members, this cooperation has begun to be effective.

The joint work on one project in East and West together with equal partners leads to mutual acquaintance, understanding, and friendship. Soviet scientists are given the opportunity of genuine cooperation in the laboratories of CERN, whereas in the Soviet Union some bureaucratic obstacles need to be eliminated in order to realize this cooperation as stipulated.

On June 8, 1972, the completion of an important stage of the CERN-Serpukhov collaboration was celebrated at the Institute of High Energy Physics. An inauguration ceremony for the CERN-built equipment (fast ejection system, beam transport, and rf separators) now in action on the 76-GeV proton synchrotron was held. A. Petrosiants, Chairman of the State Committee of the USSR for the Utilization of Atomic Energy, and W. Jentschke, Director General of CERN Laboratory I, together cut a ribbon giving access to the new building where fast ejection equipment is installed.

The two ribbon cutters then moved to sign a protocol recording the date of coming into operation of the equipment. As specified in the initial documents signed in July 1967, the Agreement remains in force for five years from this date (taken as June 8, the day of the inauguration ceremony) and is renewable from then on.

Such genuine international laboratories *with several equal* partners exist today only in Eastern and Western Europe. The United States has never attempted to establish comparable laboratories on the American continent or with Europe. Obviously, it is difficult to have such laboratories, because the United States would almost automatically become the principal partner, and small states generally do not like to take orders from larger ones. It is also clear that a common historical and cultural heritage is a strong foundation for a joint working place with a sizable technical staff of permanent collaborators.

However, the United States has developed a fellowship system that probably does not find its equal anywhere else in the world. These fellowships, as well as the positions of senior physicists, are awarded with great generosity to foreigners of all nationalities. This has resulted in a veritable stream of visitors to the universities and research laboratories in the United States, which must be called unique.

As a typical example for this type of international cooperation in the United States, one may cite the Carnegie Institution in Washington, D.C. This institution awards about 50 fellowships each year. "During the period of 1953 to 1971 about 40% of the scholars in the Geophysical Laboratory came from abroad and about 60% from the United States." Because of its different geographical environment and its historical past, the United States has promoted a more individual variant of "International Cooperation," in contrast to Europe where such individual fellowships do exist but where more funds are allocated for bilateral and multilateral institutes.

In the early 1960's there was talk about a joint laboratory for high-energy physics of the Soviet Union and the United States. Nothing tangible has resulted from these talks and plans. One could consider that it is more important for the two countries to practice collaboration in other fields for awhile before they establish a joint laboratory. The UN Conference on the Human Environment in Stockholm in 1972 set itself a considerable amount of important worldwide tasks, but it also uncovered some of the difficulties of joint international action.

Nevertheless, the discussions surrounding the idea of worldwide cooperation for the construction of a very large accelerator have resulted in the more modest suggestion for the establishment of a center for theoretical physics. After several years of intensive discussions under the aegis of A. Salam from Pakistan, it was possible to gain enthusiastic support for this plan from a number of states. Agreement was obtained on the location, namely, Trieste—where Eastern and Western Europe meet. This place was offered by the Italians and had been strongly advocated by Professor Budini. It was crucial that Sigvard Eklund, Director-General of the International Atomic Energy Commission (IAEA) in Vienna was interested in becoming the sponsor of the Center. This International Center for Theoretical Physics has indeed evolved into a truly worldwide Center since its official opening in 1964. It is necessary, however, that a better understanding of the financial needs of the Institute is generated among the member states of the IAEA, especially so because the financial needs of a theoretical institute are incomparably smaller than those of an experimental laboratory. One year after the opening, the Director, Professor Salam, could report on the success of the Center. The Institute had a staff of 52 composed of people of 28 nationalities. It had already conducted extensive seminars in the field of plasma physics and high-energy physics. Professor Salam expressed himself in the following words: "My hope for the proposed center was that, besides providing a venue for collaborative research, it might also resolve this frustrating problem of isolation for other active scientists in poorer countries. Such men could come fairly frequently to the Center to renew their contacts and engage in active research fields like nuclear theory, high energy physics, theory of plasma, and solid state physics."

I would like to enforce these thoughts, because they appear especially helpful to me. Such an international center must be a home for talented theoreticians from all countries, especially the poorer ones. At the same time, the members must have tenured positions in their respective universities and must teach there regularly. In this way, the center can surely have a stimulating international character, especially when the members consider it a meeting place instead of a permanent home.

A center for theoretical physics, similar to the one in Trieste, had, of course, been founded earlier in Copenhagen by Niels Bohr. Anyone who has ever worked there considers himself a member of a clan. Those who have been there keep in contact even though they live under different political systems, and many of them have actively committed themselves for international cooperation, international understanding, and mutual help in troubled times. Even today, long after the death of Niels Bohr, Copenhagen has remained a wide open place of great hospitality. Without any formal international structure Copenhagen has remained an example for the world and for the school of theoretical physics.

This variety of international solidarity is much harder to achieve in the field of experimental physics. I would like to explore these difficulties a little further, because our experiences are valuable for any future developments. Again, I shall call on my own experiences at CERN.

Although it is true that only a small proportion of the researchers at CERN have tenure, there exists the danger that they lose contacts with their countries of origin and that they begin to constitute themselves as a new "Nation" of their own. This danger has been somewhat corrected by the formation of "mixed teams," but it has not been removed. It is desirable that the scientists at CERN acquire even better connections with the universities. Perhaps the situation could be improved if every scientist identifies himself with a university of his choice in one of the member states with which he is in permanent, close contact. Just as university professors have a sabbatical year, which they spend elsewhere doing research, one should give a sabbatical teaching year to the scientists at the international institutions with the clear-cut understanding that they will teach for one year at a university of their choice. During this year, they should participate fully in the academic life so that they will better understand the problems connected with research at the universities. Life in the ivory tower of international organizations entails the danger of isolation from the daily life at the universities. There is a distinct risk that a permanent staff will create a new caste within the member states. The result would be that we will have a thirteenth member state in addition to the twelve already at CERN—namely, the international scientific staff.

Another aspect of the contact of a member of the permanent staff with his country of origin is that of the education of his children. They in particular are in danger of losing all connection with their nation of origin, and they may never learn their own language properly. EURATOM has developed exemplary schools that prepare the children within the common market countries for entry into the universities. At the same time, the children are given instruction in the language and literature of their respective home countries, either in normal or in separate classes. When CERN II was founded, the agreement with France included provisions for the school problem. The French will build an international school in the vicinity of the CERN II laboratories. This school will open next year.

It is far more difficult to provide the necessary contact with the country of origin for the engineers and physicists with technical tasks, who have permanent jobs. But in these cases, too, it should be possible to arrange for them to lecture at universities or colleges from time to time. As they often are specialists, their courses could be particularly useful for the older students in the engineering disciplines.

The danger of a club of international research people has shown up in the laboratories of the six common market countries. In contrast to CERN, where we have always attempted to commit the permanent staff to aid our visitors, this was not done in the EURATOM laboratories. In the latter, a permanent team of scientists became established much too quickly, without the benefit of a substantial program of visiting scientists, such that the permanent staff mainly had to work on topics or problems handed down from the headquarters in Brussels. As the original topics are no longer of interest, the member states have difficulties in finding new ones. However, the staff has tenure and rightfully claims that they cannot be dismissed. Another mistake was that the laboratories had been expanded much too quickly, and thus the age distribution of the staff has become unsound. Errors of this kind usually show up only after several years, when the time of expansion is past.

Only those laboratories of the community of the six that had a clear-cut goal that remained of interest to the member states for a longer period of time could stay clear of the crisis. However, when scientific goals of an institute are combined with applied and technical ones, one must always expect that work will be terminated sooner or later, and then the quarrel of the member states begins. For the continuation of tasks, the ministries or politicians will analyze the needs of the national industries. The most significant industries are usually in different fields in the various countries; hence each state has an industrial lobby with its own characteristics, and now the international laboratory finds itself in the midst of a violent crisis. This phenomenon has occurred particularly in those international laboratories with strong ties to industry, such as the ones engaged in technological development of reactors for nuclear energy or of rockets for satellites.

The most secure future seems to exist at an international laboratory that is centered around the construction and operation of a large and expensive piece of equipment for basic science. At CERN, the central problem has always been the construction and operation of a gigantic accelerator. Even in the agreements between CERN and Serpukhov one finds that the aspect of instrumentation is in the foreground. The same is true for the bilateral agreements between France and West Germany. For example, the two states have cooperated in the founding of an institute with a high-flux reactor and have reached agreement on the implementation of a joint laboratory for high magnetic fields. The same is also true for the agreement between the French Atomic Energy Agency and Serpukhov for the operation of a bubble chamber. In each of these cases a major element in the agreement is the construction and operation of an expensive apparatus; and in every case, the prime users will be "visiting teams" of scientists. Accordingly one can keep the permanent research staff relatively small. The scientific life is not threatened with the danger of stagnation because of the coming and going of new teams. As in all international laboratories, the permanent staff has to sacrifice an appreciable amount of time to the visitors or visiting teams.

Finally, I would like to make some conclusive remarks of a general nature. Since the foundation of IUPAP 50 years ago by a small group of prominent physicists, international cooperation has developed unexpectedly fast. There is no special field in physics that has not been dealt with at international congresses or in international journals. The time schedule of worldwide summer and winter schools, symposia, meetings, general assemblies, etc. is so tight that a physicist can hardly follow the development in his own speciality. This has made the physicists of today the modern globetrotters; and congresses of a general nature are hardly feasible without them. The international airports have become the meeting points of the physicists, and the work in international organizations has grown so immensely that it can hardly be managed. The limits to growth has become a serious question in view of travel time, preparations, and administrative work arising in this connection.

This, however, does not refer to the European Physical Society (EPS), which only a few years ago was founded on the initiative of Gilberto Bernardini and thus has closed a gap in this respect. I think that within the framework of the EPS vital questions of international science policy can be discussed, since in this organization physicists from East and West come together. And this, I hope, may also bring about better mutual understanding. We all know that questions like physics and society, which are discussed at this conference together with international science policy will become evermore acute in the future. So far, those questions have been merely considered at Pugwash meetings, the importance of which, however, ought not to be underestimated. I did not consider these vital problems here. I therefore confined myself to the functioning of international laboratories. Their growth is still modest, and we are in the state of learning. I suppose that these international laboratories are important for the future of scientific life, because they are the places for joint work on difficult scientific and technological problems and thus offer a better opportunity for intensive understanding than is the case at congresses and in the offices of international organizations.

H.B.G. CASIMIR *has just retired from the post that he has held for many years as Director of the Philips Research Laboratories in Eindhoven, where he has also been a member of the Board of Management of the N. V. Philips' Gloeilampenfabrieken. The esteem with which he is regarded in the international community of scholars is well illustrated by a partial list of learned academies that have honored him with membership: The Royal Academy of the Netherlands, the American Academy of Arts and Sciences, the Royal Flemish Academy of Science, Letters and Arts, the Heidelberg Academy of Science, the Royal Society of London, and the U.S. National Academy of Sciences of which he is a foreign associate.*

15 *Physics and Society* / H. B. G. CASIMIR

Human civilization did not begin when man started to make and to use tools, it began when he found time to decorate and to embellish his tools. The essence of culture is always in those things that from a purely utilitarian point of view are unnecessary, superfluous, or even wasteful. This holds for tools, food, and clothing; for palaces, temples, and cathedrals; but also for games and for ritual, for empirical knowledge and theoretical speculation. Certainly, in many cases such nonmaterialistic cultural activities may have been prompted by a fear of supernatural power; they may even be regarded as attempts to placate or to dominate such powers—but that does not mean they are utilitarian in the current sense.

If we take this view of civilization, then the problems of the relation between physics and society should be discussed along the following lines. To what extent can society afford to support physics as a spiritual adventure, as a search of knowledge for knowledge's sake? How much should it spend on physics, how much on other sciences, how much on humanities and on music and on art? On the other hand, what can physics do to make its achievements less esoteric, to let more people share the thrill of discovery and the satisfaction of understanding? The Chinese in the old days used gunpowder primarily for fireworks on festive occasions, thus providing innocent, though short-lasting entertainment for the people. Today, we use enormous accelerators to produce fireworks of new particles, but they last an even shorter time, and the number of people uttering groans of admiration on seeing a beautiful bubble chamber picture is small compared with the crowds audibly admiring fiery stars bursting into multicolored gerbes high above their heads.

But that is not all. Physicists should also consider to what extent their results and their methods might further other sciences that are also seeking knowledge for the sake of knowledge. And might natural philosophy not be a valuable weapon in mankind's struggle against fear and superstition and prejudice?

There have been—there still are—physicists who look at their subject in the way I indicated. I recently had the privilege to listen to Dirac talking about the early days of quantum mechanics, and he made it clear that for him the beauty of equations had always been a main guiding principle. Let me quote another instance. Shortly after the end of the Second World War and the liberation of Holland, I wrote a letter to Pauli, who was then in the United States. I explained to him that we had been rather isolated during the war years, that we would soon receive a huge pile of back numbers of journals, and that I would be grateful if he could indicate to me what he considered the main developments in physics in the United States during the war years. His answer was characteristic. He wrote: "This country is ideal for large projects like radar, or nuclear weapons, but nothing of any importance has happened here. So prepare yourself for a great disappointment when the journals arrive." On second thought, he slightly corrected this judgment and he sent me a reprint of Onsager's paper on order–disorder transitions in a two-dimensional lattice. Let us be grateful that there have existed, that there still do exist, people who rate the theory of order–disorder transitions higher than nuclear weapons or nuclear power.

I suppose that by now some of you are getting impatient, you may feel that what I have been saying so far has no bearing on the problems of physics and society at all, that I have been talking nostalgic nonsense about irrelevant futilities, and that the essential problems are quite different and are connected with the relations of physics and technology and of technology and society. Physics has become an essential element in our technological structure, and the fact that you may rightly regard my introductory remarks as irrelevant reveals the extent of this involvement.

Our Western society is an industrial and technological society. Our food and our clothes, our means of transport and of communication, our entertainment and our hobbies, and even our sexual behavior have become to a large degree dependent on industrial products, and industrial production is based on knowledge acquired from science.

This technological society does not regard physics in the first place as a noble spiritual adventure, as a real calling for a few and as a source of enrichment of intellectual experience for many, but as an activity contributing to the growth of the Gross National Product. Society does not support physics because of the things we consider essential, not for beauty and depth, not out of gratitude for "objects and knowledge curious," as Walt Whitman put it. The German poet Schiller was already well aware of this when he wrote the famous lines (excuse my imperfect translation and remember that the German word for science is feminine): "To some she appears as a glorious, heavenly goddess, to others she's a reliable cow, giving us butter and milk." On the other hand, physics has become more and more dependent on the products of advanced technology. One has only to pay a visit to one of the prominent laboratories in any part of the world to become aware of this. Milking cow or goddess, physics needs a well-equipped stable, or temple, as the case may be. I have on several occasions described this interactive system of physics and technology as a science technology spiral. Let me briefly summarize its main features.

Progress in basic physics was nearly always made by physicists doing physics for its own sake, studying the phenomena of nature out of curiosity. Of course, also the phenomena occurring in empirically developed crafts belonged to the phenomena studied by physicists. Beams had been under loads, water had been pumped, ships had been sailing, and windmills turning before the theory of elasticity and before hydrodynamics and aerodynamics were developed. But this situation has changed. Thermodynamics is probably the last example of fundamental theory partly based on empirical technology—in this case, the technology of the steam engine. Perhaps one should point out that the discovery of the ionosphere was a consequence of Marconi's practical application of Hertzian waves and that there are more cases, also in recent technology, where superciliously negative theoreticians were put to shame by enterprising practical engineers. But, although in such cases faulty conclusions may be corrected and incompletely worked out theories may be completed, no really new ideas are developed. There can be no doubt that technology has become more and more dependent on results of basic physics obtained many years previously by physicists doing physics for its own sake. Many branches of industry could even not be imagined without this basic research preceding it. Electromagnetic induction and magnetic forces on currents, on which dynamos and electric motors are based, were not invented by industrialists who were dissatisfied with belts for transmitting power. The equations of the electromagnetic field and electromagnetic waves were not invented by managers of postal services dissatisfied with the speed of horses or even trains. Radioactivity and the existence of the atomic nucleus were not discovered by boilermakers or manufacturers of hydroelectric power stations looking for other sources of power nor by the military looking for convenient means of wholesale destruction. The electron was not discovered by industrialists who wanted to bring entertainment by radio and television into every home nor by politicians who looked for new ways for getting their messages across to the public. Nor was quantum mechanics developed in order to find a substitute for the radio tube, although there can be no question that the transistor was discovered by physicists who were thoroughly familiar with the quantum mechanics of electronics in solids. Even in the older industries, empiricism is largely replaced by science-based development. It may be true that thermodynamics partly arose out of a study of existing steam engines, but no modern thermal engine could be designed without a thorough knowledge of advanced thermodynamics. That holds for steam engines and turbines and also for the Stirling engine, a pet child of the Philips Laboratories (and a rather expensive one that stubbornly refuses to come of age).

Technology does not use all the results of physics. There are many beautiful and intellectually important results that so far have found no application and may never find application. On the other hand, the results of fundamental physics may not always be sufficiently detailed for technical use. It is here that the academic scientist and the scientist working in industry meet. Most academic scientists are not working at the very front of science. They are filling gaps, studying details, they are simplifying or generalizing. This is exactly what physicists in industry are doing. Although motivation and reward systems may be different, there is not much difference in their methods or in their way of thinking once they have chosen their problem.

A further remarkable feature is that technology never uses results at the very front of science. There has always been a lead time of 15 or 20 years, and there is no indication that this time is getting shorter. On the contrary, it may get longer. It seems to me that this point is rather essential. So let me repeat a few examples. The existence of the electron was reasonably well established in 1900; before 1910, the first triodes had been made. The existence of electrons and holes in semiconductors was established before 1930; it lasted more than 15 years before the notion of positive holes was made use of in the transistor. The notion of stimulated emission was introduced by Einstein in 1917, and the idea of inverted population of energy levels must have been current among people dealing with gas discharges in the 1920's. It took 30 years before the maser was invented, and from then on it took quite a number of years before lasers found their way into technical devices. Superconductivity was discovered in 1911; economically important applications are still around the corner. No applications are in sight for particle physics nor for general relativity.

Physics, on the other hand, uses the products of technology and also the most recent ones, for instance, large computers and advanced electronics. In this respect, astronomers are even more enterprising than physicists. As soon as the first artificial satellites had been launched—if not before—they began to think of programs using satellite-borne equipment. Sometimes, however, basic physics invents for its own use technological products or methods that can be used by an industry not at all interested in the basic work as such. That happened with optical devices designed by astronomers and with the electronic circuits designed by nuclear physicists for particle counting. In this general picture, the influence of wars becomes obvious. In general, war leads to a stagnation at the very front of science—compare Pauli's letter—but in some fields, it leads to a considerable shortening of the time lag between basic discovery and application. The technology thus created may later be used in basic research. Radar, nuclear power, and to a certain extent digital computers are examples.

If we want to speak about today's problems of physics and society, we have to start by admitting that most of us, if not all of us, are part of this spiral structure. And now we have come to a parting of ways. If we believe that technological progress is essentially a good thing and that increasing knowledge is a good thing and that whatever hardships or sufferings may be caused or at least condoned by technocracy are only passing childhood diseases that will be cured by science and technology itself, then we should rejoice because of the astounding efficiency of the science–technology spiral. The problems of physics and society are then mainly problems of improving this spiral mechanism still further. Then we will aim at an even better matching of academic research and industrial research. Then one cannot avoid considering a really advanced project in fundamental research also from the point of view of the role part of it might play in this spiraling process. Now most of us have been conditioned to think this way. Even if we speak scathingly about motorcars and airplanes, we would not like to miss their comfort. Few of us do *never* look at television, and all of us use the telephone. We may still remember, be it only from stories, the weary toil of a housewife and of a working man less than a century ago, and we should not wish to do away with electrical power and mechanical tools. We cannot help being fascinated by the ingenuity of new devices. In answer to criticisms related to the obnoxious effects of technology, we may justly remark that science and technology themselves harbor many possibilities to combat such effects. Too much energy consumption might be avoided by designing equipment with higher efficiency, traffic congestion might be relieved by better telecommunications. Polluting effluents might be removed and even turned into useful products.

But we may take a different point of view. We may also feel that this curious alliance of one-sided philosophy and engineering drive and competence is so efficient that it is outstripping all other human capabilities, that it is breaking away from the restraints imposed by wisdom and charity. Then we must regard this spiral as an ominous monster, as a force that is getting out of control and that will lead to havoc and destruction. Then we may remark cynically that although a very low value of GNP goes together with abject poverty and misery, there is ample statistical evidence that increase beyond certain limits leads invariably to an increasing crime rate and therefore to a decrease in personal safety and peace of mind; to an accumulation of dirt and litter in our cities; to a deterioration in public transport, postal services, and other public services; to a growing feeling of discontent among the younger generation; to an increase in the number of suicides, be it rapid ones or slow ones by means of drugs. And once we take this point of view, we see dangers threatening everywhere. A pure scientist may admire laser physics for the insight it provides into the relations between macroscopic electrodynamics and quantum field theory, for the possibility of studying the curious details of nonlinear optics, or simply because of the beautiful demonstrations of optical interference that it provides, but we cannot forget that technology, industrial and military, is not particularly interested in these features, and that lasers are being used to assist in the callous destruction of life and property and in the engineering of irreparable damage to the human habitat. Lasers and modern electronics and many other achievements of science-based technology are the tools of political objectives that apparently justify the most ruthless measures in the eyes of some but that are loathsome in the eyes of many others and at least questionable to most of us. Then the issue of supporting particle physics is no longer a gentle farce played by both parties with their tongues in their cheeks, but it assumes much more frightening proportions. Let me explain.

Society may claim to support particle physics and accelerators because it is interested in knowledge as such, but at the back of its mind it believes that something—that is, something of economic value—will ultimately come out. Scientists, while at heart only interested in the basic results, point out that in the past research that looked equally impractical gave rise to entirely new industries. As I said, it might be a gentle farce, but if we look at the ominous power of the science–technology spiral then we may be thoroughly alarmed. Since we cannot at this moment see any application of high-energy physics, therefore, when such an application should come, it must be entirely beyond our imagination. And since we have so far been unable to keep existing scientific knowledge under control, we may well be scared about what mankind would do with entirely new possibilities.

Incidentally, such misigivings are not confined to physics. Many of you may remember an incisive article by Watson in the *Atlantic Monthly* titled "Is This Really What We Want?" where he discusses the possible impact of molecular genetics on mankind. I believe such dangers are real enough, and yet, if in some committee on science policy in my own country I make a statement that I would be happy to give more support to high-energy physics, if I could only have a guarantee that it will have no possible industrial application for the next 25 years, I am not even taken seriously but regarded as a naughty *enfant terrible* who likes to make paradoxical cracks.

I have briefly sketched two different attitudes. You may now want to know, where do I stand? And my answer will be hesitant and evasive. Like most of you, I love physics, and after more than 45 years of study, I find its achievements and its problems not less admirable and challenging—though considerably more difficult—than when I was a young student. Also, I have been closely associated with several branches of technology, I am fascinated by many of its aspects, and I do have some confidence in the ability of technology to find adequate solutions for many of the problems that technology itself has created. Yet I have in recent years, probably again like many of you, become increasingly alarmed about the influence of science-based technology not only on the waging of war but also on the destruction of the environment and on the nature of human relations. I am increasingly inclined not to speak about the efficient, the fruitful, the glorious spiral but about the ominous, the inexorable spiral. The word inexorable brings to mind a passage in C.P. Snow's essay on Einstein:

> The bomb was made. What should a man do? He couldn't find an answer which people would listen to. He campaigned for a world state: that only made him distrusted both in the Soviet Union and in the United States. He gave an eschatological warning to a mass television audience in 1950:
>
> "And now the public has been advised that the production of the hydrogen bomb is the new goal which will probably be accomplished. An accelerated development towards this end has been solemnly proclaimed by the President. If these efforts should prove successful, radioactive poisoning of the atmosphere, and, hence,

annihilation of all life on earth will have been brought within the range of what is technically possible. *A weird aspect of this development lies in its apparently inexorable character. Each step appears as the inevitable consequence of the one that went before.* And at the end, looming ever clearer, lies general annihilation."

That speech made him more distrusted in America. As for practical results, no one listened. Incidentally, in the view of most contemporary military scientists, it would be more difficult totally to eliminate the human species than Einstein then believed. But the most interesting sentences were the ones I have italicized. They are utterly true. The more one has mixed in these horrors, the truer they seem.

As I said, I am unable to clearly define my position. Still less can I present a solution. I should like, however, to formulate a few simple and straightforward recommendations:

1. A physicist should realize that being a physicist does not put him in a position beyond all responsibility, even if he is dealing with abstract academic subjects. He cannot sit in a corner, pull out the plums he likes and say: What a good boy am I. Little Jack Hornerism is no longer a justifiable attitude.
2. Since we are unwilling and probably unable to stop the development of science and technology, and since we are quite obviously unable to design a comprehensive masterplan for science and society, our best chances for gaining some control over the ominous spiral lie in a plurality of controls, in an independence of opinion of the several participating groups, and in openness.
3. In particular, universities should maintain their independence versus industry, and industry should respect this independence.

I want to discuss this aspect in more detail. In my presidential address at the annual meeting of the European Industrial Research Management Association at Dublin in 1970 I formulated a number of rules of behavior for the dealings of industry with universities. I should like to repeat them here.

(a) Do not try to influence program of basic work. Even if we are unsuccessful in trying to do this we may create a lot of ill will. If we are successful we might do a lot of harm to human culture.

(b) Do not try to determine the curriculum as long as it provides a sound training in fundamentals and gives the student a broader view and sense of perspective; do not object if it also includes subjects that are irrelevant from a purely industrial point of view.

(c) Straightforward development tasks are best done in your own laboratories. It is improbable that the university would be willing and able to do such work in accordance with requirements of industrial practice.

(d) Since much of the work in academic institutions is of a similar nature to work in industry, even though the motivation is different, I see no harm in industry suggesting certain themes or supporting work it is interested in, in a general way.

(e) A university should never accept conditions of secrecy and industrial security. Free discussions and free publication are inviolable privileges of the academic community. Economic necessities make it impossible for industry to emulate these principles in their entirety, but industry should respect them in universities. There is also a practical side to the question. If you really want to keep a thing secret, do it on your own premises.

(f) Do not insist on exclusive rights to inventions. This leads to complicated legal negotiations and usually the game is not worth the candle. If the university wants to pursue a patent policy of its own, then arrange for a nonexclusive license beforehand.

(g) Avoid under-the-table arrangements with one professor or scientific staff member. He can come to your laboratories as a consultant, that is all right, but he should not set his students to work on your problems without their knowing what it is all about.

Speaking about conditions of secrecy, let me relate a little anecdote. Many years ago, Ernest Lawrence came to visit the Philips Research Laboratories at Eindhoven. Visitors in those days had to sign a form with many stipulations in small print. People usually signed without reading, and I do not remember any single case where these rules made any difference. Lawrence, however, read them carefully and said: "I cannot sign this. It says I shall not communicate what I have seen in your laboratories to third parties and that is just what I will do. I come here to see things I can tell to my students and my collaborators. I understand that you may not show everything, that you will not answer all my questions but I want to be free to use whatever I have seen and heard."

So I struck out the offending sentences, put my signature beside it, and that was that. I did not even have much trouble with our security people afterwards: they did not really believe in their own small print.

But let me come back to my recommendations.

4. Relations between the military and the universities—if they exist at all—should *at the very least* follow rules similar to those between industry and universities.

5. It would be desirable to have also a more clear-cut separation between industry and the military. That is hardly a matter for physicists, although physicists in industrial research laboratories might have some influence. For instance, they might insist on being informed about military aspects of their work.

6. The time lag between fundamental research and technological application makes it difficult to assess possible consequences and to define responsibilities. If all academic work in physics were to stop today, this would have little influence on industrial and military technology for quite some time to come. Neither can we influence the birthrate during the next nine months by stopping all sexual intercourse today.

Yet I want to make one more recommendation. It is my personal opinion that, at the present time and in view of the alarming uses of science-based technology in warfare, no scientist in an academic position should of his own free will be active in or advise on military technology. As I said, this is a personal opinion, but I know that it is shared by many of my colleagues and an impressive number of students. Such an opinion implies a condemnation of the powers that be or, rather, a lack of confidence in the powers that be. Yes, I distrust any very large, very comprehensive, and partly secret organization, and I have scant confidence that government and the military branches of government will deal wisely and charitably with the immense power that has been made available by science and technology. In a world of brutal violence and clashing patterns of living, where contending ideologies breed murderous fanaticism, the ideologies themselves may be no more than a cloak for short-sighted selfishness. In such a world a decision of university physicists to refrain from military work can be no more than a symbolic gesture. Some may even call it a hypocritical gesture, because we continue to enjoy the privileges and the amenities and the comforts that government protects. But a symbolic gesture can be meaningful and even hypocrisies show at least some sense of values. As La Rochefoucauld put it: "l'hypocrisie est un hommage que le vice rend à la vertue" (hypocrisy is an homage that vice renders to virtue).

I have come to the end of my paper. I have tried to analyze to the best of my abilities some of the bewildering intricacies of the relations between physics and society, but I can hardly hope to have made clear to you what is confused and confusing to myself. If, however, some of my remarks have been sufficiently provocative to make you reexamine your own ideas, reassess your own individual choice of position, then I shall be more than satisfied.

ROBERT F. BACHER *is the President of the International Union of Pure and Applied Physics. He is professor of physics and provost of the California Institute of Technology, past president of the American Physical Society, and past chairman of the United States Atomic Energy Commission, for which service he was awarded the Atoms for Peace Award.*

16 *Looking to the Future*/

ROBERT F. BACHER

During the past few days, we have heard some fine papers presented before the Scientific Sessions of this Fiftieth Anniversary General Assembly of our Union. There have been reviews by distinguished physicists of most of the main areas of physics and some of the areas in which physics impinges on other sciences. These talks have shown us the present status of these many fields and some idea of the most important areas of future growth. The accomplishments of the past 50 years are very impressive. An enormous, coherent body of knowledge has been put together with many interlocks in fundamental concepts as well as in quantitative observation and calculation.

It would be easy looking at these accomplishments of physics to be too satisfied. But, in fact, some of the most basic areas of physics are understood very imperfectly. One might say that our understanding is exceeded only by our ignorance. Looking back at one large field over a period a bit longer than the 50 years of our Union, the introduction of the nuclear atom led to the Bohr theory and, almost 50 years ago, to the development of quantum mechanics. Meanwhile, properties of nuclei, especially the radioactive ones, were being studied and isotopic species were being elucidated. The whole subject of nuclear physics blossomed out in the early 1930's with the discovery of the positive electron and the neutron and the advent of nuclear accelerators. By 1940, after the discovery of fission, a sizable amount of knowledge about nuclei had been assembled, and this was increased by intensive efforts on the large-scale release of nuclear energy during World War II. This itself was a happening that had been singled out many years before by Rutherford himself as being so much moonshine. But physicists have more than once underestimated nature both in predicting what will come to pass and how complex the real world is. I can recall a distinguished theoretical physicist saying in the mid-1940's that if we just understood the nucleon–nucleon interaction up to 10 MeV a bit better, we would have the key to nuclear forces. It soon became clear that the pions and probably other unstable particles created in much higher energy interactions were involved in nuclear forces. This provided the impetus for studying nucleon interactions at high energy and for the development of high-energy accelerators. The last 20 years have been spent unraveling the various excited states and unstable particles that are created at much higher energy and that are intimately connected with an understanding of the subject. The richness of nature has always been underestimated, and nuclear and particle physics are no exceptions.

In many areas, but especially in the understanding of the very small and the very large, physics has been in the forefront of our understanding of nature. Some people—apparently an increasing number at present—are content to accept the behavior of nature as they see it with their own eyes. But many others, including all scientists, have a yearning to know more about why nature is the way it is or how various observed phenomena are related. This yearning prods us to dig always deeper into the atom, the atomic nucleus, excited states of nucleons, and unstable particles. This urge also drives physicists and astronomers to learn more about what exists in those vast reaches of space in which we play so small a part and that contain such a wide variety of unusual objects. Many of these objects were unknown or at least unrecognized relatively few years ago. Scientific work in these and other areas has greatly affected our realization of the scope and complexity of nature, while giving us some confidence that, in the extraordinary way in which some of these observations fit together, we have a significant understanding of some parts of nature. In turn, this work has added greatly to our culture and affected some of the most basic ideas of our philosophy.

As our understanding has increased, the difficulty, complexity, and cost of adding to this understanding has increased very greatly. Financial support in increasing amounts especially from public funds has been essential to the development of fundamental physics in almost every area. This, of course, has brought a greatly increased scrutiny by those who authorize the expenditures, as well as those who look at them with the thought that they could be used "more practically" or more helpfully for society, in other ways.

The real increases in support for basic science came about, however, mainly for the reason that the applications of science were vital to our technology. Technology became of major importance to industry, to the military, to transportation, to communication, to health care, to agriculture, and to a host of other developments that affect our everyday lives. Then came space exploration with its enormous costs and its dependence on very highly sophisticated instrumentation. Space missions gave a fine opportunity for exciting new scientific observations, and indeed some people reached the conclusion that this was the reason for space programs. Actually most of the missions could hardly have been justified even in small part entirely on their scientific contributions. These space missions were explorations on a new and very grand scale—not scientific experiments. The technical developments in this area often used some of the most recently discovered scientific findings, and this technology forged an even stronger connecting link between science and technology. Along with all of these links came greater support for fundamental science because it formed the base for technology and because some scientific work led directly to new technology as a part of its own development.

Inevitably the great development of technology has had an enormous impact on our society. Not all of this has been beneficial. Much of it has led to an increase in the complexity of life. For example, the great developments in communication and transportation, which have increased both the amount of exchange and the speed of exchange of ideas and news throughout the world, have in some ways degraded our environment. It may happen that a technological development that saves endless toil may have some effects that must be curbed in order not to become a nuisance or even a danger to society. As a result, there are some people today who damn technology thoroughly and science too because of its close connection.

Things have not always been this way. I recall a story about Hilbert, who, commenting to a group of students on the then often discussed hostility between science and technology, remarked that this could not possibly be true because they have nothing at all to do with one another. Indeed the separation of science and technology has played some part in the development of our Union. From the first, it was named a Union of Pure and Applied Physics, but historically there has been much greater emphasis on pure physics than on applied physics. This was probably both correct and inevitable many years ago, but it certainly does not represent the situation today. Pure physics and applied physics are so tied together that no attempt to consider them completely separately can be successful. In looking to the future, we must be increasingly aware of this close connection, whether the application of physics is to another branch of science or whether it is to technology. In the former application, there will be an immediate enrichment of pure physics itself, and in the development of technology there may be major effects on our society as well as increases in our capabilities in pure physics.

The involvement of technology in the problems of our society and in particular the blame attached by some people to technology as the origin of many of our difficulties, pose some major problems for the future. This blame has probably been enhanced by some well-meaning scientists and engineers who contrariwise believe that most of the problems of present-day society can be solved by technology itself. This has sometimes been referred to as the "technological fix." It is rather unlikely that very many of the present-day problems of society can be fixed by technology alone. It will usually require the solution of many other economic, social, and political problems in addition. To most physicists, this appears to be a journey into unknown and often disliked territory. Indeed, it is true that most scientists are untutored in these areas and should realize that their views are those of laymen not experts. But this does not change the basic fact that science-based technology may be necessary but surely not sufficient for the solution of many of society's pressing problems.

Those who say "Let us turn the clock back and live with the simplicity which our forefathers enjoyed" underestimate the magnitude of our present-day problems and the impossibility of society doing without many of the advances that are rooted in technology. We have no chance of supporting the world population today without employing our technology, and the difficulties in the near future will be much worse. Even many of those parts of presently utilized technology that are not absolutely essential for existence would be given up with the greatest reluctance if at all.

Recently, a sociologist at Harvard, Irene Taviss, has questioned whether the negative attitude of people toward technology, which we hear about constantly, is a reality. She has conducted a pilot study and contends that "the anti-technology spirit that so many commentators assure us is rampant in the land is probably a myth." Of course, this is only a preliminary study in a small part of the United States, and it may not be confirmed elsewhere. Whether the antitechnology and antiscience spirit is a myth or not, it is getting a lot of attention from people who like to attribute many of society's problems to the development and widespread use of technology.

Various writers, including Sir Peter Medawar and Murray Gell-Mann, have pointed out that antiscience and antitechnology comments are only part of the picture. These sentiments carry still further into antirationality and sometimes to a growth in pseudoscience and mysticism. The astronomers have bemoaned the fact that there are many more astrologers than astronomers. The increased attention given to astrology, palm reading, and various forms of magic and superstition is an evidence that the antirationality of those who criticize the effects of technology on our society, while probably a minority view, is a potent one, and this view is shared by a significant number of educated and presumably intelligent people, especially young people. It is true that inadequate attention has often been paid to the full impact of new technical developments on society. Too little consideration has often been given to the serious effects on our environment. Too little value has been placed on preservation of the quality of our environment.

Scientists have often pointed out that they cannot be responsible for unforseeable uses of the technology that come from scientific discoveries. Such cases are not imaginary. The early use of radio telescopes as an interferometer did not anticipate the extraordinary success of the phase-lock system that was used. This system was further developed in the long-distance telemetry of the space program. Now it is possible to use a base line of roughly the earth's diameter. This leads to very high resolution of radio sources and, with a point source, to distance determinations on earth to 10 or 20 cm. Not only does this lead to new and exciting results in radio astronomy, but it may permit the measurement of fault motion and continent drift as never before. This same development also has major possibilities for extremely accurate navigation and may have important military applications. None of this could have seemed credible at the beginning.

Another example is the laser. The discoveries of the maser and the laser came out of an investigation to see if stimulated emission could really be observed as predicted by Einstein. The first laser was operated only a little more than ten years ago. Now we have lasers used in eye surgery, in accurate surveying, and for many other purposes. Powerful lasers emitting short bursts of radiation are being applied to thermonuclear reactions. Most recently, we hear that they are being used to produce laser-driven implosions with the aim of producing thermonuclear energy releases in small pellets. Where this will lead is not easy to predict today, but the utilization of thermonuclear energy in the future seems to be almost certain. The energy demands due to increasing population and increasing energy use per person, even if restrained, will not be able to be met indefinitely from other sources.

The problem is and will continue to be: How can the benefical effects of science-based technology be maximized and the bad effects be minimized? This depends principally on how the technology is utilized and what care is taken by society in its exploitations. The problem will not be solved by trying to turn the clock back for science and technology. We need our science and technology desperately to solve the inevitable problems of the future.

It seems to me that physicists are beginning to think much more about the nature of the technology–society interaction and will try to understand it better. In the introduction to a report of an extended study of this subject at Harvard, Emmanuel Mesthene has summarized some of the studies that were conducted to ascertain the "degree to which technological change determines social change." He concludes that the technologies that are developed and applied depend on institutions and values prevalent in society at any given time, but that technological innovation provides society with new capabilities and not all of its consequences can be foreseen. He refers to social change as a second-order effect of technological change, and one might extend this to make it a third-order effect of scientific change or discovery. These higher-order effects are still significant. It is interesting that he rejects the idea that the technological developments determine the nature of society.

Others in the social sciences take a much more negative attitude toward all technology and are critical of Mesthene. They lose sight of the dependence that we have today on technological developments for the continued existence of our present-day society. Rather, they take the wholly negative view that the impact of technology on society has such disagreeable results that technology and science too should be curbed and not developed further. While this is certainly a minority view, scientists would be wise to pay more attention to the applications of science in the future.

If we look at our Union, this means that we should take applied science seriously. In the relations with other fields of science, we will gain by broadening and enriching pure physics. In recent years, some of the exciting interdisciplinary new fields have grown up in special committees outside the Unions. Some of these special committees involve the subject matter of many Unions and have built up a body of knowledge of their own too. In our Union, we have not always carried out our obligations in keeping in close communication with the work of these committees. This has led to some cases in which research under some of our own commissions is no longer fully representative of the subjects with which they are concerned. Some physicists find that the only way in which they can communicate with other physicists who work in new fields is to do so outside the Union. I am not proposing that fields of research covered by special committees should be allocated to one Union or another. Special committees have often been able to get support for their work more easily just because they were set up around a particular project. Partly for this reason, the trend for many years has been for much of the interdisciplinary work to go on without very close ties to related work in the Unions. The result is that small areas of science become isolated from the more general fields of which they are a part. Some of this separation may be inevitable with the large number of new interdisciplinary fields that have sprung up. It would be helpful to our Union and to the special committees if we put a greater effort into the communication and liaison with related special committees. The use or application of physics to other sciences is increasingly important to science as a whole.

The applications to technology and in particular the effects that these may have on our society are much more difficult problems and these are today imperfectly understood. We physicists need to understand these effects and their reaction back on science more clearly. We cannot be responsible for all the unforeseen consequences of science and technology, but we surely must be more sensitive and understanding of possible consequences. There will inevitably be vigorous arguments as to whether the net effect of a particular development is beneficial or injurious, and criteria for judgment will vary. It is not an easy task. But it is one in which physicists have both an obligation to society and a considerable amount of self-interest as well. The first problem is to improve understanding. Physicists do have something to contribute in working with these problems because of their technical content. There are also many examples of the application of scientific methods achieving success outside the fields of science. But to understand some of these problems takes knowledge in other areas as well. Perhaps the first step is to recognize this and to try to correct it.

Some of the Unions representing the sciences are by nature connected to applications that affect society. In physics, this is not often the case, although a large fraction of the work in the solid state and semiconductors has a close relation to industrial use, and much of the work in these areas is supported by industry. Very likely there will be closer ties to industrial use in the future. It is accordingly somewhat anomalous that the relations of our Union with international engineering organizations continue to be as tenuous as they are. We do not need to incorporate technology and engineering into IUPAP, but we do need to keep closer ties with the organizations that represent these areas. Once again, this is increasingly made necessary by the involvement of physics and applied physics in technology and engineering advances and the reaction that society has to this connection. We must know more about the applications of science and be able to understand better than we do today both the positive and the negative reactions of society to technological developments. Perhaps in time physicists will be able to exert more influence on how science is applied to technology and to help to maximize some of the beneficial effects and minimize some of the harmful consequences.

During these past 50 years, and especially in the past 20 years, we have seen a great increase in the international exchange of ideas in physics. Our Union has had a major part in this growth, especially through its sponsorship of international conferences by the various commissions. Ease and speed of travel have greatly increased the number of international contacts that the average physicist has. These contacts and the opportunity to exchange ideas in person inevitably lead to better international understanding, and this is an important result of our Union's principal activity—sponsoring international conferences.

We have now reached the point where our Union may be able to take additional steps. With the building of large new facilities for research, it has become increasingly common for research groups to travel long distances to carry on an experiment. It has been found to be much more effective for a group going to another country to collaborate in the work with physicists from the host country or the host laboratory. This international collaboration usually results in extended periods of residence and close association of the collaborating scientists. I have spoken to many physicists who have participated in such collaborative efforts, and each one has been enthusiastic, particularly about the increase in international understanding that results. I hope that in the future our Union can play as important a role in the fostering of international collaboration in research as it has in the promotion of international conferences. There is still far too much red tape involved in collaborative efforts, and perhaps this is an area in which we can help.

We think of our Union as being truly international, and there have been many efforts to make it so. There are, however, some major gaps in our membership, and, in addition, we are poorly represented among the developing countries. We need to pursue every opportunity to fill the major gaps in our membership and to make efforts to have an active participation by the physicists from important and influential nations whose role in IUPAP has been small or completely absent.

For the developing nations, it is not surprising that their interest in IUPAP has not been very great. Our main activities have been in sponsoring research conferences on the latest work at the forefront of the various fields of pure physics. This is not of primary interest to the developing nations, and there is no reason why it should be. The developing nations *are* interested, however, in many aspects of technology as it applies to agriculture, health care, transportation, communication, and certain areas of industry, just to mention a few. To follow these needs effectively, education is essential, particularly in the fundamentals that underlie these developments. In most cases some understanding of the fundamentals of physics is essential. Our ties to developing nations should be centered much more around the education that is needed in the course of their development. This is probably not the same physics education that has evolved in the more fully developed nations. We should, however, be able to utilize our experience in physics education and, with a study of the needs of the developing nations, be able to produce some very tangible assistance. This will not be an easy task, and it may go slowly, but it seems to me to be a real challenge for our Union and especially for our Commission on Physics Education.

Looking back over our past history we see some remarkable achievements in promoting international communication and exchange in physics. This has clearly resulted in an increase in understanding among the world's physicists. As we look to the future, there are many more opportunities ahead for our Union, and I have tried to indicate a few of these to you. I am confident that our Union will be able to continue its good work of the past and to extend it in the future. We can all hope and expect that our contributions to international understanding among physicists will be a growing part of that more universal understanding that is essential for world peace.

Index